Doctoring Traditions

Doctoring Traditions

Ayurveda, Small Technologies, and Braided Sciences

PROJIT BIHARI MUKHARJI

The University of Chicago Press
Chicago and London

Projit Bihari Mukharji is the Martin Meyerson Assistant Professor of history and sociology of science at the University of Pennsylvania, the author of *Nationalizing the Body*, and coeditor of *Medical Marginality in South Asia* and *Crossing Colonial Historiographies*.

The University of Chicago Press, Chicago 60637
The University of Chicago Press, Ltd., London

Printed in the United States of America

25 24 23 22 21 20 19 18 17 16   1 2 3 4 5

ISBN-13: 978-0-226-38179-4 (cloth)
ISBN-13: 978-0-226-38313-2 (paper)
ISBN-13: 978-0-226-38182-4 (e-book)
DOI: 10.7208/chicago/9780226381824.001.0001

Library of Congress Cataloging-in-Publication Data
Names: Mukharji, Projit Bihari, author.
Title: Doctoring traditions : ayurveda, small technologies, and braided sciences / Projit Bihari Mukharji.
Description: Chicago ; London : The University of Chicago Press, 2016. | Includes bibliographical references and index.
Identifiers: LCCN 2015050980 | ISBN 9780226381794 (cloth : alk. paper) | ISBN 9780226383132 (pbk. : alk. paper) | ISBN 9780226381824 (e-book)
Subjects: LCSH: Medicine, Ayurvedic. | Medicine, Oriental.
Classification: LCC R605 .M954 2016 | DDC 615.5/38—dc23 LC record available at http://lccn.loc.gov/2015050980

♾ This paper meets the requirements of ANSI/NISO Z39.48-1992 (Permanence of Paper).

To Manjita
For coaxing the ideas, hidden
On my crowded bookshelves,
Into my hungry hands.
For watering the book, hidden
On my dank fingertips,
Into a colorful mushroom
On the computer screen.

CONTENTS

PREFATORY NOTES

On Language

This book is about therapeutic change in Ayurvedic medicine. In order to map this change accurately, I will begin by clarifying the categories of analysis. Two major categorical confusions usually cloud our understanding of Ayurvedic medicine. First, there is no stable terminological distinction between modern and traditional Ayurvedic medicine. This is in stark contrast to Chinese medicine, where most scholarly works today distinguish between the unreformed, plural medicine of the past and the increasingly codified, standardized, and modernized version, by referring to them respectively as "Chinese medicine" and "TCM" (or "Traditional Chinese Medicine"). Such a distinction is immensely useful in studies of modernization and is also reflective of the radically refigured reality of the modern version. Learning from the Chinese scholarship, I clearly and consistently designate the specific form of Ayurveda emerging from the late nineteenth century through its embrace of small technologies as "modern Ayurveda." For me, this is a distinctive category separate from Ayurveda in general or, more accurately, "classical Ayurveda."

One might object that terms such as "refigured," "neo-traditional," or "reformed" would have served as more accurate prefixes than the clichéd and variously instantiated term "modern." I concede that terms such as "neo-traditional" or "refigured" would have had greater analytic power, but these were not emic categories. None of my actors, had they miraculously lived long enough to read this book, would have recognized the term. The category "modern Ayurveda" (*adhunik Ayurveda*), however, was one they had themselves used and thus would have readily recognized. It is this historical cogency that informs my preference. Moreover, since I already recognize the plurality of forms that the term "modern" necessarily adopts, I do not feel that its use as a prefix will be misleading in any way.

The second categorical confusion goes deeper still. When referring to pre-colonial "Ayurveda," the overwhelming tendency is to reduce the entirety of Ayurvedic practice to the operationalization of a corpus of Sanskrit texts. Yet there was much that was neither Sanskritic nor even necessarily textual that went into the everyday practice of Ayurveda. Even renowned and erudite scholarly practitioners included a lot of non-textual and non-Sanskrit material in their repertoires. To add another layer of linguistic complexity, in Bengal, this form of "practiced Ayurveda" was generally referred to as "Kobiraji medicine." Its practitioners were called Kobirajes.

The very process of modernization that this book describes and analyzes also promoted the conflation between "Kobiraji medicine" and the Ayurveda derived exclusively from Sanskrit texts. It is thus analytically necessary to occasionally distinguish these different forms of medicine. In this book, I therefore refer to precolonial medicine as either "Kobiraji medicine" or simply as "Ayurvedic medicine." When I wish to refer specifically to the Sanskrit text-based medicine, however, I write of it as "classical Ayurveda."

This conflation of classical Sanskrit and vernacular Bengali also has other ramifications. One of the complex problems it presents is in the domain of transliteration. While much of the vocabulary of Kobiraji medicine (that is, practiced Ayurveda) derived from the Sanskrit texts, their meanings were often distinctive. Moreover, Bengali words—though frequently etymologically derived from Sanskrit—have distinctive meanings. The question then arises as to whether to render these in English according to the well-established Sanskrit transliteration protocols and thereby erase all historically real distinctiveness, or to treat them according to colloquial Bengali and in turn erase their connections with the larger pan-Indic systems of knowledge.

I have chosen to abide by a middle path in my choice of transliterations. I have rendered most words according to their Bengali colloquial forms. Thus, "Baidya" and not "Vaidya." But for the very well-established and core terms, I have stuck to the Sanskritic transliteration that reflects the continuities with the pan-Indic forms. Thus, "Ayurveda" and not "Ayurbed," or "Vayu" and not "Bayu." My hope is that, though this may be a somewhat inconsistent system, it will convey something of the actual historical tension that exists between the pan-Indic Sanskritic heritage and its specific Bengali inflections. This has also meant that other than in a very small number of cases that are usually indicated, I have used my own translations from Sanskrit, Bengali, and Hindi throughout the book.

On Structure

Structurally, at the core of the book are its five case studies constituting chapters 2 to 6. Each of the first four of these five chapters describes the incorporation of a specific small technology ranging from the pocket watch to organotherapy, and from the thermometer to the microscope, into Ayurvedic practice. Each of these chapters also describes how the embrace of these small technologies resulted in the weaving of particular strands of "Western" and "Indic" sciences and the production of a new body image. The fifth case study is distinctive. It focuses on the body of the physician itself as a form of technology and charts how this body-as-technology changed in the course of the modernization. At its two ends, two independent chapters bracket these case studies. Chapter 1 provides a sociohistorical map of the people who engaged with the modernization process, while the concluding chapter provides both a potted summary of the arguments of the foregoing chapters and an account of all the styles of thinking about the body that became impossible because of the rise of "modern Ayurveda." The introduction serves to situate the entire argument of the book within a broad intellectual space that intertwines the histories of medicine, science, and technology within a larger non-Western geography.

Notwithstanding the emergent transnational scientific networks that the book describes, its geographic heart lies firmly in British Bengal between approximately 1870 and 1930. Bengal was the cornerstone of British power in South Asia and was therefore deeply touched by British and European influences. It was also the home to the very first "Western" medical college in South Asia. All of these factors positioned Bengal in a unique position and it clearly took the lead in modernizing Ayurveda in the late nineteenth century.

The period after 1870 was particularly productive as it witnessed an immense printing boom that contributed directly to the developments studied in this book. In fact, one of the key texts that I engage with in this work—namely, Gopalchandra Sengupta's *Ayurveda Sarsamgraha*—appeared in 1871. Similarly, 1930 provides a useful cutoff date since after that date the impact of the Indianization of the medical and scientific establishment, inaugurated after the Great War, begins to take effect and ushers in a new level of competition from scientifically trained Indians toward Ayurveda. This, in turn, changes much of the dynamics of modern Ayurveda.

INTRODUCTION

Braiding Science: Refiguring Ayurveda

Everyone has a slightly embarrassing hobby they don't want to make public. I am no exception. My hobby is to collect old medicine bottles. I find the sparkle, the brilliance, and the stories of old glass incredibly attractive and have spent way too much time and money hoarding hundreds of old bottles I have picked up at junk shops across at least three continents. Hence, it was no surprise that four years ago, having just relocated to Philadelphia, I was inevitably attracted to a stall selling old bottles at the local flea market. As I looked through the bottles, a predictable cache of mostly nineteenth-century American and some British patent medicine bottles, I was suddenly struck by one curious bottle. What had caught my eye was that the embossed label on the glass, under the dirt and the grime, was in Bengali! I could barely believe my luck. After years of collecting and several futile attempts, I had long despaired to find a nineteenth-century Indian medicine bottle with an Indian script embossed on the glass. Even now, every time I am in India, I try to look into a few junk shops on the off chance of finding one. But, alas, India has a much more thriving culture of recycling old bottles and, as a result, I have never been able to find a nineteenth-century Indian medicine bottle with an Indian-language inscription. This was my holy grail. But what was it doing all the way across the world in Philadelphia? I asked the seller where he had found it and he said that everything he had on sale had recently been dug up by him and a friend in neighboring Pittsburgh. The site, he further mentioned, had been a landfill or rubbish heap dating from the late nineteenth century.[1] He did not think much of the bottle since, unlike the rest of his trove, he could not make out what the bottle said on its body. Having quickly purchased, brought home, and cleaned my prize possession, I discovered to my further surprise that the bottle had once contained an Ayurvedic preparation. A well-known Ayurvedic family firm called "C. K. Sen & Sons" had sold the medicine and the bottle dated from around the 1890s.

It has been a while since I had been working on the history of the modernization of Ayurveda and I knew both the firm and their history reasonably well, but the last place I had expected to find this last material trace of their once thriving business was in Pittsburgh, Pennsylvania. It was this fortuitous find that made me radically rethink the way I was approaching the history of modern Ayurveda. For the first time it dawned on me just how "transnational" the story of modern Ayurveda was. Extant historiography on the modernization of India's various indigenous traditions have done well to point out the links between these "modernized traditions" and various nationalisms.[2] But modern Ayurveda was much more than simply a nationalist project. Undoubtedly, a lot of the energy, support, and political resonances of modern Ayurveda in the colonial era derived from anticolonial nationalism. But it was also a subject fostered at the intersection of various late nineteenth-century transnational trajectories. Increasingly globalized commodity cultures, unprecedentedly large global markets, a variety of scientific universalisms disseminated through equally particularistic and parochial regimes of governance, and above all the emergent regime of an inchoate, heterogeneous and radically fissured global technomodernity, had all shaped the modernization of Ayurvedic medicine.

The bottle of Ayurvedic medicine that was used by some anonymous Pittsburgher did not simply attest to the reach of Ayurvedic entrepreneurs, but also raised other questions. Through what kind of shared or discordant frameworks did the physician-entrepreneur in Calcutta and the patient-consumer in Pittsburgh comprehend the material object—that is, the medicine, which connected them? That C. K. Sen & Sons were selling medicines in America and the dialogue that this implied, meant that the Sens in turn must have been exposed to nineteenth-century American and European science and medicine. How did this impact what they were doing? Most importantly, how did the transnational flows of knowledge, materials, and people produce modern Ayurveda?

My project, through a serendipitous encounter with an empty bottle, evolved from one about the moderization of tradition in India to one about braiding knowledges. I became keenly aware of how much of modern Ayurveda was produced through the braiding of "West" and "East" around material objects and ideas. Wanting to retain something of the sense of scale that had first surprised me about the bottle, I honed in on small, non-spectacular, and everyday technologies. Like the bottle, I thought these humble material anchors could move between places, traditions, and times more easily and yet reveal some of the most ubiquitous sites of braiding.

What Is Modern about Modern Ayurveda?

David Hardiman has rightly pointed out that no matter how radically refigured, or even how badly disfigured, we cannot dismiss modernized versions of tradition simply as "invented traditions."[3] Besides the necessary bad faith that such allegations of "invention" often carry, it is also true that nothing can be "invented" out of nothing. The invention must mobilize intellectual and material resources to its ends, and these resources in turn must come from specific sources. In the case of modern Ayurveda, these resources were gathered widely from across the world. Therapies, ideas, material objects, and expertise were brought together from the diverse flows of men, ideas, and things that straddled the late nineteenth-century globe. Thus, ironically, the more traditional and national modern Ayurveda tried to be, the more transnational and modern it became.

These are not merely stories from a dead past. Modern Ayurveda remains a vibrant and viable therapeutic alternative in India. Launched at the end of 2012, India's Twelfth Five Year Plan both raised the funding for public health and made a concerted recommendation for the mainstreaming of "traditional" medicine. The larger global market in such "traditional" medicines has already been growing for sometime now.[4] Such a growth has further underlined the need to interrogate the characteristic modernity of traditional medicine. Not only medical sociologists and anthropologists, but historians too, have increasingly been drawn to this question. Until the end of the twentieth century, the main accounts of the history of "modern Ayurveda" as opposed to longer, more classically rooted histories had been rare. In 1999, when I first encountered modern Ayurveda as an object of historical analysis as a master's student at the Jawaharlal Nehru University in New Delhi, our reading list was limited to a single monograph by Poonam Bala and a handful of essays and journal articles by a few pioneering authors such as Charles Leslie, Gananath Obeysekere, Brahmananda Gupta, and K. N. Panikkar.[5] In 2006, Kavita Sivaramakrishnan's *Old Potions, New Bottles* became the first monograph-length, purely historical study of "modern Ayurveda."[6] Two years later, Dagmar Wujastyk and Frederick Smith published an edited volume of essays exploring the "modern" and increasingly globalized histories of Ayurveda.[7] The very next year, Madhulika Banerjee published *Power, Knowledge and Medicine*, on the pharmaceuticalization of Ayurveda.[8] Four years later, 2013 witnessed the publication of Rachel Berger's *Ayurveda*

Made Modern.[9] In the meanwhile, half a dozen or so PhD theses have also been written at various universities across the world on the subject. Some of these, such as Burton Cleetus's important account of the entanglements of caste and Christianity in the construction of modern Kerala Ayurveda, make valuable contributions to our knowledge of the subject but unfortunately remain unpublished.[10] These strictly historical accounts apart—certain ethnographic works such as Jean Langford's brilliantly imaginative ethnography of contemporary Ayurvedic practice, published as *Fluent Bodies* in 2002; or Maarten Bode's ethnography of contemporary marketing practices of pharmaceutical firms dealing in "traditional" medicines, *Taking Traditional Medicine to the Market* (2008)—also cast sidelong historical glances.[11]

Pari passu with this welcome efflorescence of historical scholarship on modern Ayurveda, works on the modernized versions of other traditional medicines of South Asia have also emerged in this period. Guy Attewell, Seema Alavi, Claudia Liebeskind, and, to some extent, Markus Daechsel have written on the histories of modern Unani Tibb.[12] Similarly, E. Valentine Daniel, Richard Weiss, and Gary Hausmann have written on Siddha Medicine.[13] Most recently, Laurent Pordie has written on Indian editions of the modern Tibetan Sowa Rigpa medical tradition.[14]

Despite the robust historiography, however, the simple question of precisely what is "modern" about modernized traditional medicine, remains far from fully settled. The historiography on modern Ayurveda has focused particularly on the changes in the institutional context.[15] Changes in patronage, professionalization, clinical interaction, medical educational institutions, state regulation, and industrialization of the pharmaceutical side of medicine have all been extrapolated. As a result, David Hardiman has called modern Ayurveda "syndicated Ayurveda." The term, borrowed from Romila Thapar's analysis of neo-Hinduism, posits that "groups—or syndicates—with certain vested interests sought through combination, organization and publicity, to establish a particular, limited notion of their practice that set it apart from other forms of practice."[16] Clearly, the emphasis here is on groups, their interests, and so on. It is silent about the actual content of the medical tradition as such—that is, the medical theories and practices.

This absence is particularly striking in light of comparable work on the modernity of other traditional medicines outside South Asia. Nathan Sivin's pioneering exploration of the conceptual mutations engendered by the state-led modernization of Chinese traditional medicine remains a benchmark for such studies.[17] Recently, others like Volker Scheid and Sean Hsiang-lin Lei have further fleshed out the intellectual and practical content of the modernization of Chinese medicine.[18] Similarly a number of historians and anthropologists have recently illuminated the complex neo-traditional modernity of Tibetan medicine.[19] In Africa, particularly in East Africa, Steven Feierman wrote extensively on the actual intellectual and practical nuances of modernized healing knowledges. Later, Julie Livingston and Stacey Langwick have further fleshed out the story.[20]

Instead of exploring the historical evolution of the content of Ayurvedic medicine since its engagement with modernity, as has been done in the case of traditional medicine in China, Tibet, and Tanzania, the extant scholarship on modern Ayurveda has focused almost exclusively on the new contexts of medical practice. It might well be objected that "content" and "context" are not independent from one another. But that does not mean any engagement with the theoretical and practical content of a medical tradition becomes redundant. Simply knowing that modern Ayurvedic pedagogy moved from the feet of the guru to the college classroom, or that the modern Ayurvedic physician charged uniform fees to all patients unlike his forebears, does not tell us exactly how these things impacted the conceptual, practical, and technical repertoire of Ayurvedic therapeutics.

To understand exactly what is modern about modern Ayurveda, therefore, I will argue that it is important to go beyond the extremely useful but limited accounts of institutional change and explore what Charles Rosenberg called "therapeutic change."

Therapeutic Change

Rosenberg's classic exploration of therapeutic change in nineteenth-century America is in many ways a model for all historians of medicine writing in his wake. In the piece, Rosenberg argued that the key to understanding the therapeutic change in the nineteenth century was focusing on "a system of belief and behavior participated in by physician and laymen alike. Central to the logic of this social subsystem was a deeply held metaphor—a particular way of looking at the body and explaining both health and disease."[21] Though Rosenberg himself explicitly limited his insight to nineteenth-century American medicine,[22] his formulation actually works splendidly in analyzing non-Western medical paradigms of the nineteenth century as well. The "body metaphor," as he calls it, remains the key to understanding the "central logic" through which therapeutic change was effected.

There are two elements of Rosenberg's formulation that are worth underlining. First, that "therapeutic change" cannot be reduced to merely how "therapies" change. All too often the words *therapeutic* and *therapy* become too closely identified. The word *therapeutic*, in its adjectival form, simply means "of or relating to the healing of disease." It is not just limited to *therapies* or medicines. This is what Rosenberg's account reminds us: that "therapeutic change" is an alteration that affects every aspect of a medical tradition. Second, Rosenberg's framework underlines the mutation of the body image as the key to mapping a change that is daunting in its many-sidedness. It is only by understanding how particular ways of looking at the body shift that the many other aspects of therapeutic change can be comprehended.

Rosenberg's insistence on shift in body image implicitly recognizes the historically contingent nature of anatomical knowledge. Unfortunately, much of the historiography of Ayurveda fails to recognize this historical contingency. An obstinate and Whiggish insistence that "Western" anatomical knowledge based upon cadaveric dissections is "true" or "scientific," whereas any other anatomical imagination, such as one based on humors or humor-like flows, mars many histories of Ayurveda. Two examples will attest to this Whiggery. Poonam Bala, citing the ancient Susruta's encouragement of physicians to acquire of a structural sense of the human body by studying cadavers, asserts that "unrefined and rudimentary as this may appear to modern physicians, it indeed was a bold venture. . . . Thorwald marks this as 'the oldest lesson in dissection in human History.'"[23] Clearly the specificities of Susruta's own anatomical imagination or cadaveric practices are entirely lost in this framework. The value of the body image that informed his therapeutics can only be judged on the basis of its proximity to modern biomedical anatomy. Jayanta Bhattacharya provides an even clearer instance of this line of historical inquiry. Writing of a much-celebrated incident in 1836 when a high-caste Bengali performed a human dissection at the Calcutta Medical College, Bhattacharya declares that

> it reconstituted "psychologized" epistemology of the Indian knowledge system in the mould of objective, value-neutral, clinical detachment of modern medicine. As dissection became the primary means to know the human body, the living body was regarded as a kind of "animated corpse." . . . The act of dissection brought Calcutta on the same footing with London.[24]

Allegedly, this momentous act, besides making Calcutta and London equal, also completely "banished the divine" and replaced a "two-dimensional Ayurvedic bodily frame with a three-dimensional anatomical space."[25] Notwithstanding mutual differences, Bala and Bhattacharya clearly see cadaveric dissection as a sign or marker of higher and better knowledge. To them, clearly a therapeutic ensemble that includes an anatomical imagination attached to the practice of cadaveric dissections is to be preferred over other forms of anatomical imaginations. In fact, it is doubtful that either Bala or Bhattacharya would even allow the term *anatomical* to be deployed to describe body metaphors and imaginations that are not derived through cadaveric dissections.

Yet there is now a wealth of literature explicitly demonstrating what is implicit in Rosenberg's framework for studying therapeutic change, namely that body metaphors are always historically contingent. Shighehisa Kuriyama's classic comparative study of the ways in which the body was understood within ancient Greek and Chinese medicines is perhaps the strongest argument against the kind of Whiggery that plagues the history of Ayurveda. Kuriyama writes that "doctors in China missed much of the detail observed by Greek dissectors and incorporated invisible features that dissection could never justify. . . . Yet indifference to anatomy didn't mean a slighting of the eyes. Not at all: ancient Chinese doctors evinced great faith in visual knowledge. Like their Greek counterparts, they scrutinized the body intently. Only they somehow saw it differently."[26]

Recent works on the history of modern Western anatomical knowledge by scholars such as Ruth Richardson, Carin Berkowitz, and Michael Sappol have provided further proof for the historically contingent nature of body knowledge.[27] While Berkowitz has shown how anatomical illustrations, rather than simply serving as mirror images of a dissected corpse, were mediated by a range of artistic and technical concerns, Richardson and Sappol have drawn attention to the multiple local political investments which promoted the hegemony of dissection-based anatomical knowledge. Together what these works prove for modern "Western" anatomy is what Kuriyama showed for the ancient world—namely that anatomical knowledge is neither neutral, nor apriori universal. Its universal validity, like all other aspects of scientific knowledge, has to be constructed rather than discovered. As Nelly Oudshoorn points out, "the naturalistic reality of the body as such does not exist, it is created by scientists as the object of scientific investigation."[28]

These insights convince me that to understand "what is modern about modern Ayurveda," we must follow Rosenberg's framework for investigating therapeutic change, and do so without any Whiggery about different types of body knowledge. In studying nineteenth-century Ayurvedic practice, speaking of a single body metaphor is unsustainable. In a context where there were few institutionalized pressures toward homogenization in the form of legislation, professional oversight, or pedagogical streamlining,[29] there was much more room for the variation of body metaphors.

Moreover, as David Gordon White and Dominik Wujastyk point out, Indic traditions of body knowledge had traditionally sustained a multiplicity of body metaphors and images.[30] This plurality of body metaphors also led to many of them being only partially or inchoately worked out. They were more like transparencies laid on top of one another. Yet it is these inchoate and plural body metaphors through which therapeutic change must be accessed. I call these partial, plural, and translucent body metaphors "physiograms."

Physiograms

Despite my commitment to Rosenberg's general model of analysis, the reason I break with a single body metaphor is to highlight the diversity of and dynamism of nineteenth-century Bengali culture. The religious, sectarian, linguistic, ethnic, and intellectual diversity of nineteenth-century Calcutta where most of the developments discussed in this book took place can hardly be overstated. Kapil Raj points out that, by the closing decades of the eighteenth century, Calcutta was "an intersection of myriad heterogeneous networks, a node in the transformative circulation of knowledge which structures and organises cultural encounter and its outcome."[31] Recovering this heterogeneity and diversity is of both historical and political import. Historically, its value lies in subverting the colonial "rule of difference," while politically its value lies in busting the myths of autochthony that inspire contemporary essentialist nationalisms.

Yet the impossibility of a single, unified image of the body need not doom us to the other extreme. The antidote to too much consensual uniformity need not be a nihilistic world of utter disagreement among all. Instead of dwelling in such extremes, I suggest that it is better to adopt a modified and somewhat less consensual version of Rosenberg's body metaphor. This modified and less homogenized image of the body is what I call a physiogram. I borrow the idea of a physiogram from John Tresch's notion of cosmograms. The latter are inchoate images of the cosmos manifested in practical and material texts, objects, and practices. They are middle-level generalities in that they are not necessarily held by everyone in a historical moment. Yet they are shared and not limited necessarily to individual actors. As inchoate, they also leave enough space for being shared differently. Most importantly, however, they can be discerned from actual objects, texts, and praxes, even when they are not clearly spelled out. They are, to use Tresch's own phrase, "materialized cosmologies."[32] I redeploy this notion of cosmograms alongside Rosenberg's notion of a body metaphor to produce my physiograms, namely materialized physiologies or materialized body metaphors.

By drawing upon Tresch's usage, I also want to signal my interest in reanimating a dying dialogue between history of science and history of ideas. As Tresch points out, of course there is no question of returning to an old, internalist, and largely Platonic conception of the history of ideas. Yet, informed by the critiques that have been made of the history of ideas over the years, a new history of ideas seems to be the only way to reconnect a pluralized view of science to a larger social, cultural, and political milieu. Without such a reanimated dialogue, asking the kind of big-picture questions that Robert Kohler and James Secord have been urging us to ask would largely be impossible.[33] In my context of colonial Bengal, this engagement with a history of ideas is even more pressing, since without it much of the temporal and geographic specificity of scientific work remains invisible.

I think of these "physiograms" as entities similar to Steven Feierman's "folk anatomy." As Feierman clarified, though the people themselves may explicitly deny having any concrete knowledge of the insides of the body, one can discern in their actions a broadly shared and fairly specific understanding of the body. It was these materialized body metaphors that Feierman collected together as a "folk anatomy."[34] My physiograms are similar, except that they are culled from therapeutic discussions and practices that are themselves not as diffident as the Ghaambo to speak of the insides of the body. But I also go a step further and posit that there might be several such physiograms that can be overlaid on top of each other like transparencies without necessarily being articulated as a single "anatomical" image or a single body metaphor.

It is both curious and unfortunate that with the exception of the few Whiggish nods to the superiority of biomedical anatomical knowledge, most of the extant historiography on modern Ayurveda has been almost entirely silent about the body or body metaphors. Whatever little exists by way of critical scholarship about Indic body knowledge is either beholden to classicists such as Dominik Wujastyk or anthropologists like William Sax, Murphy Halliburton, Harish Naraindas, Helen Lambert, and Robert Desjarlais, to name but a few.[35] Of these, the most influential have been the works of Joseph Alter and Jean Langford.[36]

While this scholarship has taught me much, it has unfortunately seldom been more than cursorily engaged with the history of (Western) science. As a result, in some cases, the accounts have contrasted a vibrant, fluid, and dynamic Ayurvedic body with a caricatured and ahistorical view of the biomedical body. Langford's espousal of a contrasting "docile" biomedical body, for instance, is derived entirely from a Jungian psychoanalyst's reading of the idealized images of anatomical atlases, though she contrasts this with her own immensely rich ethnographic data for Ayurveda.[37] Yet, as Annemarie Mol and others have shown us, the biomedical body cannot be simply derived from anatomical atlases. It, too, is performatively constituted and therefore locally enacted in myriad different ways.[38]

One of my main endeavors in this book is to present both Ayurveda and biomedicine as dynamic, historical realities. While keeping my primary focus on Ayurveda, I constantly try to give a sense of contemporary developments in Western science and avoid anachronistic retrospection based on present-day sciences. By thus flitting between Ayurveda and "Western" science I also want to undermine any notion of discrete "medical systems," "epistemic fields," or "cultural logics." Following Scheid, I want to see medical systems as porous: "Rather than possessing clearly defined boundaries, medical systems are permeable to all kinds of technological and ideological influences effecting systemic change and local adaptations."[39]

Recognizing such pluralities allows me to follow Sandra Harding's lead in "rais[ing] new questions for histories, sociologies, epistemologies, and philosophies of science. How have scientific projects and traditions around the globe interacted with each other? What did each borrow? How has the West invented and maintained the notion of static, timeless, 'traditional' societies"?[40] I want to use physiograms—those incomplete and plural body metaphors—like Alter did for yoga, to "force . . . Ayurveda out of classical texts and locate its history in the body of practice, as the body of practice is a fact of everyday life."[41]

These physiograms are then a species of middle-level inchoate generalities embedded in everyday forms of practice. Langford described how Ayurveda became a "strategic sign evoked in political and cultural maneuvers" and came to be "framed as an ethnomedicine."[42] As a historian of medicine, I recognize the importance of such semiotic analysis, but my own focus is more on therapeutic change (in the larger sense of the term). I want to explore how understandings of the body changed and how such changed understandings operationalized in everyday practice and applied theory.

My project is thus admittedly somewhat descriptive. But I argue that not only is an accurate description of the nature of therapeutic change in Ayurveda sorely missing, but that a critical description is in itself a mode of analysis. For one, such descriptions, as I will show, constantly run up against the brute fact of transnationalism: of intellectual impurity. "Culture," "system," and other such closed totalities are indeed recognized as the "strategic signs" Langford said they were, but they do not overly detain me in this book. My quest is to keep my nose close to the messiness on the ground, map the shift in body knowledge through it, and eventually get a handle on my basic question: what is modern about modern Ayurveda?

Ayurvedic Technomodernity: Causes

One of the most obviously novel aspects of modern Ayurveda, both in the period I am studying and since, has been its embrace of a range of modern technologies. Discussing similar developments in the context of yoga, Alter calls these the "*materia scientia*, that is in many ways, the materialization of a technocentric modernity."[43] Pocket watches to check the pulse, thermometers to check the temperature, stethoscopes to hear the patient breath, microscopes through which germs and blood cells could be seen, injections of endocrinal extracts to rejuvenate the patients, and even glass phials (rather than carpet bags full of crushed herbs) in which medicines were stored are some of the mundane small technologies that made up the *materia scientia* of modern Ayurveda in the late nineteenth and early twentieth centuries.

Unlike the imported X-Ray machines and complex laboratory setups that Alter studies in modern yoga, the technomodernity I explore here is humdrum, mundane, and nonspectacular. But it is precisely their everydayness that makes them so potent a motor of change. These small technologies and the technomodernity they engender interest me not simply because they were so conspicuously new in Kobiraji practice in the late nineteenth century. They are of interest also because they in turn became the catalysts for further change.

It would be cogent to clarify what I mean by calling these small technologies "motors" or "catalysts" of change. Words such as "motor" or "catalyst" undoubtedly have a causal ring to them. Yet it is worth noting that there are many types of causes. The ancient Greeks, for instance, recognized four distinct types of causes: *causa materialis*, *causa formalis*, *causa finalis*, and *causa efficiens*. Heidegger explains that in the ancient Greek world, each of these causes were thought of as a way of being responsible and "the principal characteristic of being responsible is this starting something on its way to arrival. It is in a sense of such starting something new on its way into arrival that being responsible is an occasioning or inducing to go forward."[44] This older, classical usage, Heidegger points out, is much more expansive than the usually popular, but mistaken, reduction of the notion of "cause" to that of the "maker" or "creator"—that is, *causa efficiens*.

Modern notions of causality have been further reduced to fit an image of causality derived from (classical) physics. Thus, causes are thought to be always preceding the effects in time, exerting a direct "pushing force" and of producing determinate results that aid future prediction. But none of these characteristics are particularly necessary in the Greek notions of causality that Heidegger sought to recover.

Heidegger's efforts are particularly cogent for historical scholarship today. With the demise of economic determinism and the mounting critiques of social history/determinism, the straightforward cause-effect explanatory models found in much of the extant scholarship have become unsustainable. Yet, at the same time, we are still haunted by the "why" question. Frequently we hear of research described (and usually dismissed) as "too descriptive," "not sufficiently analytic," etc. The problem here, I argue, is not really with analytic failure, but an overly rigid and unworkable notion of causality.

While Heidegger's resurrection of the more expansive original notion of Greek causality is extremely useful in inspiring us to rethink a more capacious, less physics-driven notion of historical causality, I argue that we need to go further still and engage with locally relevant models of causality as well.

Bimal Krishna Matilal points out that causality was extensively debated within Indic philosophical traditions throughout history and right up to the modern period. These debates only partially mapped onto the fourfold Aristotelian notions of causality. The Navyanyaya school that was dominant in early modern and modern Bengal clearly rejected any notion of causality as a "force" or "power" (*shokti*). Instead, the Navyanyaya position partially coincides with David Hume's view of causality as constant conjunction of two events. What Navyanyaya philosophers added to this notion of constant conjunction was a notion of "not-otherwise-determined-ness" (*ononyothasiddhotwo*), as well as an insistence that the constant conjunction was not merely "perceived to be" (as Hume would have it) but a real conjunction.[45]

Small technologies thus were motors of therapeutic change in modern Ayurveda not because they exerted some direct physical force on body metaphors, but because they were constantly present and "occasioned or induced a going forth" that cannot be otherwise fully explained. It is worth clarifying here that Navyanyaya notions of causality are explicitly plural and resists the determination of single causes. Thus, to assert that small technologies were a cause of therapeutic change does not rule out other possible causes of the same change as well. The rich cornucopia of more "externalist" causes of change, such as emergence of multiple public spheres, nationalisms, and mutations in patronage structure, described by previous authors such as Kavita Sivaramakrishnan, are not at all excluded by the small technologies.

Alongside the issue of multiple causes arises another problem—namely that of the relationship between the cause of the cause and the effect. In the present context, this is an extremely tricky issue. Even while accepting my argument that small technologies caused therapeutic change, readers might wonder why Kobirajes adopted small technologies in the first place. I answer this question at some length in chapter 1, but, put briefly, my answer is inspired by the recent of insights of scholars like Douglas Haynes and others who argue that class identities in South Asia were premised on patterns of consumption.[46] Sanjay Joshi goes further and argues that the South Asian middle classes not only self-defined their identity through consumption practices, but also took the lead in framing local modernities for all classes.[47] In fact, Haynes has also pointed out that technological objects formed an important part of South Asian middle-class consumption patterns.[48] My argument then is simply that the Kobirajes, as members of the colonial middle class, were engaged in class-identity formation by consuming small technologies.

These discussions of middle-class identity and its role in framing local modernities naturally also raise fundamental questions about the nature of modernity itself. Following the work of Partha Chatterjee, Sudipta Kaviraj, and others, the multiple modernities thesis has come to be widely accepted among South Asianists.[49] There is unfortunately less acceptance or even awareness of the plurality and multiplicity of modernities in medical history circles. When such awareness is there, it tends to be "misrecognised, in a recuperation of sovereignty, as a novel process of laminar global flow."[50] Given its enormous variability, some have even begun to wonder whether the notion of "modernity" (or the even more chimeric "early modernity") any longer has any heuristic value.[51] Lynn Hunt has baldly stated that "nothing is gained" by deploying the label "modernity" and it has become redundant to historical analysis.[52] While recognizing the heuristic chaos to which historians such as Hunt are responding, I have still retained the notion of "modernity" in this book—albeit in its multiple, shifting, and inchoate sense. My reason for its retention is simple. The notion of "modernity" or *adhunikota* had tremendous cache for my historical actors. They ranted against it, they embraced its possibilities, they argued over its true nature, and sometimes even sought to undermine it by claiming that the truths of Ayurveda were eternal and transmodern, but throughout all this they always spoke of it. Whatever their positions toward it, they recognized it. Late-nineteenth century Bengal, for me, was certainly a "modern" time.

This obviously did not mean everyone was equally modern or seen as such by either themselves or their peers. Religion, caste, gender, urbanity, the ability to consume commodities (including books and other modernist paraphernalia), and much else all combined to make Bengali modernity patchy, partial, and selective. Nor did it ever mean that those who embraced modernity were able to actually engender the kind of total rupture certain extreme versions of modernity seemed to call for. Senses of selfhood, which have been given a pride of place in many historical debates over modernity, too, seemed interrupted and differentiated. Yet, beneath and beyond all this plurality, incompleteness, contradiction and failure, *adhunikota* was a term that seemed meaningful to everyone. Even though those who opposed it took up a political position against it, they did not dismiss it as redundant.

Though I elaborate upon these arguments further in chapter 1, within the larger scheme of the book, I see these issues of modernity and consumption as a "cause of the cause" and, in the tradition of the great Nyaya philosopher Gangesa, choose to bracket it off from the more putative discussion of the immediate cause: the small technologies. This is because I do not want to lose sight of my central question of what is modern about modern Ayurveda. Ayurveda's technomodernity is unquestionably new. But what is it really constituted of? What was the therapeutic (again in Rosenberg's expanded sense) consequence of the presence of these small technologies?

I think of the small technologies as the proverbial speck of sand on the back of the oyster. They instigated and induced new intellectual and practical ferments. These ferments crystallized into the new physiograms that made up modern Ayurveda. Small technologies were the causes of therapeutic change that brought about modern Ayurveda, in the same sense as the speck of sand is responsible for the birth of the pearl.

Non-Biomedical Technomodernity

Historians of medicine working on a variety of contexts have recognized the inauguration of medical technomodernity at the turn of the nineteenth century. Stanley Joel Reiser states that the "physician has become the prototype of the technological man."[53] Joel Howell's study of American medicine connects this technomodernity to the rise of hospital.[54] Margarete Sandelowski has related biomedical technomodernity to the professionalization of nursing.[55] In contrast to these stories of centralized organization, Graham Mooney points out that, in the case of TB patients, a techno-material culture focused on the domestic space and personalized consumption.[56]

All the extant work, however, has been focused on biomedical technomodernity. Anthropologists have alerted us to the widespread uptake of medical technologies in non-biomedical settings.[57] But with the exception of Alter, very few scholars have looked at these non-biomedical technomodernities historically. In the case of the early Ayurvedic technomodernity that I am interrogating in this book, when compared to the extant scholarship on biomedical technomodernities, there are interesting points of departure. For instance, Ayurvedic hospitals did not emerge till the mid-1910s and did not become popular till much later. Hence early Ayurvedic technomodernity did not have the same relationship with hospitals that Howell has found in America. Likewise, Kobirajes for the most part did not use professional nurses in the period under study. Finally, while elements of Ayurvedic technomodernity did target domestic spaces, the technologies I focus on here were targeted mainly at the physicians rather than the patients.

Ayurvedic technomodernity was fundamentally dependent upon the individual choices made by individual Kobirajes. In a presidential address delivered at the first annual meeting of the All India Ayurvedic Physician's Conference in 1916, Jaminibhushan Ray, founder of one of the earliest Ayurvedic hospitals in Calcutta, asserted that it was the duty of every individual Kobiraj to look favorably upon certain key modern technologies. Among these, he included the stethoscope, the microscope, injections, X-Ray machines, Oxygen Inhalers, alongside two specific medicines—Emitin and Diptheria Antitoxin—and the Colli Vaccine.[58] There were clearly no hospital or professional nursing association to support the growth of Ayurvedic technomodernity. Modernizers like Ray had to appeal directly to the individual Kobiraj.

Ayurvedic technomodernity, like any other technomodernity, cannot be reduced simply to the technological objects. No doubt these were the most conspicuous elements of it, but as Marie-Noelle Bourgeut and others point out, "Instruments are not to be solely identified with wood, metal, or glass machines. A human body whose walking pace and perceptual skills have been trained and disciplined is also functioning as an instrument."[59] As I will show in the penultimate chapter of this book, Ayurvedic technomodernity radically refigured the way the body of the physician itself was instrumentalized for therapeutic purposes.

These general features of technomodernity receive an interesting twist when implicated within an Ayurvedic milieu. Most extant work on the circulation of instruments have tended to see them as part of the elaboration of a singular "Western" science. For instance, Bourguet et al. point out that "instruments and experimental apparatus retain their meaning if, and only if, they can be displaced or replicated" and therefore despite all variations that global circulation subjects instruments too, "what [remains] at stake is the construction of a shared experience, a common knowledge, forms of intersubjectivity, trust and consensus."[60] Yet non-biomedical medicines—be it Ayurveda, Unani, Sowa Rigpa, TCM, etc.—were constantly under pressure to foster precisely the opposite—namely, shared experience, common knowledge, forms of intersubjectivity, trust, and consensus.

Ayurvedic technomodernity is fabricated in the cauldron of these two distinctly opposed propensities. On the one hand, global medical instrumental culture sought to engender a shared, singular scientific modernity; on the other, non-biomedical medical traditions were predominantly interested in fabricating a distinctive modernity. The concatenation of these alternate and mutually opposing forces clearly shows us that facile distinctions between the "global" and the "local" are inadequate for making sense of Ayurvedic technomodernity.[61]

The rejection of the binarism between the "global" and the "local" does not, and indeed cannot, mean the erasure of all difference and the fabrication of a singular scientific modernity. Following Alter, I will argue that we need to recognize the many post-Western universalisms that have emerged in the colonial and postcolonial contexts. Ayurvedic technomodernity was resolutely not "local"—neither in its constitution nor in its aspirations and theoretical claims. Yet it was global in a way that off-staged the centrality of the "West." It did so not by rejecting the "West" entirely, but by provincializing it.

While this book is certainly limited to exploring Ayurvedic technomodernity, the problem of how the planetary circulation of instruments interacts with multiple post-Western globalisms is a theme with much wider resonances. I am confident that the insights gleaned from this study have the potential to illuminate the processes at work in non-biomedical technomodernities elsewhere as well.

Ayurvedic Technomodernity: Agency and Difference

David Arnold's fascinating and brilliant new book, *Everyday Technologies*, which remains an inspiration for the present work points out that, "since the late nineteenth century in particular, our ideas of time, space, of body, self, and 'other,' have been profoundly transformed by technological innovation and by the incorporation of new and ever-changing technologies into our daily existence." Yet what is crucial is that "the functions and meanings assigned to modern technology are not everywhere the same. Identical technologies can take on vastly different meanings between one society and another, even when that technology shares a single point of origin and its physical form remains fundamentally the same."[62] Negotiating these opposing pulls was not easy. "Much of the scholarship on the history of technology in colonial India," writes Gyan Prakash, "has focused on large projects—the textile industry, railways, telegraphs, bridges, dams, and modern manufacturing."[63] There was thus no fixed rubric within which to study the circulation of small technologies.

Arnold crafted a framework to study such everyday technologies by drawing upon a diverse array of approaches. He drew upon the Social Construction of Technology (SCOT) approach developed by Wiebe Bijker and others, which sought to align sociopolitical interests with internal histories of technological design, the "material culture" approach of Frank Dikotter, the user orientation of David Edgerton, Ruth Schwartz Cowan's exploration of the gendered patterns of technology use, and, to a lesser extent, Dipesh Chakrabarty's critiques of the singular histories of capital.[64] Combining these diverse interests and analytic frameworks, Arnold constitutes a unique analytic framework that is attentive to cultural and political resonances of technology-in-use. Jettisoning the dissemination paradigm, he looks anew at the creativity and ingenuity involved in adapting technological objects, not just in terms of design and operation, but also culturally and socially to successfully embed them in a new context. Such an approach, while fully teasing out the analytic possibilities and extending the transformative aspects of the extant approaches to technology, also serves to sober us in our newfound enthusiasm for the "agency of the non-human."

Bijker, drawing on Langdon Winner,[65] has pointed out that the basic starting point of his analysis is "that things could be otherwise"—namely, that at every stage of a technology's development and uptake there are a plethora of choices and these could be determined in different ways.[66] The final outcome is therefore never a matter of teleological determination. Arnold's emphasis on technology-in-use extends this further to show how, even after basic design elements have been stabilized, a number of cultural, social, economic, and political choices continue to shape the trajectory of the technological objects.

Such choices also demonstrate that an overemphasis on the newly fashionable creed of the "agency of the non-human" can in fact lead to the obscuring of the ingenuity and range of choices open to the users and adapters of technology. The insistence on the in-built affordances of a technological object naturally serves to limit the range of uses and meanings that might be derived from it by adapters. This in turn would then once again restage the superiority of the act of invention/innovation and through it the global hierarchy of the "West" over the "rest." Aditya Bharadwaj, referring to works of Bruno Latour, Donna Harraway, Marilyn Strathern, and Paul Rabinow, points out that

> the insights of authors writing on the mutual influences of natural and cultural domains have enriched the social sciences conceptually and methodologically. Yet they privilege the existence of one among many other possible cultural biographies of human biology. . . . The very act of demolishing the hegemonic formulations of nature/culture oppositions in the Euro-American worldview has unwittingly led to the "anthropologization" of an equally dominant model of the biological and the social that continually bleed into each other.[67]

This is not to dismiss the "agency of the non-human" altogether. Far from it. Of course, all human ingenuity operates within historically imposed limiting values, including material limitations appropriate to a particular historical context. Yet, at the end of the day, the meanings, uses, and politics of a technological object, so far as I am concerned, are still decided by a combination of creative thought and embodied practice, both of which in turn are historically and culturally conditioned.

In fact, there are other efforts to render the history of science more materially sensitive without undermining the culturally fashioned distinctive histories of technologies-in-use. The efforts to investigate the "mindful hand"—a conceptual framework for interrogating culturally specific embodied practices that blur the line between use and thought—is one possible way of accessing technologies-in-use in ways that are neither purely ideational nor wholly culture-blind. Lissa Roberts describes the "mindful hand" as the "embodied locus of a practitioner's competence."[68] Such embodied locations of competence literally create the material world through distinctive ways of seeing, feeling, understanding, and acting upon it. The nonhuman world does not exist prior to such "cunning ways of world making" by the "mindful hand." As a result, Ian Inkster points out "the purported distinction between the world of the hand and the world of the mind is a social construct of the clerical masters of the early modern Western world."[69] Another related approach that also tries to render the history of science less idealized (pun wholly intended) without erasing all distinctions between human and nonhuman agents is the "mangle of practice" described by Andrew Pickering. Pickering confessed that he was "forced to regard [the intentional structure of human agency] as differing from non-human agency in its temporal structure, through its orientation to goals located in the future," yet at the same time he insisted, following Giles Deleuze and Felix Guattari, that such goals "be seen in the plane of practice . . . rather than as controlling practice from without."[70]

In the case of Ayurvedic technomodernity, the meanings and identities that both the technological objects themselves and the data they produced came to acquire were highly historically contingent. They were shaped both by the bifocal ideational resources of the colonized bourgeois (of which more in chapter 1) and their embodied forms of practice. In the subsequent chapters, it is those contingent uses, meanings, and identities that I recover. It is through those peculiar and particular meanings and identities that we can begin to map the shifting forms of the body metaphor that underwrote modern Ayurvedic therapeutics. These identities and meanings were not at all predictable or derivable merely from the nature of the objects themselves.

In fact, the very genealogies or "histories" of these technologies mutated in the course of their reimagination and redeployment. These small-scale technologies often acquired new genealogies or pasts. It was part of the process of reframing their identities. Such new pasts frequently sought to indigenize these technologies by locating local antecedents. Chapter 4, for instance, attempts to portray the enhanced vision of the microscope as an inheritor of the supervision of ancient seers, while chapter 3 attempts to identify thermometric data with the kind of information premodern Kobirajes were able to obtain via the pulse. Such claims might well be seen as such pure "inventions." But to reject such genealogies would miss the point behind their invention. We would end up insisting on a "Western" genealogy of the technological modernity that by the nineteenth century was increasingly global. Instead, we must see in these alternate genealogies creative efforts to embed the technologies in locally relevant cultural practices, social, and bodily habits, recognizable uses and familiar cultural memories. Such embedding naturally also meant aligning available tools, techniques, practices, and descriptions with incoming material objects and practices. Eventually it was the success of such embedding and alignment practices that allowed the technologies to be creatively redeployed to new functions and acquire new meanings. In a sense, therefore, these new genealogies were real insofar as they made new and unforeseen futures possible for these technologies. The eyes of ancient seers were indeed endowed with microscopic power, because it was this that allowed the assimilation of modern microscopic regimes of vision to be assimilated into modern Ayurveda.

This is not to suggest that the ingenuity and creativity of usage seen among Ayurvedists were entirely unlimited. It was limited by the culturally and historically conditioned limits of the embodied capacities of their "mindful hands" and the material culture in which they were immersed. It was also to some extent guided, if not limited, by the nature of the objects themselves. But there was sufficient freedom of interpretation and imagination to make the final identities and meanings of assimilated technological objects unpredictable and open-ended.

What made these identities and meanings even more unpredictable were the unstable and heterogeneous traditions of knowledge to which the early developers of modern Ayurveda had access. As Kapil Raj points out, colonialism in South Asia created in Calcutta fertile "contact zones" where multiple peoples and knowledges jostled for space and domination.[71] In these "contact zones" the possibilities of imaginative redeployments of small-scale technologies were increased manifold. It was not simply a matter of re-creating the identity of an object passing from one monolithic culture to another equally monolithic one. It was rather reimagining the possibilities of a technological object at the interface of myriad cultural, intellectual, and practical traditions.

Hence, while these small-scale medical technologies initiate therapeutic change and the birth of modern Ayurveda through the mutation of the underlying body metaphor, I do not claim for them their own "agency." I am happy to see agency rest with human beings, that is, the Ayurvedists, all the while recognizing that human beings exercise agency within historically defined cultural, social, political, and material constraints. Ayurvedists embraced and redefined these small-scale medical technologies for their own ends and by their own wits used their historically situated mindful hands; but once the objects had thus been redeployed, they also required other changes.

Para-Sciences and Braided Knowledge

Recent advances in the history of science have evinced a new and welcome sensitivity to the non-Western contributions toward the emergence of contemporary global techno-modernities. Three distinctive trajectories have engendered this new sensitivity: first, the critiques and visible inadequacies of the old dispersal models of science championed by George Basalla and others;[72] second, the increasing global presence of non-Western doctors, scientists, and technicians;[73] and third, the demise of epistemic and ontological justifications for holding on to any single, coherent, and homogenous definition of "science" in the singular.[74] As we learn to accept a wide variety of scientific rationalities and acclimate to the ubiquitous presence of the non-white scientist/technician even in the minority world, there has emerged an all too obvious need to understand the historical processes through which we arrived at the present globalized yet heterogeneous moment of technoscientific modernity.

Alongside this new sensitivity has also arisen the related issue of how localized, embodied, and embedded practices were able to generate and sustain the universal claims that are the bases of the sciences. Bruno Latour's work has been immensely influential in framing the discourse around these questions. He has in fact created an entirely new vocabulary to talk of such matters. The concepts of "immutable mobiles" that "circulate" across networks of actors and are calibrated in "centres of calculation" usually based in the "West," have all now become familiar to historians of science, as indeed to many others.[75]

What Latour's actor-network theory has obscured, however, is the place of radical alterity in general and alternate rationalities in particular.[76] The "immutable mobiles" are extracted from embedded alternate rationalities, circulated, calculated, and thereby re-embedded into an emergent set of practices seen to constitute "Western science." But there is no account of what the Sakhalin Islander, whose map forms Latour's classic example of the "immutable mobile," made of Laperouse and his practices. Did the encounter only enrich Laperouse or did it leave some trace on the Islanders as well? Did the Islander take on any of the European practices? If so, how did they adapt and adopt it? Did they creolize European mapmaking in turn? To use Helen Tilley's pithy formulation, the main inquiry here is focused on "just how 'Western' Western science really is."[77]

At the heart of these questions, however, is an elision: the elision of any interest in the traditions of non-Western knowledge. The interrogation of just how "Western" "Western science" is covers up the silence around a more fundamental question: does the Sakhalin Islander's traditions of knowledge have any place in histories of science? Should we, as historians of science, be interested in the Islander's science beyond the limited role it played within a larger apparatus of "Western" science?

Even today, when the din arising from a multitude of paeans to the "global" have reached almost a point of distraction, history of science as a field remains dominated by Laperouse's side of the story. It is true that there is increasing attention to what he absorbed from the Sakhalin Islander, but the Islander's forms of knowledge themselves have no room in the histories of science. The non-West exists mainly as "practice" or "intelligence" within newly globalized histories of science. Not as para-rationalities. Not as coherent forms of thinking about and acting in their own versions of "nature." Even the limited, and in itself in some ways problematic, rubric of "ethnoscience" has hardly had much play within the larger field of history of science as such. It is indeed unfortunate that nearly a decade after Robert Kohler had aired his hopes in the pages of *Isis* that the study of "ethnoscience" would illuminate aspects of what he—turning the Otherizing logic of the prefix "ethno" upon itself—had called "our own ethnoscience," the number of articles engaging with non-Western "science" (as opposed to "practice" or "intelligence") in that very eminent journal is less than ten.[78]

It may well be opposed that all forms or traditions of knowledge do not constitute "science." Though such a position seems increasingly untenable as scholars become more and more suspicious of there being any single or fully coherent "scientific rationality," yet, even if we hypothetically admit the validity of such a position, a determination on this could only be legitimately made after the careful study of non-Western intellectual traditions of "natural" knowledge. Not before that. Particularly throughout Asia—from West to East and South Asia—there are numerous traditions of textualized, formalized abstract knowledge that demonstrate most of the generic characteristics of "science" sought out in Europe, and yet seldom are these studied independently by historians of modern science; rather, they are usually left to classicists, linguists or area studies scholars. Book series and journals devoted to the history of science, even while vigorously pursuing a globalizing agenda, therefore remain largely inoculated against para-sciences or their para-rationalities.

In the South Asian context, classicists have given us enough evidence of the existence of a rich seam of knowledge that was undoubtedly scientific and that remained vibrant up to the eve of colonialism. David Pingree's and, more recently, the work of Kim Plofker on mathematics, Sreeramula Rajeswara Sarma's work on astronomy, Jim McHugh's fascinating explorations of scents that touch upon the histories of chemistry, David Gordon White's unorthodox reading of the history of Indian alchemy, Gerrit Jan Meulenbeld's enormous corpus of writings on Indian medicine, and, recently, Daud Ali's exploration of medieval automata have clearly evinced the unquestionable existence of rich and robust bodies of scientific and technical knowledge in precolonial India.[79] It is but natural to assume that some of this survived in some form or the other into colonial times. Yet, there has been little interest among historians of modern science to delve into these. As Dominik Wujastyk, speaking particularly of medicine, points out,

> European colonialism established itself decisively in the Indian subcontinent in the period from 1770 to 1830. This period, and the century following it, have in recent years become the subject of much creative and insightful work by medical historians working on the colonial period. . . . [But] there is . . . a historiographical gap concerning the period preceding that studied by colonial historians. Little attention has thus far been paid to what was taking place in Indian scholarly—including medical—circles immediately before the colonial period. The lacuna is a particularly striking one because the two centuries from 1550 to 1750, just preceding the European colonial establishment, constitute a strikingly creative era in Sanskrit intellectual history.[80]

As a result, the best of the global studies have moved somewhat away from knowledge traditions as such and focused instead on the *people* doing science. They have sought out "brokers" who moved between communities and their traditions of knowledge thereby fashioning "global intelligence." These go-betweens have thus emerged as the figures who reformatted and translated parallel rationalities.[81] Such a move has been successful in gaining some space within history of science for non-Western knowledges through the rehabilitation of the biographies of these translators.

Laudable as this exercise is, there remains one drawback. The majority of studies look at how non-Western knowledge is transformed into globalized intelligence useful to the "West." They do not seek to explore how "Western" scientific intelligence might have conversely fitted into other, non-Western traditions of natural knowledge. The only serious exception to this trend is Marwa Elshakry's fascinating new account of the fate of Darwinism in Arabic.[82] Similar works in South Asia remain rare. One partial exception pertaining to South Asia is Simon Schaffer's brilliant and sensitive study of Tafazzul Hussain Khan's Arabic translation of Newton's *Principia Mathematica*. At a time when most historical scholarship is focused on how "intelligence" traveled westward, Schaffer artfully asks, "What happened to some versions of Newtonian sciences in Calcutta in the last decades of the eighteenth century?" Schaffer is interested in how "go-betweens worked indispensably within and across learned cultures."[83] Unfortunately, however, this emphasis on the figure of the go-between prevents Schaffer from following through with his earlier query of "what happened to Newtonian sciences in Calcutta." Yet, despite the lack of follow-through, (to use a Cricketing expression), Schaffer's brief tryst with the Shi'ite appropriation of Newton shows us the rich rewards awaiting anyone willing to systematically pursue the "non-Western" scientific traditions through their tryst with "Western" sciences.

It is from Schaffer's model that I draw inspiration. But for me the go-betweens are the small medical technologies. It is these that cross over and simultaneously remain within learned traditions. It is around these that their users are forced to develop new forms of knowledge. But I also want to modify Schaffer's model in another way. He avoids rigorously defining the entities he calls "learned cultures." These cultures, regardless of how we might affiliate them—"Western" or "Islamic" or "Indic" or "Bengali"—are always already internally variegated. Every "knowledge tradition" or "learned culture"—two phrases I use interchangeably—is already internally plural. In the case of Bengali Ayurveda, I argue such pluralities were often defined with reference to creedal filiations with different streams of religious piety such as Tantric, Shaiva, and Vaishnava. Some of these internal differences could even be—and indeed frequently were—extremely antagonistic and mutually irreconcilable.

By thus conceptualizing learned traditions as always already internally pluralized and seeing the technological objects in the figure of the go-between, I see how different strands of knowledge drawn from different "learned cultures" were braided around the technological objects. By thus refusing to call the people doing the braiding, namely the Ayurvedic scholars, go-betweens, I avoid any implication that any of the knowledge traditions were already clearly formed. Instead, I signal toward a situation wherein the Ayurvedic physicians of colonial Bengal could and did access multiple strands of knowledge from within multiple learned traditions. What led them to draw together particular strands and braid them together were the new technological objects that had come to populate the material culture of their everyday lives.

My thinking here is shaped to some extent by Nicholas Thomas's work on the material culture of the colonial Pacific.[84] Thomas uses the notion of "entanglements" to emphasize the polysemic, mutable, and historically situated use of material objects. But rather than focus on systems of exchange and value as Thomas does, I trace the particular strands of broadly Ayurvedic and broadly "Western" scientific knowledges, that were pulled together and braided around the technological objects.

The braided knowledge framework is reminiscent of Volker Scheid's fascinating exploration of modern Chinese medicine.[85] But in focusing on objects through which to access medical knowledge and practice, it also draws on the works of Nancy Hunt and Stacey Langwick in Africa.[86] From Hunt, I draw her aspiration for a "science of the concrete," while, from Langwick, I draw her openness to the radically historicizable understanding of "materiality" itself.

The most significant difference between Hunt's and, to some extent Langwick's, historical situation and mine, however, resides in the social identities of the people doing the transcoding. For Hunt, these are the classic colonial "middles." Trapped between colonial elites and the "lows" or *basenji* ("uncivilized"), the "middles" for Hunt were "leading translators . . . the central midwives of colonial mutations." She described their identities as "hybrid." Defending the term against postcolonial critiques of essentialist visions of biologically pure parents being written into the word, Hunt argued that metaphorically such purity and intermixture was exactly what the "middles" stood for.[87] The colonial Kobirajes who undertook to refigure Ayurveda in the late nineteenth and early twentieth centuries were not "middles."

For starters, they did not putatively work within the colonial medical or bureaucratic apparatus. They were all in independent practice, and direct governmental regulation of Ayurveda was minimal. In their own society, they hailed from one of the highest castes and were usually firmly middle class. In chapter 1, I give a detailed picture of their social and cultural milieus. Their identity, I argue, is best described as a hyphenated Baidya-Bourgeois one. While the Baidya caste identity came with social status as well as access to long intellectual traditions, the bourgeois identity came with its own status, investment in coding modernity, and a shared culture of consuming modern artifacts. These men were firmly upper-middle class and though they performed the task of transcoding just as the "middles" did, there were important distinctions.

Their world was a world of double-barrelled elites accessing both "Eastern" and "Western" intellectual and material traditions with a degree of autonomy that was much more constrained for the "middles." Even more importantly, for these Ayurvedists, many of the intellectual lineages and textual sources of tradition were still putatively accessible. These learned traditions, still accessible through unbroken lineages of intellectual guruship in the late nineteenth century, constituted what is best described as para-sciences. They were systems of codified, abstract systems of thought and practice that exhibited most of the formalized characteristics of the so-called sciences, and yet developed along a largely parallel and independent pathway. As a result, the Ayurvedic scholars who accessed these traditions developed a full and rigorous understanding of these traditions and their esoteric vocabularies. Language, for them, had not been reduced to mere lexicon. When they learned new "Western" sciences, they added further linguistic competencies to their existing competence and therefore enriched their linguistic and practical repertoires.

None of this is to suggest that such intellectual continuity was to be seen everywhere in South Asia. In any walk of South Asian life, including in the realm of *daktari* medicine, traditional intellectual lineages had indeed been undermined and language transformed to rough-and-ready lexicons of colonial rule. Employed in the colonial medical establishment, *daktar*s were akin to "middles"—though perhaps enjoying slightly higher status owing to the earlier introduction of full-fledged medical degrees in South Asia. In my previous book, *Nationalizing the Body*, I have explored the history of these *daktars* and the vernacularized form of "Western" medicine they practiced. But the world of Ayurveda was different. Here intellectual traditions were much more alive. The Baidya-Bourgeoisie who transcoded the material culture of modernity did so with far greater autonomy and with access to much more coherent sources of tradition. These sources had not yet degenerated into mere "remains and debris."

This is why, despite the shared interest in "recovering the science of the concrete" by focusing on "entangled objects," instead of the linguistic metaphor used by Hunt, I find the metaphor of braiding much more useful in my historical situation. If we think of cultures as spools of numerous diverse and different threads, we might envisage their interaction as the braiding of certain threads taken from different reams. My inspiration for the metaphor comes from Ranajit Gutha's and Gautam Bhadra's discussions of the "braiding" of elite and subaltern politics, but I draw the specifics of this metaphor from a combination of two sources: first, a brief introduction by the art historian Monica Juneja; and second, a programmatic statement on *histoire croisee* by Michael Werner and Benedicte Zimmermann.

Juneja struggles with many of the same challenges I have outlined above, though she does so from within the disciplinary world of art history. In speaking of "braided histories," a notion that she unfortunately does not explicitly define, she refers to both the problem of totalizing notions of culture and its indispensability, while also noting its historically contingent materialities. She proposes that the ideas of "transculture" and "transculturation" are useful in bringing out the "cultural braidedness." By transculturation—a notion borrowed originally from the Cuban anthropologist Fernando Ortiz—Juneja signals a view of culture that forestages "contact, interaction and entanglement" without reducing all of culture to simply regimes of "spatial mobility, circulation and flows."[88] Expressly rejecting the narratives of biological purity written into the notion of the hybrid, Juneja describes the transcultural as "something always in process and not a thing with an essence."[89]

As my second source of inspiration for the metaphor of braiding, Werner and Zimmermann advocate the use of "crossing" as the figure through which to think of historical configurations where objects and ideas are in motion. They cite four specific advantages in favor of such thinking. First, "the notion of intersection precludes reasoning in terms of individual entities, considered exclusively in themselves, with no external reference point." There is, hence, no chance of once again regressing to the "pure" parental entities that "hybrids" evoke. Everything is always already crossed and relational, including objects. Second, "relational configurations and active principles also requires paying particular attention to the consequences of intercrossing." Such histories do not stop, as a lot of studies tracing the global circulation of objects do, merely by tracing the movement of an object or idea. It has to follow through and interrogate the consequences of such inter-crossing. Third, "to cross is also to crisscross, to interweave, that is, to cross over several times at a tempo that may be staggered." There is, thus, no original departure point and no final destination. Multiple movements constantly braid the objects and ideas into inchoate and forever emerging regimes of value and meaning. Finally, "the entities, persons, practices, or objects that are intertwined with, or affected by, the crossing process do not necessarily remain intact and identical in form."[90]

"Entangled objects" are therefore the figure of crossing through which we might glimpse the cultural interactions of Ayurveda and "Western" science via a transcultural perspective. Neither Ayurveda nor "Western" science are static or fully defined entities in this crossing. They are multi-stranded spools selectively braided together by the presence of the "entangled objects." But this braiding in turn also transforms the objects themselves, giving them new meanings and transforming the relations between the two cultural entities being crossed.

The "entangled objects" themselves do not have agency as such. But they are what Christopher Pinney describes as zones of "affective intensity," or what Jane Bennett calls "vibrant matter"—that is, they have the capacity to pull together and cross cultures in ways that give rise to new identities and figures.[91] The "entangled objects" indeed as spindles—perhaps pearly ones (apropos my earlier figure of the speck of sand on the oyster's back)—pull rich threads from different spools and braid them into new patterns.

Science or Medicine?

In a now familiar public spat between two of the founding fathers of history of science and history of medicine in the mid-1930s, George Sarton mocked historians of medicine who proceeded without knowing the history of science as staging *Hamlet* without the eponymous Prince of Denmark. Henry Sigerist retorted that what Sarton imagined to be *Hamlet* was really *Othello* and that his suggestion to include Hamlet in it was absurd.[92] Where would modern Ayurveda fit into this debate between history of science and history of medicine?

While the ancient history of Ayurveda has been more ambivalently addressed, the historiography on modern Ayurveda has clearly chosen to position itself within the history of medicine, rather than the history of science. Yet, numerous popular and academic works continue to resuscitate the nineteenth-century translation of Ayurveda as the "Science of Life." This nomenclature, whatever its linguistic merits, was a moniker adopted as a self-designation by many of the founders of modern Ayurveda such as the redoubtable Bhagavat Singhjee.[93] Why were they so interested in calling Ayurveda a "science" rather than "medicine"? And how did that choice functionally play out?

It is relatively easy to answer the first question. Gyan Prakash has enlightened us about how colonial ideology invested power in the very sign of science.[94] Irrespective of cognitive content, practical utility, etc., the mere framing of something as "science" came to command prestige and power. Prakash also shows us how Indians then sought to reappropriate this sign of power and thereby reposition it. But this does not fully answer my second question. How did the choice of the term *science* rather than *medicine* affect modern Ayurveda?

Once again, a simple perusal of the archive of modern Ayurveda will immediately make it clear that the choice had very real functional consequences for modern Ayurveda. The Ayurvedists, whom I have called braiders above, having defined their subject as a "science," felt a legitimate obligation to look for parallel theories and practices not simply in "Western" medicine, but along a much larger terrain of "Western" sciences. As the subsequent chapters illustrate, these braiders drew their intellectual and practical threads from a heterogeneous and wide-ranging number of Victorian and Edwardian sciences.

Particularly popular were theories of electromagnetism. Such theories had flourished in metropolitan Britain throughout the long nineteenth century and had themselves taken on many diverse forms. The modernizing founders of modern Ayurveda drew eclectically and diversely upon these electromagnetic theories in order to constitute their own traditions of electromagnetic thinking which in turn shaped the newly emergent body metaphor. Briefly, these electromagnetic theories existed cheek and jowl with still later ideas about protoplasm. While in Bengal, at least in modern Ayurvedic circles, these protoplasmic ideas never acquired the kind of iconic proportions it seems briefly to have acquired in Europe, it was nonetheless an important moment in the process of the birth of modern Ayurvedic notions about the body. The nearly ubiquitous use of a notion of chakras within contemporary Ayurvedic discourses almost certainly has its roots in the brief and early tryst of modern Ayurveda with protoplasmic theories. Finally, what did overwhelm the strong electromagnetic and protoplasmic ideas were ideas about endocrines. Building on the early proto-biochemistry of the protoplasmic ideas, hormones emerged by the 1920s as a powerful and dominant framework within which modern Ayurveda was discussed. While hormones were part of certain strands of "Western" medicine as well, the Ayurvedic sourcing of it was not exclusively from medical contexts. It was rather from emergent physiological and biochemical experimental milieus. Moreover, the tenor of modern Ayurveda's discussion too was as much therapeutic as it was cosmological. Endocrinal ideas became an avenue through which the very nature and character of life itself were described.

This account of a gradual move from a nervous anatomy to an endocrinal one is not unique to India. Nor are the resonances that endocrinal anatomy develops with humoral models of the body. Indeed, Chandak Sengoopta's insightful inquiry into European science in the late nineteenth and early twentieth centuries describes a very similar transition: "The discovery of internal secretions displaced olidest theories of bodily regulation, bringing about a humoralist orientation."[95] The transition I map in this book is redolent of Sengoopta's roadmap. But there are also interesting differences. To begin with, the "solidist," mainly neural, anatomies were never as solid as they were in Europe. The pressure to reconcile them with a para-humoral framework, which Ayurvedists had inherited, meant that the emphasis was always on the "fluid" nature of the electromagnetism that underwrote the solid, neural anatomy. This in turn meant that the break between the solidist, neural anatomy and the more fluid, hormonal one was much more gradual.

It is clear, therefore, that when Ayurvedists sought intellectual and practical traditions to draw upon, they looked beyond strictly the clinic. In many cases, they looked wholly beyond medicine. Ayurveda's translation as the "Science of Life" was not merely a discursive strategy to acquire power and prestige, but also an intensely practical guideline about where to look for threads to braid their version of modernity.

John Harley Warner comments on how the rise of social history from the 1960s made historians of medicine look more toward sociologists, urban historians, labor historians, students of politics, etc., rather than toward historians of science. Indeed those still interested in the overlaps between science and medicine were stigmatized and the study of the "cognitive content of medical science" diminished.[96] Since histories of Ayurveda continued to be largely social and occasionally political histories, they continued the trend toward disregarding or at least severely downplaying modern Ayurveda's connections with a diverse array of "Western sciences."

Warner has been also among the first to point out that "science" was neither unitary nor simply of the sort done in laboratories. From the late nineteenth century onward, the term *science* has progressively been applied to an increasingly varied set of practices, spaces, projects, and rationalities. In fact, Marwa Elshakry has recently argued that it is only around the turn of the twentieth century—precisely the period I study in this book—that a singular identity of "Western science" begins to emerge.[97] John Tresch has gone a step further and argued that heterodox sciences often nourished alternate modernities within Europe as well as regions connected to Europe through trade, empire, or otherwise.[98] To resituate modern Ayurveda within a larger historiography on science, rather than just medicine, is therefore not merely to insist on a larger canvas of historical inquiry. It is also an insistence on a recognition of a much more heterodox imperial tryst with "sciences" and a possibility of recovering oppositional modernities that, while antagonistic to empire, are neither autochthonous nor authentic.

One of the central political anxieties of many who study the histories of science in South Asia and particularly those who seek to engage with para-sciences such as Ayurveda is that in the process they might end up unwittingly feeding the fires of xenophobic cultural nationalisms.[99] Once we embrace the pluralized view of science and scientific rationality, however, the opportunity opens up to locate oppositional modernities that do not lend themselves to be fodder for the proponents of cultural purity. We can trace braided knowledges that trip up dominant imperial binaries and the politics that underlie them without claiming that such braided knowledges are purely or authentically indigenous.

The reappraisal of such heterogeneity is also one way of getting beyond the mistaken, problematic, and dangerous homogenization of the "West" that so often unwittingly emerges in postcolonial histories. In a bid to account for the violence, disruption, and injustice of colonial forms of knowledge, recourse is often had to a monolith called "Enlightenment Rationality." Science, suitably capitalized, naturally becomes the main repository of this driving force of all that is evil and inequitable. Yet, such homogenization of "Western" rationality, not to mention "science," runs counter to postcolonialism's own valuable insights about hybridity, plurality, and the determinations of the self-in-the-other. Moreover, actual historical analysis has long since disproved these cardboard caricatures of "Enlightenment Rationality."[100] Unfortunately, however, the intimate enmities of the imperial contact zone are forgotten in the witch hunt for a single villain.

Once we look beyond the stifling depictions of "Enlightenment Rationality," we notice in Victorian Britain, perhaps above all other contemporary European locales, a thriving culture of counter-rationalities. Spiritualism, animal magnetism, occult chemistry, cell psychology, protoplasmania, and much else that was radical and subversive emerged out of precisely the imperial encounter where "Enlightenment Rationality" was supposedly performed as a monolithic enterprise.[101] Largely owing to a lack of conversation between postcolonial scholars working on South Asia and historians of science, the former, even while noticing the widespread engagement of reformers and revivalists in India with these heterodox sciences, have continued to call them "pseudo-sciences."[102] Such descriptions have trivialized the engagement and continued to render imperial science monolithic and hegemonic. Yet, historians of science have long since dispensed with the strict dichotomy between "real science" and "pseudo-science," seeing them as historically and contextually developed political labels that obscure a number of significant continuities between what in retrospect seems like totally disparate camps.[103]

In fact, many of the ideas about electromagnetism, protoplasm, and eventually endocrinology that entered modern Ayurveda had initially entered British India through Spiritualist and Theosophical networks. Many of the stalwarts of modern Ayurveda were directly involved in the Theosophical Society. A noted Tamil writer on modern Ayurveda, Captain Srinivasa Murti, for instance, was an active member of the Theosophical Society in Madras. Likewise, Hemchandra Sen, who introduced both protoplasmic and endocrinal ideas into modern Ayurveda, was also known to attend meetings of the Theosophical Society in Calcutta. Numerous other lesser-known authors directly or indirectly quoted ideas that were prevalent in Theosophical and Spiritualist circles.[104]

Once we escape the restrictive binarisms through which colonialism engendered its rule of difference—namely science/religion, East/West, medicine/quackery, tradition/modernity, rational/irrational, etc.—and begin to see these terms as performative and fluid boundary makers presaged into action by interested parties trying to deny a more plural, heterogeneous, and interpolated world, we begin to notice a range of rationalities linked to both "science" and "non-Science" that mutually engaged each other at the colonial "contact zone" to form new assemblages of discourse and practice such as modern Ayurveda. The guiding question remains, as Lawrence Cohen had put it: "Can a hermeneutic of other subjugated knowledges [scrambling geographic boundaries] be constructed without reducing them to Romantic visions of epistemic alterity"?[105]

Colonialism and Alterity

Finally, there is the vexed issue of modern Ayurveda's alterity or Otherness vis-à-vis biomedicine. Is modern Ayurveda still sufficiently distinct from biomedicine, or has the modernization process incrementally reduced the distance between the two medicines and rendered them different versions of the same? Some like Langford contend that despite all mimicry and hybridity, modern Ayurveda remains sufficiently different from contemporary biomedicine.[106] Lawrence Cohen and Harish Naraindas, on the other hand, are much more circumspect.[107] Both see Ayurveda's alterity severely compromised by its increasing proximity to biomedicine. Dominik Wujastyk feels that the random hybridization has produced a cognitive incoherence.[108]

There are resonances of some of these debates beyond the history of Ayurveda. It has long been a matter of debate among postcolonial historians as to precisely how much genuine alterity survived the colonial encounter. The historians of the Subaltern School had famously argued that Indian society had two distinct levels and while alterity among its elites was mere window dressing, more genuine alterity had survived in the political and intellectual culture of the subaltern classes.[109] The matter has remained contested ever since and the question of difference and alterity continues to surface in South Asian histories repeatedly.

In the Ayurvedic context, I think the debate will be well served to eschew any search for "authentic" or "pre-colonial" difference and to think instead of shades of difference. It is clear that Ayurveda today is not biomedicine. Every practitioner and most patients insist that it is something distinctive from not only biomedicine, but also Unani Tibb, Sowa Rigpa, etc. Yet, as this book along with ethnographies such as Naraindas's show, the cognitive difference of Ayurveda has been progressively reshaped.[110] By repeatedly refiguring Ayurveda in light of "Western" science, the object that has been produced as modern Ayurveda is not utterly dissimilar to biomedicine. Indeed, there is a lot that is explicitly defined along biomedical lines.[111]

What we need then is a way to acknowledge levels or shades of difference: to open up a distinction between simple difference and radical alterity. So that we can distinguish the sorts of difference that have been maintained and cultivated so as to give Ayurveda its distinct contemporary identity as well as take stock of the more radically subversive alterity that has had to be compromised to make Ayurveda compatible with biomedicine and modern Western sciences. Jean Baudrillard and Marc Guillaume point out that radical alterity is utterly unrepresentable. Its presence is spectral and can only be surmised from traces. Unlike simple difference, radical alterity is unintelligible and subverts the very premise that reality is rational.[112] It is only by disaggregating these levels of difference, by distinguishing what is intelligible difference from that which is utterly untranslatable, that we can grapple with the diverse enactments and deployments of difference under colonialism. Partha Chatterjee has spoken convincingly of the "rule of colonial difference."[113] But all forms of difference are not alike. Some forms, as I have been arguing, are intelligible and cultivated, while others are radically subversive, untranslatable, and total. The former produced colonial hierarchies and the latter fed into the constitution of a range of para-modernities and even non-modernities.[114]

Of the former, tamer version of difference, Homi Bhabha has written that colonialism produced a "desire for reformed, recognizable Other, as a subject of difference that is almost the same but not quite."[115] Beneath its protestations of dissimilarities, there often lurked a sea of sameness. In a biting critique of this kind of tame difference now masquerading as "multicultural science," Lawrence Cohen points out that it risks "mapping difference onto an underlying hegemony." This underlying hegemonic sameness is nothing but the "unexploded Occidentalism of 'the West' as synonymous with modernity."[116]

A much more radical, subversive, and unintelligible alterity haunted the margins of colonial modernity. The consciousness of peasants, outcasts, and unreformed Others engendered this radical alterity. These degrees of difference had first been noticed in South Asian historiography by the Subaltern Studies project. The collective's later works, however, frequently conflated these forms of difference by emphasizing an undifferentiated notion of "difference."

Especially in the context of the history of medicine, distinguishing shades of difference will allow us to take better stock of the politics of colonialism. As Rachel Berger points out in her recent work, one of the major issues with tracing colonialism's impact on Ayurveda is that the former had so little to directly say about the latter.[117] Since the colonial state remained largely aloof from Ayurveda or its fate until after the Great War, it is difficult to draw any direct conclusions about the role of colonialism in shaping modern Ayurveda in the pre-War era. Yet, colonialism cannot be reduced merely to the overt machinations of the colonial state.

Colonialism operationalized an equally important and incisive cultural politics organized around subtle and not-so-subtle ideological apparatuses that promoted particular values, attitudes, and styles of thinking. One of the most pervasive impacts of colonialism was the constant need perceived by colonized intellectuals to justify or demonstrate the validity of their statements in light of European standards: to make non-colonial knowledges translatable into the idioms of "Western" science. Very few intellectuals, if any, escaped this perceived need to cite European validation. It was this need that led modern Ayurveda to progressively divest itself of elements that did not conform to European notions of modernity. But neither did modern Ayurveda simply submerge itself within the growing body of "Western" medicine in India. It played on the levels of difference in creative ways. It retained an identity by maintaining a certain amount of distinctiveness, while surrendering its more stringent modes of alterity.

In the conclusion to this book, it is these levels of difference that I interrogate and juxtapose. For the sake of clarity and accessibility, I divide the conclusion into two distinct parts. In the first part I briefly pull the strings together from the foregoing case studies and provide a summary sketch of the new body image that underwrote technomodern Ayurveda. Having done this, in the second part I briefly explore the radically different therapeutics that were utterly unrepresentable in modern Ayurveda. In order to draw these contrasts without undertaking a full-blown study of premodern Ayurveda which would be well beyond the ambit of this book, I will instead piece together particular elements of therapeutic engagement that, while clearly present in pre-modernized Ayurveda, were progressively marginalized through the modernization. I also braid these elisions together in order to conceptualize an approximate figure of the therapeutic modality that would have permitted these elements to coexist in a single repertoire. I argue that these elements of an earlier therapeutic modality—one I call cosmo-therapeutics—come to seem absurd and unrespectable in modern Ayurveda because of the latter's progressive investment in the late-colonial state's biopolitical apparatus as well as its rampant pharmaceuticalization.[118] What looks, to modern eyes, like some of the most absurd and irrational elements of cosmo-therapeutics, was also what made it utterly unsuited to the calculability necessary for biopolitics and pharma-capitalism. Neither can one make population-level calculations necessary for forming public policy in biopolitical states based on the possibility of miraculous interventions by saints, nor can public hospitals that are run according to codes of public accountability and mass efficiency abide by the possibility of serendipitous miracles. By exiling this incalculable rationality based on singularities and unpredictability, modern Ayurveda constituted itself as a therapeutic alternative to biomedicine that could still function within the dominant political architecture of the modern state.

ONE

A Baidya-Bourgeois World: The Sociology of Braided Sciences

Gangadhar Ray (1798 [1205 BE]–1885 [1292 BE]) is a name to conjure within the history of modern Ayurveda. Practically every history of the subject assigns to him a prominent role as one of the founding fathers of the modernization or revival project. Yet it is surprising how little is actually known about him. Besides dates and places of his birth and death, we hear of his publication of the *Charaka Samhita* in 1868 for the first time, of the long lineage of students he trained, and at best the partly apocryphal story of his having left Calcutta in a huff the day Madhusudhan Gupta became the "first modern Hindu to dissect a corpse." We never hear that he was a precociously early leader of a caste movement who wrote energetically and extensively to forward partisan claims for his caste. Nor do we ever hear of the fact that having written treatises disproving the authenticity of the *Bhagbat*—the canonical religious text of the Vaishnava sect—he was deeply hated by Bengali Vaishnavas. We remain similarly ignorant of the fact that despite his staunch antimodernism in many ways, he owned a printing press. Or, for that matter, that, besides being an erudite scholar and physician, he was also a talented artist and sculptor who used to personally sculpt the elaborate idol every year for the annual Durga Puja.

The one-dimensional, cardboard image of Gangadhar is not unique. Nearly all the major actors in the history of modern Ayurveda suffer the same fate. Thus, Gananath Sen, the first dean of the Faculty of Ayurveda at the Benares Hindu University and a towering stalwart of the whole process of the modern transformation of Ayurveda, is seen merely as a modernizer and advocate of syncretic Ayurveda. We hear nothing of his tireless leadership of Baidya caste—a specific caste only found in Bengal to which most Bengali Kobirajes or Ayurvedic physicians belonged—associations or his efforts to get Baidyas to change their surnames to the Brahmin patronym, "sharma." Nor indeed do we hear of his mother's family's intimate connection with the medieval Vaishnava saint, Chaitanya Mahaprabhu. Thus, we are unable to make sense of the caustic language that this otherwise mild-mannered, amiable man occasionally used toward Gangadhar in print.

The extant scholarship on modern Ayurveda does not even mention the conflicts and attacks by Gananath on Gangadhar. They, and a slew of other characters, are simply recounted en passant in what is effectively a linear narrative of progressive "modernization." Moreover, in most of these histories, these figures are merely Ayurvedists. Their lives are entirely encapsulated by their brief role in the teleological story of modernization. They have no further social, cultural, or political interests or identities beyond the narrowly Ayurvedic. Their cultural, social, and intellectual lives have been either wholly ignored or reduced to a few signposts of dates and places. The vacuum that has resulted from the clearing away of thick contextualization has been filled either by hollow census statistics and homilies about caste, or indeed by anthropological and cultural theory. In the former case, individual lives, intellectual traditions, personal identities, etc., have disappeared under generalizations generated by the colonial state. In Poonam Bala's pioneering "sociohistorical" account, therefore, what we are given by way of social background are neither numerical charts of students at colonial medical schools, enumerated according to colonial categories like "Hindu," "Mohammedans," "Eurasians," "Christians," etc., nor elaborate graphs based upon the numerical tables.[1] At best, we get occasional blurbs uncritically lifted from colonial ethnographers telling us, for instance, that "the *vaidyas* (*vaids*) of Bengal . . . were the *ambasthas* of Manu."[2] Colonial categories, in such instances, not only hijack the "social," but also suffocate the physician-intellectual's individuality by somehow implying that these numeralized generalizations will enlighten us about the way a historically situated person will act. Such approaches to mapping the social background of historical actors are a particularly acute case of what Pierre Bourdeiu has aptly described as the "intellectualist illusion that leads one to consider the theoretical class, constructed by the sociologist, as a real class, an effectively mobilized group."[3] What makes this particular variant even more misleading is that the sociologist builds upon categories that were constructed under a colonial regime and were tempered by the agendas of imperial rule.

By contrast to the approach of scholars such as Bala, we have the approach of anthropologists such as Jean Langford. Here numeralized fictions are replaced by rich contemporary contextualization. But the context that emerges is structured by the anthropologists' choices rather than by the choices of the actors. Thus one major drawback of this approach is an extremely limited engagement with intellectual traditions. Ayurvedists are seen to be responding to contemporary social, political, and cultural forces, but doing so in a way that is not shaped by their individual histories to any considerable extent. What unites the cast of characters brought together in a study such as Langford's is not intellectual proximity or sectarian profile, but simply the fact that the same anthropologist studied them all. And their individual actions seem to all follow a common logic that the anthropologist is eventually able to discern.

As would be obvious by now, I wish to move away from both these approaches. Instead, my intention is to use individual biographies to build a historico-sociological map: a map that will reflect the emic social groupings of the actors engaged in the transformation of Ayurveda at the cusp of the nineteenth and twentieth centuries. I take my cue here from Kavita Sivaramakrishnan's rich and insightful study of Ayurvedic modernization in the Punjab. Sivaramakrishnan's study demonstrates the importance of sectarian affiliations in constituting patronage networks in precolonial Punjab. This in turn resulted in colonialism and the loss of patronage it ushered in affecting different groups of physicians differently. As a result, these groups also participated in the later modernization process in distinctive ways.[4]

While it is unlikely that sectarian affiliations in Bengal were as closely linked to patronage networks as they were in precolonial Punjab, sectarian affiliations not only shaped social solidarities but also influenced later participation in distinctive forms of nationalist politics. My intention, therefore, is to see the social field, so far as possible, through the eyes of my historical actors. Moreover, I want to connect these emic groupings to actual therapeutic change. This is easier said than done. Naturally, all my actors did not explicitly state what social identities they embraced and what social worlds they inhabited. But here it is useful to follow Marshall Sahlins in tracing the "structures of the conjunction" by which he avowedly meant "the practical realization of cultural categories in a historical context, as expressed in the interested action of the historical agents."[5] Reinserted into my project, Sahlins's program translates as the determination of the historico-sociological map of modernization of Ayurveda, not through the colonial categories of identification, but through the categories through which the historical actors self-identified. In the following sections, I highlight how widely sometimes the colonial categorization of social identities and the categorization of self-identities were at variance.

The Small World of Ayurvedic Technomodernity

Steven Feierman points out a crucial problem with using Sahlins's formulation. It does not define "historical agent" with any precision.[6] Whose actions should we look at in recovering the "cultural categories" in a given historical context? Surely, it would be impossible to look toward each and every individual involved in a historical drama. In the present instance, however, this is a very limited problem for the simple reason that I am actually dealing with a very small group of people. Despite the fact that at its height the Ayurvedic developments in Bengal had an influence throughout South Asia—with its leaders and students acquiring important positions in places as far away as Benares and Colombo—the actual group of people involved in the transformation was remarkably small.[7] Moreover, ties of caste, neighborhood, kinship, and factional membership related many, if not most, of the main actors to each other. It is this remarkably small group and its relative cohesiveness that permits me to use Sahlins's method without risking the imposition of the cultural categories of select historical agents on all the others.

There are several reasons that operated to make the effective group of modernizers small and cohesive. It is impossible, and I daresay somewhat unnecessary, to flesh them all out in detail. Yet I outline three preeminent factors that restricted the social base of the modernization process.

The first of these reasons is the impact of familial and "preceptorial" kinship. The real kinship ties are fairly straightforward and easily documented, though they remain significantly under-appreciated. A mere three families have played an enormous part in the modernization of Ayurveda in Bengal. In 1884, a leading medical periodical, the *Chikitsa Sammilani*, published a list of eminent Ayurvedists who had founded a new association in the city of Calcutta. The first five names on the list, comprising of a total of twenty-eight names, were all from a single family. These were Gangaprasad Sen, Durgaprasad Sen, Annadaprasad Sen, Bijoyratna Sen, and Nishikanta Sen.[8] The first three, Gangaprasad, Durgaprasad, and Annadaprasad, were siblings; Bijoyratna was their sister's son, who had been raised by his uncles in their home, while the last, Nishikanta, was the son of the eldest brother, Durgaprasad. Even relatively lesser-known figures such as Bhagabatiprasad, Gangaprasad's son, later went on to play an important role by editing the very first Ayurvedic periodical, the *Ayurveda Sanjiboni*, in 1921.

Another equally important family group was the sons and nephews of Chandrakishore Sen. Brahmananda Gupta's account mentions the entrepreneurial exploits of Chandrakishore Sen who established one of the first

Ayurvedic pharmaceutical companies under the name of C. K. Sen & Sons.[9] The empty Ayurvedic medicine bottle dug up in Pittsburgh, with which I began this book, had been manufactured by this very company. Chandrakishore's nephew was Binodlal Sen, one of the earliest and most successful authors of the modernization. Not only was Binodlal's two-volume *Ayurveda Bigyan* (The Science of Ayurveda) (1887 [1294 BE]) a classic, but his efforts even earned him a reference on the Calcutta stage during his lifetime.[10] Binodlal's name was to be found on the list published in 1884. Binodlal's youngest brother, Nagendranath Sengupta, authored several books in Bengali and English including such works as the two-volume *Sohoj Kobiraji Siksha* (Simple Manual of Kobiraji Treatment) (1897 [1301 BE]), and the two-volume English work titled *The Ayurvedic System of Medicine* (1909). Nagendranath's cousins, Debendranath and Upendranath Sengupta, the sons of Chandrakishore, similarly wrote numerous books including the four-volume jointly authored work titled *Ayurveda Samgraha* (Ayurvedic Compendium) (1892 [1299 BE]). Each of these works is a major intervention and deservedly has been recounted as part of the history of the transformation of Ayurveda, but recognizing them all as the works of a single family demonstrates just how narrow the social base of the transformation could at times be.

Slightly later, the family of Shyamadas Bachaspati also emerged as a powerhouse. Shyamadas and his son, Bimalananda Tarkatirtha, were both deeply involved in the transformation of Ayurveda. Both the father and son were equally active in establishing and running the Vaidyashastrapith Ayurvedic College in Calcutta in the 1920s. In 1933, Shyamadas served as the President of the Nikhil Bharotiyo Ayurvediyo Mohasommelon (All India Ayurvedic Congress). Shyamadas's elder brother, Dharmadas, was also an eminent Ayurvedist based in Benares. Both of Bimalananda's sons, Brahmananda and Krishnananda Gupta, also played an important part in the later history of Ayurveda in Bengal.

Just as three families dominated the modernization, two preceptorial lineages served to further consolidate this domination. Brahmananda Gupta and Subrata Pahari have both documented how practically every well-known Ayurvedist of the nineteenth century worked within either of the two preceptorial lineages. The first derived from Gangadhar Ray, while the latter derived from Gangaprasad Sen.[11] Given that until the late 1920s, all Ayurvedists trained through the master-disciple tradition (*guru-sishyo porompora*) by apprenticing and living in the home of the guru for years, these preceptorial lineages began to function as a pseudo-kinship. Ronald Inden and Ralph Nicholas have documented the existence of a set of pseudo-kinship terms to refer to preceptorial kin.[12] An eminent mid-twentieth century Ayurvedist, Shiv Sharma, described how preceptorial kinship worked by close analogy to familial kinship.[13]

These ties of preceptorial kinship operated in two ways. On the one hand, they reduced diversity of opinion within the preceptorial lineage by embedding intellectual activity within a close-knit kin group. On the other, it became a handy constituency that could be readily mobilized in the cause of particular modernizing projects that one person within a lineage undertook. Thus Jaminibhushan Ray and Gananath Sen, for instance, worked together in establishing and running the Astanga Ayurveda College. Both of them were students of Gangaprasad's nephew Bijoyratna Sen. Similarly, when Shyamadas Bachaspati opened the Vaidyashastrapith, he invited others who were part of the preceptorial lineage of Gangadhar. Haranchandra Chakraborty, who taught surgery at the Vaidyashastrapith, and Bijoykali Bhattacharya, who served at its principal, were both from the preceptorial lineage of Gangadhar. Bijoykali Bhattacharya's younger brother, Shibkali Bhattacharya also taught at the Vaidyashastrapith.

The second major factor that contributed toward the restriction of the social basis of reform was geography. The geographic influence operated at two levels. First, the dominance of the city of Calcutta served to exclude anyone who was not immediately situated in the city. This is both reminiscent of and distinctive from Sivaramakrishnan's account of the Punjab. While urban centers clearly dominated in the Punjab too, there were multiple urban centers. By the 1880s, besides the major Punjabi cities like Lahore, even minor towns such as Bhasaur, Firozepur, and Tarn Taran were mobilized in the cause of modernizing Ayurveda.[14] By contrast, even Dhaka lagged far behind Calcutta in its participation in the modernizing project in Bengal. Second, urban neighborhoods served as important sites for mobilization behind various modernizing initiatives. Thus, the associational politics that Sivaramakrishnan finds dispersed across the urban geography of Punjab, in Bengal came to be embedded in a condensed form in the neighborhoods of colonial Calcutta.

The earliest effort at forming an association for modernizing Ayurveda was most likely the Chikitsa Sobha formed in 1871. The association encouraged and financed the publication of Gopalchandra Sengupta's three-volume work, the *Ayurveda Sarsamgraha,* of which we will hear more in the subsequent chapters. The work eventually had fifty-two subscribers, but several were simply individual corresponding members residing in isolated villages or towns without any interaction among themselves. A handful of others subscribed to the work on nonmedical grounds, such as the aristocrats, Raja Jyotindramohan Tagore and Rani Swarnamayee, who patronized learning of all sorts without any direct input into the project. Of the twenty-odd physicians who actually attended the meetings of the association and hence probably had some putative input into the work, nearly all lived on or around Sookea Street.[15]

The long-term fate of the Chikitsa Sobha unfortunately remains unknown. A little over a decade later, in 1884, another effort was made. This time the association was called the Dhormo Mondoli and "all eminent Ayurvedists" were said to have joined.[16] This association, however, seems to have been dominated by Gangaprasad Sen's family and their extended networks. Much later, around 1913, recalling these early associational attempts, Gananath Sen recalled some more failed attempts. The earliest attempt he spoke of was one instituted by Gangaprasad Sen, Kaliprasanna Sen, and Gananath's late father, Bishwanath Sen. Subsequently, around 1904/1905, Gananath mentioned another attempt by Surendranath Goswami. Both of these attempts failed before one of Gananath's teachers, Bijoyratna Sen, was successful in establishing an Ayurveda Sobha. Around 1907, a second Ayurveda Sobha was formed. By 1913, a third association called the Brahmin Ayurveda Sobha had also come into existence.[17] All of these associational attempts were based in Calcutta and often involved the same set of characters or others closely associated with them. Gangaprasad Sen, for instance, was part of both the Dhormo Mondoli and the latter attempt along with Kaliprasanna and Bishwanath Sen. Bijoyratna, who eventually succeeded in establishing the Ayurveda Sobha, was Gangaprasad's nephew.

Some sense of the level of exclusivity around these efforts can be gathered when this small set of characters resurfacing in different combinations is contrasted with the fact that a census of the city of Calcutta taken in 1891 found 177 Kobirajes, including fourteen women.[18] Hardly a tenth of that number ever participated in the modernization process. Yet, despite this small participation, Calcutta was able to dominate the early history of Ayurvedic modernization. By contrast, outside of Calcutta, the modernizing projects, which relied heavily upon publishing success, failed repeatedly. Hence, despite eastern Bengal having historically had a high concentration of Ayurvedists,[19] we find that two Ayurvedic periodicals, the *Ayurveda Hitoishini* and the *Ayurveda Bikash*, that appeared out of Dhaka, the preeminent cultural center in eastern Bengal, both wound up within just two years.

In terms of localities, Sookea Street and, slightly later, Bhabanipur, remained the two dominant areas within the city that shaped the modernization process. The Chikitsa Sobha that met for its meetings at No. 6, Madan Mitra's Lane, was, as I have already noted, largely made up of Ayurvedists in the Sookea Street area. A commemorative souvenir published by a local heritage group and authored by Brahmananda Gupta recounts thirty-two major Ayurvedists throughout the nineteenth and twentieth centuries having resided for at least a part of their active professional lives in the relatively small area between Maniktala (Sookea Street) and Shobhabajar (Shyampukur). A glance at this list of past residents reveals the names of Gangadhar Ray, Gangaprasad Sen, Bijoyratna Sen, Bishwanath Sen, Dwarakanath Sen, Shyamadas Bachaspati, Bijoykali Bhattacharya, and many more. It is almost the entire cast of characters who shaped modern Ayurveda in Bengal. They all lived at some point or another in an area that is at best two kilometres squared.[20]

Later, another concentration of Ayurvedists developed in the Bhabanipur area of southern Calcutta. Evidence of this concentration of Ayurvedists in the Bhabanipur area is available from the proceedings of Gananath Sen's Baidya Brahmin Somiti. Though the association was a caste association rather than a medical one, the traditional connection of the Baidya caste with the practice of Ayurveda meant that a large number of the members were practicing Kobirajes. In 1925, the Baidya Brahmin Somiti opened a special Bhabanipur Centre to spread its message among the Baidyas of the area. Within the first year of the center's existence, it had acquired eighty-six members. This was clearly only a fraction of the total number of Baidyas in the area since the annual report of the center on its first anniversary emphasized the need for more aggressive outreach.[21] While it is impossible to tell how many of these Baidya residents of Bhabanipur were practicing Ayurveda, there is every reason to believe that a significant number among them would have been associated with it. Such an assumption is partially validated by the occasional lists of the attendees at meetings of the Somiti that specifically mentioned the occupations of the members. For instance, eighty-three members attended the first annual general meeting of the Bhabanipur Centre. Some of these attendees did not belong to the Bhabanipur Centre as such, but were members of the parent body and had traveled from north Calcutta, such as Kobiraj Haripada Sensharma, Kobiraj Jadunath Guptasharma, and, of course, Gananath Sen himself. Yet, the majority was members of the local center. And though a large number of them did not mention their occupations, happily for me, some did. These included Kobirajes Sukumar Dassharma, Ashutosh Dassharma (Ray), Naliniranjan Sensharma, Jogendramohan Sensharma, and Sashibhushan Dassharma.[22] In fact, even without these specific names, the very fact that the association first opened a "branch" (*shakha*) in the Bhabanipur area and then, within a year, upgraded it to an almost autonomous status as a "center" (*kendro*), proves the area's importance to the caste. Moreover, since Kobirajes were still overwhelmingly (though not exclusively) Baidyas, it is clear that Bhabanipur, by the mid1920s, had a large concentration of Kobirajes that was comparable to the Maniktala-Shyampukur area.

Finally, what limited the social basis of the modernization was its heavy dependence upon two exotic languages: namely, Sanskrit and English. As I demonstrate throughout this book, the modernization proceeded by braiding Sanskritic and European (accessed mostly through English) sciences together. This meant that any physician wanting to participate in the process had to master two languages beyond what he used in his daily practice. While Sanskrit had long been nominally the source of Ayurvedic knowledge, few practicing physicians actually knew or read it. In this regard, the situation in Bengal was not dissimilar to that in the Punjab, where "Ayurveda" (or more accurately for the period "Baidak") throughout the nineteenth century was mainly studied in the local vernacular: Braj Bhasha written in the Gurmukhi script.[23] William Adam's report on education in Bengal in the 1830s provides a striking picture of how limited a role Sanskrit played in medical practice prior to the modernization. In Natore, Adam reported a total of 123 practicing physicians. Of these, eighty-nine were Hindus and thirty-four Muslims. Adam estimated that a maximum of eleven to twelve physicians out of these 123 knew some degree of Sanskrit. The remaining one hundred odd physicians derived the "only knowledge they possess of medicine from Bengali translations of Sanskrit works which describe the symptoms of the principal diseases and prescribe the articles of the native materia medica that should be employed for their cure and the proportions in which they should be compounded."[24]

What made this even more remarkable was that Adam confessed that Natore was one of the few places in the province where there was a devoted medical school maintained by two scholar-physicians salaried by eminent local families. If the level of Sanskrit competency in the vicinity of one of the few medical schools was so abysmally low, one can surmise how much lower it would be in other, less well-provisioned regions. Though Adam, reflecting perhaps local prejudices and hierarchies, continued to call the Bengali-literate physicians "uneducated," he confessed that "the only difference that I have been able to discover between the educated and the uneducated classes of native practitioners is that the former prescribe with greater confidence and precision from the original authorities, and the latter with greater doubt and uncertainty from loose and imperfect translations."[25]

Beneath this class of physicians was another that Adam called "village doctors." Of these there were, by Adam's reckoning, 205. These people were largely unlettered and treated using a combination of simple herbals and some incantations. Lower still in social rank was a motley crew of specialists ranging from smallpox inoculators (21 in Natore), midwives (297 in Natore) and "conjurors" specializing mainly in snakebites (722 in Natore).[26] As is clear from Adam's own vocabulary, it was only the Sanskrit-literate physicians who were recognized as genuinely "educated." Yet they formed a microscopic minority. In the case of Natore, a place with a higher than normal chance of acquiring a Sanskrit education, twelve or so Sanskrit-educated physicians formed a measly 0.86% of the total strength of medical manpower—that is, 1,389 people in all.

It was this enormous gap between the total numbers of people working as physicians of one sort or another and those accorded the social respect as scholar-physicians that is reflected in Subrata Pahari's list of eminent nineteenth-century Ayurvedists in Bengal. Despite the fact that his list covers all of Bengal and the entirety of the long nineteenth century, it consists of a mere forty-one names. Of these again, for most, there is little but a name and a date. It was only for twenty-three such physicians that Pahari was able to gather some limited personal information.[27] That a single small region, Natore, had over a thousand medical practitioners in the mid-1830s and that a historian studying the entirety of the region throughout the century yet struggles to find biographic information on more than twenty-three people testifies to the enormous gap in the archive.

The modernization process, heavily textualist and based upon the braiding of Sanskritic and English sciences, operated to marginalize all those people whom Adam had thought of as "uneducated." Effectively this meant—accepting the Natore numbers to be erring on the higher rather than lower side for the "educated"—over 99 percent. The modernization was therefore engendered by a tiny, relatively privileged and closely knit elite group of urban physicians at the turn of the nineteenth and twentieth centuries. It is this relatively small size and cohesiveness of the modernizing elite that allow me to create a common historico-sociological map for them.

Cast(e)ing Modern Ayurveda

I have already hinted at the importance of caste for two of the stalwarts of the Ayurvedic transformation, Gangadhar Ray and Gananath Sen. But two further important caveats are necessary before discussing the issue further. First, though both these figures are larger than life in the history of modern Ayurveda and their interest in caste alone would have been enough to make the issue important, they were by no means the only ones to be interested in both caste matters and Ayurveda. It is therefore a more general "structure of conjuncture" that I seek to unravel here and not just flog the accidentally common hobbyhorse of two founding fathers. Second, pace what Bala has stated and what historical actors explicitly assert, Baidya caste identity and their monopolistic claims to practice Ayurveda were not a settled issue in the nineteenth and twentieth centuries.

The entire rich and complex historiography of the phenomenon of caste in South Asia has taught us to deeply distrust the putative conflation of ancient labels and contemporary identities that Bala engages in.[28] Careful archival work and a serious engagement with a wider array of sources—particularly beyond the narrow cache of colonial governmental records in English—plainly shows why such conflations are illegitimate. Writing almost forty years before Bala made her assertion, D. C. Sircar had pointed out that throughout history several completely disconnected groups had claimed to be the "ambasthas." The *Aitreya Brahman* of ca. 500 BC mentions an Ambastha King named Narada. Greek and Roman authors mention the Ambastha people living around the Chenab during the invasion of Alexander in 327–25 BC and having a republican government. The Mahabharata also mentions an Ambastha tribe in the Punjab. The *Brihaspati Arthashastra* placed them in the region between Kashmir and Sind. A Pali work entitled *Ambattha Sutta* said they were a group of Brahmins, while the *Jatakas* referred to them as agriculturalists. It was the *Manusmriti* that propounded the myth of their descent from a Brahmin father and Vaishya mother and asserted their identity as physicians. Around 140 CE, Ptolemy's *Geography* as well as the *Markandeya Purana* and the *Brihat Samhita* spoke of an Ambastha settlement in the Mekala country, perhaps near the central Indian Maikal hills. In more recent times, a wide variety of groups claimed descent from Manu's Ambasthas. In Bihar, they are a subcaste within the Kayastha caste. In southern India, they tend to be barber-surgeons, performing some slight ritual functions and minor medical tasks. Their women are often midwives. In Bengal itself, apart from the Baidyas, the Mahishyas—a dense and low-caste group involved in an aggressive Sanskritizing movement since the nineteenth century—also claimed descent from the Ambasthas.[29] The title, in effect, was little more than a label that different groups claimed as they reinvented or transformed themselves. This of course is also in keeping with contemporary scholarship on castes in general which notes that the caste groups of India, far from static, ahistorical, or biologically discrete entities, are in fact dynamic social corporates that continually fuse, split off, reinvent, and renegotiate their identity and status within local milieus.[30] The operative unit in this dynamic system is not the fourfold varna categories usually referred to as "caste": namely, Brahmin, Kshatriya, Vaishya, and Sudra, but rather more localized units called *jatis*.[31]

Writing in 1971, Hitesranjan Sanyal pointed out that the "Baidya" as a *jati* emerged in Bengal sometime between the twelfth and the fourteenth centuries in caste lists found in texts such as the *Brahmavaivarta-puran* written at that time. There is no mention of such a group prior to this. Sanyal also points out that the Baidyas, along with their nearest rivals, the Kayasths, had been successful in ritually and socially elevating their status considerably by the sixteenth century.[32] Though Sanyal did not explore the specific case of the Baidyas beyond the sixteenth century, his broader argument in the article was that there were strong continuities from the precolonial to the colonial period in terms of the tendency toward upward mobility exhibited by some *jatis*. In order to go beyond the literal truth of what the Baidyas were saying in the nineteenth century, we must look toward this longer, deeper continuity.

A fuller history of the Baidya *jati*'s upward mobility will have to wait for another opportunity, but suffice it to say that the trend did not buckle after the sixteenth century. In fact, by the eighteenth century, Raja Rajballabh, a proud Baidya, had acquired enormous wealth and power under Bengal's post-Mughal nawabs. He used his power and influence to the fullest to formally convince a large body of Brahmins to accept the right of the Baidyas to wear the sacred thread.[33] This act admitted Baidyas into the uppermost fold of the social and ritual hierarchy in Hindu society. For a *jati* that in the earliest hierarchic lists from the twelfth century had ranked among the Sudras (the lowest of the four ideal-typical Varnas), this was a massive ascent. To what extent medicine played a role in this ascent is open to question. While the Baidyas certainly had long-standing associations with medicine, their ascent had clearly been enabled by wealth and influence accrued through government service. In fact, there is no hint that Rajballabh, the eighteenth-century leader who enjoys almost mythic status among Baidyas, ever had anything to do with medicine.

Be that as it may, Rajballabh had clearly set the stage for the claims of the nineteenth century. The Baidyas were, as a matter of fact, the very first Bengali *jati* to attempt to form a caste association. In 1831, Khudiram Bisharad, a teacher at the erstwhile Native Medical Institution, founded the Baidya Samaj, the first modern caste association. Even before this, a "continuous pamphlet warfare" deploying the newly available print technology, had erupted between Brahmins and Baidyas from 1822 onward. Whereas Brahmins now resented the Baidyas for wearing the sacred thread, the latter, using the thread as evidence, sought equality with the Brahmins.[34]

After Bisharad's early attempts, it was Gangadhar who emerged as the champion of Baidya caste claims. Nagendranath Basu, staunch Kayastha-propagandist and author of the multi-volume Bengali encyclopaedia, the *Bishwokosh*, mentioned that "he [Gangadhar] had attempted to make the Ambastha-jati [used here as a synonym for Baidya] into Brahmins. Many Baidyas followed his reasoning, performed ritual penances and took to wearing the sacred thread."[35] A Baidya pamphleteer also stated that "there was none who worked harder than him toward the amelioration of the condition of the Vaidya caste and it was he who started the theory of the Brahmanic origin of that caste."[36] One Baidya author celebrated Gangadhar's role by saying that

> though many highly capable Baidyas have from time to time undertaken various investigations [into the history of modern Baidyas], none have been fully successful in their endeavours. The only exception was the Berhampore-resident, polymath (*sorboshastroparodorshi*) and foremost scholar (*ponditagrogonyo*), the late, lamented and sage-like (*rishitulyo*), Kobiraj Gangadhar Ray. It was he who, through great care (*jotno*) and effort (*cheshta*), collected a large number of ancient scriptural quotations (*shastriyo bochon*) and extrapolated their true meanings (*prokritartho nishkashon*) before publishing these in the form of several small booklets that included commentaries and exegesis.[37]

I find two aspects of Gangadhar's involvement worth underlining. First, his claims of equality with Brahmins were clearly building on the achievement of Rajballabh. Second, according to Gangadhar's grandson, as per family lore, the traditional leadership of the caste had rested with one of Gangadhar's ancestors until Rajballabh usurped it.[38] So by stepping up as the champion for his *jati*'s claims, he was also attempting to reassert his family's perceived rights to the leadership of the *jati*.

Another prominent Ayurvedic modernizer to be actively involved in the Baidya caste mobilization was Binodlal Sen. Binodlal Sen, one of the most important early figures of the transformation as well as a senior member of the powerful family of Chandrakishore Sen, also contributed to *jati* politics. He never enjoyed the kind of leadership roles that Gangadhar and Gananath did, but he was clear about his commitment. He undertook to publish the two earliest genealogical works of the Baidyas, the *Chandraprabha* and the *Ratnaprabha*, both authored in the seventeenth century by Bharatamallik. In the foreword to the *Ratnaprabha* that appeared first, Binodlal stated that the text "described in detail the origins, lineages, segmentary lineages, marital alliances and original places of residence of all the Baidyas." He also mentioned the difficulties he had faced in locating the original handwritten manuscripts of Bharatamallik, but declared that if the publication of this work proved even remotely helpful to his caste fellows (*swojatiyo mohodoygon*) then he would consider his labors worthwhile.[39] Genealogical works, better known as *kulajis*, were not simply textual aids for remembering a family's past. In Bengal's volatile *jati* society, these texts had always had an ideological function and were frequently rewritten to articulate shifting ideological requirements. In the nineteenth century, their insertion into the domain of print was not devoid of further ideological operations. As Kumkum Chatterjee points out, "Kulajis [by the latter half of the nineteenth century] were . . . being viewed as artefacts from the past that could be pressed into the service of late-nineteenth- and early-twentieth-century agendas."[40] In Binodlal's case, not only did he press these early modern artifacts into a new, print-dominated domain of caste mobilization, but also cleverly braided his professional, entrepreneurial interests together with his caste-related ambitions. Thus, when he published the *Chandraprabha* the year after the *Ratnaprabha*, he advertised on its back cover both his pharmacy and his medical works such as the *Ayurveda Bigyan* as well as his two translations, the *Astangahridayam* and the *Nidan*. Going further, he declared that anyone who bought medicines of a certain value could have any of the medical or genealogical works for free.[41]

In subsequent years, advertising became more and more prominent as a way of intercalating caste and medical interests. Medicines and medical works were advertised in caste journals and caste matters were furthered through medical networks. On the pages of the caste journal *Baidya Protibha*, for instance, there are advertisements for medicines such as the *Chaulmugra Molom* (for all skin diseases), the *Prosonno Botika* (for malarial fevers), the *Shokti Botika* (for nervous weakness), etc., all advertised by Prasannakumar Sen.[42] There is nothing unusual about the medicines per se and they seem like any other set of patent medicines of their time that could be obtained from a specific postal address. What is interesting, however, is that from the same postal address readers could also order a set of books by Kobiraj Shyamacharan Sen. The three of the five titles offered were the *Bongiyo Baidyajati* (Baidya Caste of Bengal), the *Ambasthabrahman ba Baidyaporichay* (The Ambastha-Brahman, or Introduction to the Baidyas) and the *Baidyajatir Utpotti* (Origins of the Baidya caste). Predictably each of the works sought to argue and prove that Baidyas were Brahmins, that they had always been so and had the right to wear the sacred thread.[43] The medicines and caste books, thus, were once again bundled together. Similarly, on the pages of another journal, the *Baidya Hitoishi*,

Binodlal Sen's younger brother, Nagendranath Sengupta, advertised his medical texts by stating that

> [the advertised texts] belong to the Baidya society as a whole. For generations it has been their lot to selflessly serve humanity with nothing but respect for recompense. The books of the Baidyas are those which tell of how they saved the lives of others in the past and how they shall continue to do so in the future. . . . The advertised texts are a testament to how even today books that are completely written and disseminated by members of the Baidya caste can be complete and free of error.[44]

The works advertised by Nagendranath included titles such as the *Sochitro Kobiraji Siksha* (Illustrated Manual of Kobiraji Treatment), *Sochitro Daktari Siksha* (Illustrated Manual of Daktari Treatment), *Drobyogun Siksha* (Manual of the Quality of Substances), and *Poricharjyo Siksha* (Manual of Caring). The caste journals and involvement in caste mobilization thus became a way to advertise medical books, medicines, and even one's own practice. Just as medical works helped subsidize the publication of caste tracts, the latter functioned as vehicles for advertising the medical works.

There were two directions to *jati* politics: an upward push and a downward push. The upward push involved competing with higher-ranked groups. This meant both disseminating new origin myths, but also the internal reform of the social, ritual, and cultural practices of the *jati* so as to conform to the practices of the higher-ranked groups. In the past, this process was dubbed "Sanskritization" and was seen to be a mimicry of Brahmin customs.[45] Today, however, "Sanskritization" has been criticized for being too plastic. There is a lot of regional variation in terms of the models of behavior that are followed. Brahmins are not universally the model. Moreover, in regions like Bengal, Brahmins themselves might have a number of different models among them depending on, for instance, theological differences between Vaishnavas and Shaktas.[46] Hence the actual practices followed for successful mobility in any context are historically specific.

While the upward push receives much attention, the downward push is often relatively neglected. Any *jati* category, despite posing to be a biologically discrete group, is not so. As a result, as a *jati* successfully rises up the social ladder, more and more individual families seek to enter it at the bottom. Writing in the early 1880s, James Wise wrote that "in Silhet (*sic*) . . . Baidyas, Kayaths, and even Sunris are at liberty to intermarry."[47] Later, in a *Report on the Census of Bengal, 1901*, Sir Edward Albert Gait wrote that "east of the Brahmaputra Kayasths and Baidyas intermarry. The Sudras of East Bengal if well-to-do can generally manage to obtain Kayasth brides and eventually gain recognition as good Kayasths. Baruis and Maghs are believed sometimes to become merged in the Kayasth caste, so also do well-to-do Carims in Rangpur."[48] This infiltration at the bottom had two consequences. First, it diluted the privileged position sought by the upward push. Second, it constantly brought in families whose ritual, cultural and social practices were further away from those of the highest groups. Thus, in short, this constant infiltration at the bottom undermined the upward push. The downward push was thus necessary to police the boundaries of the *jati,* to try to keep out new entrants and assert the distance of the *jati* from everyone ranked lower than it. Genealogical projects, such as Binodlal's, were aimed at this downward push, while the others attended to the upward push.

These nineteenth-century efforts were, ironically, forcefully rejected by the next generation of leaders. Though Gananath Sen was by far the most prominent, energetic and devoted among the next generation of leaders, others such as Jaminibhushan Ray, Haripada Shastri, Satyacharan Sen, and many more were active to varying degrees in the Baidya movement. Some such as Nagendranath Sengupta, Binodlal's younger brother, did not partake of the leadership, but openly financially supported the efforts by placing advertisements in the movement's mouthpiece. Gananath, as the unquestioned leader of the *jati* politics of the Baidyas in the twentieth century, claimed that Gangadhar and his generation had been wrong to suggest that the Baidyas were equal to Brahmins. He contended that the Baidyas were in fact originally Brahmins—that is, that the relationship between Baidyas and Brahmins was not one of equality but rather identity. To that effect he founded the Baidya Brahmin Somiti and among other moves vigorously propagated the adoption of Brahmin surname "sharma" by all Baidyas. At one point Sen even dismissed those who held positions akin to Gangadhar's as "tales of a marijuana-addict" (*ganjakhuri*).[49]

Despite the rather harsh language, it was clear once again that Gananath was building upon Gangadhar's legacy just as he in turn had built upon Rajballabh's, if not Bisharad's, before him. Gananath's claims to Brahmin status were based upon a whole host of reasons including both the right to wear the thread that had been won by Rajballabh in the eighteenth century and the right to read Sanskrit-texts that had officially been admitted in the 1820s when Baidya boys were allowed to enter the *Native Medical Institution* to study under Bisharad. But crucially, around the time of Gananath, the Baidya's foremost right, and occasionally even a monopoly, over Ayurvedic practice began to be asserted. Thus, even as Gananath himself stopped short of making such claims and limited himself to exhorting Baidya boys to take up Ayurveda rather than enter the rat race for salaried employment, his colleagues went much further. One such author, Khagendranath Choubey, writing in the mouthpiece of the *Somiti*, wrote that "in this great city of Calcutta today, besides the Baidya-Brahmin Kobirajes, we find Brahmin Kobirajes, Kayastha Kobirajes, Napit (barber caste) Kobirajes, Tontubaye (weaver caste) Kobirajes, Chandal Kobirajes, and even Muslim Kobirajes. Many today claim that they have been practicing medicine for generations. In ancient times, however, all and sundry did not have the right to practice medicine. Only Baidya Brahmins had that right."[50]

It is unimportant to delve further into this *jati* politics of the Baidyas at this point. What is important, however, is to take careful note of two related facts. First, that those who shaped modern Ayurveda in Bengal had an equally avid interest in *jati* politics. Second, the cultural legacy of this was an assertion of both a legacy and a right to classical Sanskrit medical learning. The classical and textual bent that restricted the social base of the reform was therefore not a mere unhappy accident, but a tendency born out of the connections of the modernizing process with *jati* ambitions that were structured around claims of an ancient Sanskritic pedigree for the Baidyas.[51]

Creedal Cracks in the Small World

Gaudiya Vaishnavism originated in the wake of the tremendous social and religious upheaval caused by the medieval Bengali saint Chaitanya Mahaprabhu (1486–1534).[52] Though later channeled—at least in its mainstream form—into more orthodox directions, the spontaneous movement that had grown up around the Mahaprabhu had had a strong message of social equality and preference for personal devotional religiosity in preference to ritual and philosophy.[53] One of the oft-forgotten aspects of this movement was the prominent and disproportionately high participation of Baidyas in this movement. Mahaprabhu's foremost biographer, Krishnadas Kobiraj, was a Baidya and so were many in his inner circle. Ramakanta Chakravarti, in his authoritative work on the movement, writes that "it is beyond question that Brahmins and Baidyas numerically dominated Chaitanya's Bhakti movement."[54] This is even more significant when we recognize that demographically Baidyas are a much smaller caste than Brahmins. It was most likely the Chaitanya movement and its aftermath that gradually encouraged Baidyas to seek the parity with Brahmins. In any case, after the demise of Chaitanya, when the movement was molded toward more conservative directions and hierarchies were reintroduced, many of the Baidyas who had joined the radical movement under Chaitanya came to enjoy high quasi-Brahmanic status as gurus.[55]

Not all Baidyas looked at the Vaishnavas positively. But a complicated structure of self-governing societies within the Baidya *jati* had allowed for a fair degree of ritual and religious variation within the *jati* without bringing these into open conflict. The Baidya *jati* as a whole, like most other *jati*s at the time, was internally divided and subdivided into a number of distinct *samaje*s or "societies." These *samaje*s were internally self-governing on matters of ritual and social etiquette and were strictly endogamous. Generally, the *samaje*s claimed either a shared geographic origin or to have resulted from a mythic schism between the medieval King Ballal Sena and his son Lakshman Sena. Both these theories seem doubtful since, among other things, they often also had ensconced within them other self-governing *samaje*s. Thus there were the major *samaje*s of *Barendra, Rarhi* and *Banga.* Underneath these larger *samaje*s were numerous *sthan*s or "places" which too occasionally had their own self-governing status.[56] One of these smaller but autonomous *samaje*s was the *Panchakut-samaj*. This latter in turn was split into two autonomous "societies": namely, the Birbhumi-samaj and Senbhumi-samaj. These latter again in turn split into nine *sthan*s and thirty-two village-caste councils.[57] Other geographically organized "societies" enjoying varying degrees of self-government included *samaje*s such as *Srikhanda, Saptagram, Satsaika,* etc. In some cases, completely eschewing geography as an organizing metaphor, the *samaje*s were organized around patronyms. Thus there were in places "societies" such as *Sen-samaj,* the *Gupta-samaj,* the *Chaudhuripada-samaj,* the *Maulikpada-samaj,* and so on.[58] These self-governing *samaje*s could allow the Vaishnava and non-Vaishnava Baidyas to remain within the caste without forcing them to conform to a single religious identity.

Alcohol use provides an excellent case study through which to see this structure at work. It is worth noting that alcohol was a substance that, apart from its religious connotations, also had distinct medicinal uses. Yet we find that Baidyas were greatly divided upon the use of the substance. The Baidyas in Jessore, who were largely non-Vaishnavas, regularly consumed alcohol as a ritual substance. Giving evidence to the *Commission on the Spread of Country Spirit* in 1883, Shibendranath Gupta, a Baidya and a teacher at the Eden Female School, reported that "I speak of Jessore, where I was born, and I have found that among the Baidyas, the caste to which I belong, owing to the prevalence of Tantric worship, the practice of drinking was formerly observed only on particular days of the moon, and in the midst of *chakras.*"[59]

However, the Gaudiya Vaishnavas, including the Baidyas in their fold, were staunchly opposed to alcohol and preached abstention.[60] What allowed these divergent practices to be simultaneously contained within the Baidya *jati* was the institution of *samajes*.

Outside the *samajes*, as individuals vied for the overall leadership of the *jati* as a whole, these conflicts came increasingly to the fore. Thus, when Gangadhar made a bid for the leadership of the *jati*, he had to tackle the question of theological differences. Gangadhar was a Shaiva-Tantric—that is, a believer in a tradition of Tantrism organized around the worship of the god Shiva—and he sought to promote the Shaiva-Tantric perspective. Sectarian differences rarely meant the wholesale denial of the gods or goddesses of the opposing sects. Instead it meant the subtle assertion of the primacy of one's own sectarian deity within the pantheon by undermining the texts and claims to the paramountcy of other deities. Gangadhar thus denied the divinity and the authenticity of the *Srimadbhagbat*, a key Vaishnava text claiming supremacy for Vishnu. Naturally, this infuriated the Vaishnavas and they dubbed him a "Vishnu-hater" (*Bishnudweshi*). In the *Gobardhanbarnan* and the *Radhakrishnabarnan* he had tried to reach out to his Vaishnava detractors. Despite this, his own loyalties remained within the broadly Tantric camp with him embracing Shaktism (a moderated version of Tantrism) late in life.[61] Vaishnavas, therefore, continued to regard him with suspicion and dislike.

Once again, Gangadhar is not alone in expressing his religious differences openly. The process, once we recognize it at work, is all too ubiquitous in modern Ayurveda. Surendranath Goswami's writings are another excellent example of how *jati*, religion, and Ayurveda could be braided together. Goswami was the founder of the first short-lived Ayurvedic school entitled The Calcutta Ayurvedic Institution and Pharmacy in 1905.[62] After its demise, he also served as one of the founding faculty at the first Ayurvedic college established in Calcutta, the Astanga Ayurvedic Bidyaloy.[63] In 1916 Madan Mohan Malaviya invited him to take charge of the Ayurvedic department at the newly established Benares Hindu University. Though he had to decline the invitation owing to ill health, he was later elected president of the All India Vaidya Sammelan's annual meeting at Lahore. He also held a parallel allopathic qualification. According to an anonymous biographer, he was the first graduate of the Calcutta Medical College to have embraced Ayurveda.[64] Besides all this, Goswami was also a direct descendant of two of the Baidya members of Chaitanya Mahaprabhu's inner circle: namely, Sadashib Kobiraj and his son, Purushottam Das Thakur. Goswami's family enjoyed quasi-Brahmin status as hereditary gurus and his father made his living exclusively from such guruship.

When Goswami set out to transform Ayurveda, he did so clearly with an intention of subtly making modern Ayurveda more Vaishnava. The evidence was there to be seen on the very first page of what was arguably his most influential publication. The publication date on *Ayurveda o Malaria-jwor* was given according to an obscure calendar called the *Chaitanyabdo* that commenced from the birth of the Mahaprabhu. It is a calendar virtually unknown among non-Vaishnavas and was clearly used to make an ideological statement. But that was not all. In the book he argued for a complete overhaul of the way Ayurvedists had classified fevers for nearly a millennia. He argued that that classification was fundamentally mistaken. In order to get the classification right, one had to reread the entire Sanskrit medical corpus from the perspective of myths contained in the *Srimadbhagbat*—the very text whose authenticity Gangadhar had questioned.

Continuing further, Goswami argued that once the medical corpus was reread through the *Srimadbhagbat*, a completely new history of Ayurveda emerged. This new history, he argued, showed that the original humoral ideas about disease had been upstaged long ago by a new view that had established "ghosts" and "demons"—which he creatively redescribed as "germs"—as the preeminent cause of fevers. This "new view" (*nobyo-mot*) he asserted was "none other than Lord Krishna's."[65] He stated that the Vedic seers who had championed the old position had feared only two things: first, the manifest/perceptible (*protyoksho*) facts, and second, the words of Lord Krishna.[66]

It is from this perspective that some of Gananath's positions also begin to make sense. It is curious how he hails Surendranath respectfully as a pioneer, but often uses language bordering on the unseemly when speaking of a much better-known and respected scholar like Gangadhar. Yet this makes perfect sense when I take into account that Gananath's mother was the daughter of a family which had descended from a son who had been miraculously born to a childless couple in the sixteenth century through the direct intercession of the Mahaprabhu after the couple met him and sought his blessings.[67] The family was devout Vaishnavas, and Gananath had almost certainly grown up with some antipathy toward the man he would later compare to a marijuana addict.

My point, however, is not to reduce theological differences to petty rivalries and dislikes. There is, of course, some of that. But it is much more than that. My point here is that colonial religious categories like "Hindu," "Mohammedan," etc., do not work precisely because these alternate designations such as "Vaishnava" or "Shaiva-Tantric" continue to be meaningful for the historical actors concerned. How precisely it becomes meaningful of course varies. For some it is merely a matter of designating the "us" from the "them," but to others it is equally important in thinking of medical matters like the classification of fevers or the use of alcohol in medicines. Colonial categories substantially alter the importance of identity categories, often making categories like "Hindu" more important than categories like "Vaishnava," but they do not wipe the latter out. At least not in the period I am discussing.[68]

Moreover, these theological concerns also fed back into the question of caste. In Srinivas's classic model of "Sauskritization," for example, temperance from alcohol is one of the key modes of rising up the caste ladder.[69] But in a Tantrism-dominated area where even the Brahmins ritually consume alcohol, this model is naturally not going to work. It is significant that Ganath's mouthpiece for his *jati* politics, the *Baidya Hitoishi*, was relatively quieter on the temperance issue than many other contemporary *jati* movements were. It is quite likely that this was a strategic silence so as to not alienate the large numbers of Shakta and Shaiva-Tantric Baidyas.

Medical Orientalism and *Daktari* Education

Despite their commitment to Sanskrit learning and the continued relevance of traditional religious ideas, the Baidyas were also one of the earliest and keenest groups to embrace "Western" education. "Western" higher education in Bengal—and South Asia more generally for that matter—commenced with the founding of the Hindu College in 1817 (later renamed Presidency College), while the first officially sponsored medical college, the Calcutta Medical College, was established in 1835. By the middle of the nineteenth century, Calcutta University had been founded and numerous colleges imparting "Western" education gradually sprung up. Baidyas took to these institutions in large numbers. Their embrace of "Western" education was reflected in the disproportionately high representation of the Baidyas in the colonial government circles as well as the liberal professions. Census Commissioner and colonial ethnographer par excellence, H. H. Risley stated that "many Baidyas have distinguished themselves at the Bar, and as agents, managers, and school-masters, while others have taken to the study of English medicine and have entered Government service or engaged in private practice as medical men."[70] The Census report of 1901 explicitly commented on the disproportionate Baidya presence in the upper echelons of the government bureaucracy. In terms of populations, the report pointed out that the Brahmins and Kayasthas outnumbered Baidyas at ratios of 34:1 and 18:1, respectively. Yet seven of the Covenanted and Statutory Civil Service positions were held by Baidyas as opposed to a mere two by Brahmins and thirteen by Kayasthas. Among deputy and sub-deputy magistrates, seventy were Baidyas,

128 Brahmins and 144 Kayasthas. It was only among sub-judges and munsifs (lower-court judges) that the margins were relatively smaller, though here too the Baidyas held the upper hand. Here Baidyas held forty positions as compared to 136 for Brahmins and 160 for Kayasthas.[71]

This exposure to "Western" education had two major consequences for the modernization of Ayurveda. First, almost all the major agents of transformation after the early disapproval of Gangadhar had some training in "Western" medicine. Second, either through medical school or, more likely, during their general stream higher education, they had all been exposed to what is best described as "medical orientalist" writing.

"Medical orientalist" writing had its roots in the late eighteenth century and displayed many of the classic features of disciplinary orientalism. By disciplinary orientalism, I refer to the academic discipline inaugurated by scholars like Sir William Jones as opposed to the heuristic category of "orientalism" popularized in recent times by Edward Said. The two are of course related, but they are not identical categories and my intention is to describe the former and its impact on medicine and the Baidyas. The academic pursuit of orientalism was premised mainly upon the study of classical languages. In South Asia, this usually meant a study of classical Sanskrit. Michael Dodson has rightly pointed out that these early orientalist pursuits and their connections with governance resulted in the institutionalization of the cultural authority of Sanskrit learning and its purveyors. Thus, orientalist knowledge production based upon the relation of the Sanskrit-*pandit* and the European orientalist, "while seeking to undermine the *pandit's* control over matters relating to Sanskrit-based knowledge, simultaneously strengthened their ascendant socio-cultural status."[72] This resonated strongly with the Baidya *pandits* and their long-standing efforts to use Sanskrit learning as a tool for advancing their claims to higher status. Pascale Haag has recently drawn attention to the long-standing and multipronged attempts by Bengali Baidyas to claim Brahmin status through the study of Sanskrit.[73]

A whole host of orientalists, including Sir William Jones himself (despite his early negative assessment of Asian medical works), had written on medicine. Among the more prominent authors were H. H. Wilson, J. F. Royle, Allan Webb, F. J. Mouat and T. A. Wise.[74] These medical orientalists shared a common framework for conceptualizing "traditional" medicine. They all agreed in valorizing the classical medical works in Sanskrit over contemporary practice. They were dismissive and disdainful of the postclassical works written in vernacular languages. They also insisted on seeing the tradition as exemplifying a single, coherent body of work that could be labeled "Hindu." Indeed all of them insisted on naming the tradition they studied as "Hindu medicine" rather than as "Ayurveda." This "Hindu medicine" lacked any internal fissures or disagreements and formed a monolithic, systemic unity. Royle introduced the additional argument of this "Hindu medicine" having been the original source of all classical medical traditions of the world, including the Greeks.

Occasionally, and admittedly inconsistently, medical orientalists also fell back upon a more racialized discourse of "Aryanism." The ancient "Hindus" who produced the Sanskrit works became a racial group called the "Aryans." Thomas Alexander Wise, comparing the Chinese and the "Hindus," was thus able to write that Chinese "literature remained devoid of that refinement and elegance, that splendour and power which, from the Ganges to the Thames, has for five thousand years characterized the production of Aryan art and science, whether in the Sanscrit [*sic*] language, or its derivatives."[75] Tony Ballantyne has pointed out that "the theory of shared racial origins was a powerful influence in nineteenth century India because of its generality and flexibility."[76] The theories had little administrative impact and were often redundant at the regional level, but at a more general, civilizational level, it served diverse purposes and often subverted the increasingly racialized divide between the colonizers and the colonized. Together with a theory of degeneration, Aryanism and orientalism could work well in asserting the glories of an ancient Sanskrit-Aryan medical tradition and yet ignore and denounce contemporary practice and postclassical literature.

It was this framework that insisted on a singular, unity of the entire tradition, underplayed or ignored internal differences, marked the tradition as "Hindu" and, above all, valorized classical Sanskrit elements over both vernacular, postclassical writing and contemporary practice, which was embraced and adopted by the Baidyas.[77] And Aryanism was far more widely and consistently deployed by the Baidya authors than the European medical orientalists had ever done. Gananath Sen, in his *Ayurveda Samhita* wrote that "in the course of time, when the Aryan-nation (*Arjyo-jati*) became weak, Buddhist professors (*bouddhacharjyogon*) (who were) the legatees of Aryan knowledge (*arjyogyanadhikari*) along with a new religion distributed Indian knowledge treasures (*Bharotiyo gyan sompod*) among many nations of the east and west."[78] In a more poetic vein, Kobiraj Brajaballabh Ray wrote that

"swagoto" boli, turjo ninande	["Welcome" was the battle cry
dakila Ognibesh,	made by Agnivesa,
Sosh-uchhase uthilo kanpiya	Waves of passion
Arjyo uponibesh[79]	Wracked by the Aryan colony.]

Such Aryanism, in its appropriated form resonated with several agendas, including the politics of retaining and strengthening upper-caste privilege. As Tom Trautmann points out, by the end of the nineteenth century a "racial theory of Indian civilization" had become widely accepted. According to this theory, Indian civilization was the product of a clash and subsequent mixture of two races—namely, a light-skinned Aryan race and a dark-skinned Dravidian or aboriginal race.[80] Authors like Sen and Ray were clearly narrating the history of medicine in a way that made it the exclusive legacy of the light-skinned Aryans of the upper caste. This was the *jati* politics operationalized in the work of the Baidya Brahmin Somiti.

The most pervasive legacy of medical orientalism, however, was the consolidation of the cultural authority of Sanskrit classical medicine. Thus, Binodlal Sen began the preface to his *Ayurveda Bigyan* by stating that "there is not even the slightest scope for doubting the veracity of the hidden truths that the ancient seers had revealed by the power of their religious austerities and their pristine knowledge."[81] In support of his contention, he cited T. A. Wise's *Commentary on the Hindu System of Medicine.* In short, then, the uptake of medical orientalism by Baidya authors on the one hand promoted the cultural authority of ancient Sanskrit medical authors, but also did so in a way that emphasized greater claim of the upper castes to that classical legacy.

"Western" education also had a second, relatively more predictable impact. Most of the Baidya physicians who undertook the transformation of Ayurveda were exposed to some kind of "Western" medical education or the other. Tarashankar Bandyopadhyay in his acclaimed novel, *Arogya Niketan,* (The House of Healing), tells of the first generation of Kobirajes who learned "Western" medicine in the late nineteenth century. Their training was often through apprenticeships to *daktar*s (often themselves trained through apprenticeship or even autodidacticism) rather than through college education.[82] T. A. Wise, for instance, mentioned one Neemchand Desgupta, the scion of an ancient Baidya family, who had traveled far from his home in Orissa to Dhaka in order to apprentice with Wise.[83] Early authors of texts transforming Ayurveda, such as Gopalchandra Sengupta and Binodbihari Ray, were possibly educated in *daktari* medicine through such apprenticeships.

Toward the end of the nineteenth century, such ad hoc apprenticeships were replaced by more formal qualifications. Usually these tended to be the lower "LMS" (Licentiate of Medicine and Surgery) degrees, rather than the full MB degrees. Surendranath Goswami, Nagendranath Sengupta, Jaminibhusan Ray, and Gananath Sen—all of whom will appear in much of this book—were all holders of LMS degrees. It was on rare occasions that an MB, such as Hemchandra Sen, took up Ayurveda. The LMS degree was the lower of the two degrees awarded by Calcutta University to medical students since the University's inception in 1857. In hindsight today, it is often difficult to distinguish those who held the LMS degree from the holders of the VLMS (Vernacular LMS) degree. The former was awarded after a full five-year study in the MB course, though it was lower than the MB. The latter, on the other hand, was given to those who went through the Bengali or Hindustani Vernacular Classes where the training was not in English and the full duration of the course was three years.[84]

The extensive familiarity with and use of "Western" medical notions were a direct and obvious consequence of this *daktari* training. The training was also prominently advertised and, in itself, became a deployable sign in the cultural politics of modern Ayurveda. Trained in *daktari* medicine, these Ayurvedists claimed to be able to neutrally judge the merits of either "system" of medicine. At a deeper level, however, the two "systems" did not remain discrete in their minds. They interpreted the *daktari* training in light of what they knew from Ayurveda and transformed the latter by reading into it notions from their new *daktari* training. Modern Ayurveda was a result of this intricate braiding of two scientific knowledges.

In sum, the Baidyas, through their immersion in "Western"-style educational institutions and disciplines, acquired a deep-seated commitment to the epistemic structures of "Western" scientific knowledge. Their participation in institutional frameworks, both educational and professional, where such knowledge and its underlying assumptions were privileged, also cultivated in them an abiding commitment and affiliation to these assumptions and traditions of knowledge. These affiliations in turn competed and intricately interacted in their minds with the filiations to Sanskritic and precolonial religious traditions of knowledge.

Bugbear of Dissection

The intimate braiding of *daktari* education and the modernization of Ayurveda raises the specter of dissection. One story, often repeated in much of the extant historiography, is that Gangadhar Ray had decided to leave Calcutta upon hearing the cannonade that celebrated Madhusudhan Gupta's dissection of a corpse at the Calcutta Medical College. Though the story's veracity has been doubted, it clearly had apocryphal value in stressing the opposition of traditional Kobirajes to dissection.

Ironically, despite noting this opposition and Gangadhar's huge influence on the modernization process, very little account has been taken of how this opposition to dissection affected modern Ayurveda. Many historians have simply assumed that modernization of Ayurveda entailed a radical refiguration of the Kobiraji body imaginary and that this must have entailed an uptake of dissection. Madhusudhan Gupta's dissection, for instance, has been described by one recent author as a "scientific breakthrough" that

> reconstituted "psychologized" epistemology of the Indian knowledge system in the mould of objective, value-neutral, clinical detachment of modern medicine. As dissection became the primary means to know the human body, the living body was regarded as a kind of "animated corpse." . . . The act of dissection brought Calcutta on the same footing with London.[85]

This is precisely the kind of Whiggish hyperbole that is refuted by the apocryphal story of Gangadhar leaving Calcutta. Thus, if Calcutta had become London because of Madhusudhan's act, modern Ayurveda, one might say, left Calcutta straight away.

More cogently, as Shighehisa Kuriyama points out, merely dissecting a cadaver does not ensure that the dissector would arrive at the same anatomical conclusions. Brilliantly comparing ancient Greek and Chinese body knowledges, Kuriyama has convincingly demonstrated that what one sees in a corpse that has been cut open is still largely influenced by the culture within which the dissector operates.[86] The role of dissection in shaping modern Ayurveda has therefore been a bit of a bugbear.

Dissection was not as important to the early history of modern Ayurveda as some authors seem to suggest. In 1916, J. Donald, the Secretary of the Government of Bengal, spelled out the crux of the matter. In a lengthy memo he wrote that the government would not officially support any Ayurvedic college that did not teach cadaveric dissections; but since no college at the time was willing to do this and risk losing its orthodox students, Donald felt there would be little scope of incorporating dissections directly into Ayurveda. Instead, he advocated the teaching of "Western" anatomical textbooks.[87] Yet, in 1922, when a private college did indeed request the government for a supply of dead bodies with a view to teaching morbid anatomy, their request was refused.[88] Later in 1928 when the Astanga Ayurveda Bidyaloy, one of the top Ayurvedic colleges, opened its Dissection Hall, the government once again asked them to ensure their own supply of corpses.[89] What is remarkable is that this was already after over fifty years since the emergence of modern Ayurveda.

Until almost the 1930s, Ayurvedic students, including those who went to dedicated Ayurvedic colleges that taught a modernized curricula in one way or the other, did not practice cadaveric dissection regularly. In the late nineteenth century, when the more foundational developments took place, the scenario was even starker. The only connection cadaveric dissection had with modern Ayurveda was a mediated and episodic one. Some of the founding fathers of modern Ayurveda, especially those who might be thought of as a second generation following the pioneering generation of Gangadhar Ray, Gangaprasad Sen, and others, were educated in "Western" medicine before switching to Ayurveda. These men, such as Surendranath Goswami, Gananath Sen, and others, would obviously have encountered dissections as students at the "Western" medical colleges. This brief encounter in early life was all.

While it is true that some form of anatomical knowledge did percolate through them into modern Ayurveda, that knowledge was not a direct and regular knowledge formed through continuous exposure to the dissection table. Rather, it was mediated, episodic, and distant. Moreover, they had to communicate this knowledge not by repeated demonstrations, but rather by writing and illustrating books and articles using their earlier knowledge. Representations of anatomical knowledge, like all other forms of representations, are mediated. They cannot be taken to be identical with actual dissections. Such representations are subject to technical, epistemic, ideological, and even aesthetic mediations.[90] For the vast majority of Kobirajes, therefore, the knowledge of dissection arrived secondhand and mediated through words and images of a handful of reformist stalwarts.

There is both a tacit confirmation and an explanation of this absence, or at best the minimal presence, of cadaveric dissections in the formation of modern Ayurveda in Tarashankar Bandopadhyay's iconic and critically acclaimed novel, *Arogya Niketan*. The novel's central character, Jibon Mohashoy, the son of an Ayurvedic physician, sought out a *daktar* to learn "Western" medicine.[91] The latter, after much hesitation, accepted the new pupil as an apprentice and soon, impressed by Jibon's intelligence and devotion, came to see him as an heir. Yet, when the teacher sought to introduce his student to dissections, Jibon fell violently ill. His visceral reaction combined with deeply held moral and religious qualms about the desecration of the dead with simple human empathy. The corpse they were to dissect was that of an young girl and Jibon could not stop seeing the image of his young daughter in the corpse. Despite the teacher's reprimands and even a threat to discontinue the lessons, Jibon could not bring himself to dissect. Eventually, the teacher—himself deeply disappointed in his student's failure—had to accept that Jibon's medical training will have to proceed without dissections and that he will combine what he learns from the *daktar* to his dissection-independent understandings of the body.[92] Though this is a fictionalized episode, Tarashankar wrote the novel itself in close consultation with his friend and political comrade, Bimalananda Tarkatirtha. The latter was one of the most eminent practitioners of modern Ayurveda, descended from a long familial tradition of physicians and a key actor in the story of refiguration this book maps. Hence the incidents depicted in the novel lie somewhere in between the purely imagined and thinly veiled allegory of the real.

It is thus extremely interesting that not only is the modern Kobiraj, despite his quest for "Western" therapeutic knowledge, shown to be deeply averse to dissections but also that this aversion was understood as being equally located in religio-moral concerns and in the seemingly overwhelming capacity for empathy that trounced the emotional distance required for dissections. John Harley Warner and Lawrence Rizzolo point out that it was precisely around this time, at the dawn of the twentieth century, that dissections in the US were reimagined in light of new experimental sciences, by strongly silencing the human affective aspects of the act. By contrast, in Jibon's case—if we accept it as an allegorical portrayal of Ayurvedic modernity—there is a surfeit of affect that resolutely subverts dissections as a way of knowing.[93] Empathy, rather than emotional distance, is presented as the basis of modern Ayurveda's authority and capacities.

In short, then, two things are clear. First, dissections—as unmediated, actual practice rather than mere representations—did not play much of a role in the early history of modern Ayurveda. Second, the strength of sentiment against dissections even into the 1920s suggests that it cannot be held to be an important cause for the shifts that had taken place in Ayurvedic body knowledge in the second half of the nineteenth century. That cause had to be located elsewhere. Cadaveric dissections are a mere bugbear: a historiographic red herring.

Consumption as Modernity

The large-scale participations of Baidyas in the liberal professions and in government service meant that the *jati* leadership had a very pronounced bourgeois middle-class culture. Those who shaped the transformation of Ayurveda were undoubtedly members of this bourgeois class. The South Asian middle class, however, has been notoriously difficult to define. Ever since the appearance of a volume of essays edited by Carol Breckenridge entitled *Consuming Modernity* in 1995, a robust new scholarship has emerged on the South Asian middle classes. The works of Peter van der Veer and Christoph Jaffrelot, Mark Leichty, Markus Daechsel, Sanjay Joshi, Douglas Haynes, and Joanne Punzo Waghorne, among others, have given us a new set of analytical tools and critical insights into the South Asian middle classes.[94] One of the key contributions of this scholarship has been to turn the traditional Marxist scholarship on its head by arguing that consumption, rather than the role in production, is the crucial site for class formation. As Jaffrelot and van der Veer point out, "The middle class is a notoriously elusive social category"; it is difficult to define. Yet there seems to be consensus around the fact that it was absent in the precapitalist era. It is, as Jaffrelot and van der Veer baldly state, a "phenomenon of the capitalist era."[95] What unites this amorphous and rather heterogeneous social group is its "rather homogenous consumption pattern."[96]

What makes the middle class important for historians, suggests Joshi, is that in India, as elsewhere, it was the middle class that took the lead in defining what "it meant to be modern." But this definition of the modern that the middle class etched out was not a stable one. Joshi argues that, rather than seeing the class as a sociological fact, we need to see it as a continuous process of self-fashioning. "The definition and power of the middle class came from its propagation of modern ways of life. But Modernity in this sense," Joshi clarifies, "represents more than a fixed set of indicators regarding economic organization, social relations or even a single set of cultural values. To be modern in colonial India but also perhaps across much of the post-Enlightenment world was also an aspiration, a project."[97]

Going a step further, Markus Daechsel argues that the self-expressionism of the middle class is crucially inscribed upon the corporeality of the middle-class body. The middle class, Daechsel argues, seeks to acquire its hegemonic role in modern societies by insinuating its own body as the national body. A historically engendered, class-specific body is therefore metonymically passed off as the idealized body of the entire nation upon which biopower is then negotiated. The power of the middle class emerges then from the fact that whatever the outcome of the rival and heterogeneous biopolitical projects, their substrate remains the middle-class body. To access the historically specific shape of the middle class, then, we must acknowledge that the middle class "selves do not exist by themselves, they dwell in bodies that perspire, digest, have sex and die" and do so in class-specific ways.[98]

Three important elements emerge from this discussion of middle-classness. First, an aspiration to modernity; second, an emphatic forestaging of a carefully calibrated middle-class body as the ideal body; and third, the class's "homogenous consumption pattern." Each of these elements can be observed among the Baidyas involved in modernizing Ayurveda.

The aspiration to modernity took on a rather curious aspect within the transformational project. Instead of simply seeking to establish the modernity of their medical ideas, the Baidyas claimed that these ideas were in themselves ancient but either matched or more commonly surpassed modern "Western" ideas. Their modernity was therefore a sort of transcendent modernity that transcended narrowly chronological contemporaneity and somehow always remains modern. Throughout the subsequent chapters I will note this tendency toward transmodernity. A characteristic statement of this transmodernity can be found in Binodlal Sen's *Ayurveda Bigyan.*

> Some people erroneously believe that the therapeutic ideas expressed in the Sanskrit medical classics are either not completely correct or not relevant to our times. It is blatantly obvious for any man of wisdom to decide for themselves how legitimate it is to hold that those who had risen to the highest pinnacle of all sciences may have remained somehow deficient in Ayurveda.[99]

Beneath the circuitous and complicated prose, what Sen is saying is simply that the ancient therapeutic ideas are eternally valid and remain as relevant in modern times as they ever were. This is transmodernity.

Their forestaging of the middle-class body as the national body, or indeed the ideal body, can also be easily discerned from a cursory perusal of the Ayurvedic periodical literature. In a lengthy essay entitled *Onukorone Amader Obostha* (Our predicament through imitation), Satyacharan Sengupta, an eminent Kobiraj, poured scorn on the tendency to give up long-established customary practices of daily regimen and adopt "Western ways." Throughout the essay he spoke of "us" as the people who were suffering deteriorating health through the imitation of the "West." In one of many statements in the essay that added flesh to the eponymous "us," Sengupta wrote, "through 'Western' education we have come to try to imitate 'Western' society."[100] Clearly, the unmarked "us" was the middle-class body exposed to "Western" routines through the embrace of "Western" education.

Finally, a satirical anti-dowry poem published in the *Baidya Hitoishi* provides an excellent testimony to the uniform consumption patterns among the Baidyas. Surendralal Sensharma, bitterly critical of dowry demands among the Baidyas, listed the kind of things people usually demanded as dowry,

> Offering in cash two thousand
> gold and jewellery about sixty *bhori*

A gold watch with *bel*-leaves and sandalwood-paste
chair, table and a bed
Tableware two sets: of silver and gun-metal
gramophone and a sewing machine
Diamond ring for the right hand
Pearls and rubies are not desired
To make travel comfortable
Motor-cycle is a must
A harmonium is a limb of a song
No one complains to have one.[101]

More than the perceived homogeneity of demand for commodities during weddings, what is remarkable in the list is the number of technological objects that are included. From gramophones and sewing machines to watches and harmoniums, this is not a "traditional" list of consumables. Rather, it is an emphatically modern list made up largely of technological commodities. Yet their demand is embedded within a seemingly traditional practice of dowry and ritualized through overt markers such as the *bel* leaves and sandalwood paste.

Given that this poem appeared in a Baidya caste journal, it amply sums up the centrality of bourgeois consumption to not only Baidya caste identity but also modern Ayurveda that, as I have shown, was intricately intercalated with these caste periodicals. By positioning a uniform pattern of consumption of technological objects couched significantly within a seemingly traditional and ritualized milieu, the Baidya caste was aspiring to its own version of modernity or, indeed, transmodernity, to be more accurate. It was this techno-consumerist transmodernity that was to shape modern Ayurveda as well. The coming together of consumption and technological commodities is central to the arguments of this book. Throughout the period under study, we find the Baidyas braiding their precolonial intellectual resources with the latest "Western" intellectual trends on and around technological objects.

The usual advertisements published in the Ayurvedic periodicals of the day also reflect this obsessive consumption of technological commodities. Hence, around Christmas, there were advertisements in Gananath Sen's *Journal of Ayurveda* for luxury pocket watches and pens offered on discount.[102] Similarly, in the caste periodical, *Baidya Hitoishi*, one of the most regular advertisements was Indumadhab Mullick's *Ik-Mik Cooker*, an automatic rice cooker.[103]

The technologically nurtured transmodernity that emerged as an aspirational goal for the bourgeois Baidyas deeply structured their medical project

1. Kaviraj Gangadhar Ray in traditional attire.

as well. Thus, in his annual speech, the president of the annual Baidya Sommelon categorically stated that "it is our duty (*kortobyo*) to look favorably upon the inventions of modern science."[104] He then proceeded to introduce the audience one by one to such recent inventions as the X-rays, oxygen inhalers, saline drips, stethoscopes, and so on.

An excellent exemplar of the way in which the consumption of technological inventions shaped both Baidya identity and influenced their engagement with Ayurveda is photography (see fig. 1). Gangadhar himself was photographed and his photograph had been widely reprinted in books on Ayurveda. Upon inspecting the photograph closely, an expert and deliberate attempt to shape the message communicated by the image can be discerned. In the photograph, Gangadhar is seated like a Sanskrit *pandit* with a manuscript laid out in front of him. More importantly, the shawl that is loosely draped over his shoulder is just sufficiently parted in the front to show the bright white sacred thread proudly hanging across his chest. From what

I have discussed of Gangadhar's politics above, it is clear that both the claim to Sanskrit learning and wearing the sacred thread were central to his politics. It is therefore almost certain that these elements had been deliberately organized to communicate exactly the political message he sought to communicate through his published works. Christopher Pinney has pointed out that though photography was quickly pressed into the service of anthropological objectification of colonized people, both anthropology and photography also gradually—and most likely through their mutual entanglement—fumbled toward an awareness of their "own uncertainty and impossibility" in the period.[105] But even as the sinews of imperial power tried to work out just how far photography could aid its agendas, colonized elites like the Baidyas began to develop enough savvy to appropriate and turn the technology to their own ends. Gangadhar's careful calibration of his own photograph is a glowing testament to the precociousness of that savvy. An even clearer example of such manipulation can be seen in Surendranath Goswami's use of photographic technology. I have been able to track two different photographs of his. Both were published in periodicals (see figs. 2a and 2b). The first one (fig. 2a) for which I have not been able to trace the original forum of publication,[106] shows Goswami dressed in a dhoti, with a formal buttoned jacket and black shoes. He has a shawl around his neck and holds a walking stick. He is seen seated on a typically ornate Victorian chair with one arm resting on another ornate and typically Victorian side table alongside a flower vase. In striking contrast, the second photograph (fig. 2b) depicts him wearing a dhoti and a shawl and sitting on the floor with his legs folded as pandits do. His shawl is draped in such a fashion as to reveal his sacred thread and in front of him is possibly yet another manuscript—though the poor quality of the publication has somewhat obscured this. In short, it is almost exactly the same pose as Gangadhar before him had struck. This photograph was published in a periodical with pronounced Vaishnava sympathies alongside Goswami's obituary. The obituary emphasized his piety and recounted two alleged miracles from his childhood. Goswami himself had been a regular contributor to the periodical and had first published his views on the Vaishnava interpretation of Ayurvedic ideas about fevers in this very periodical.[107]

The most striking example of the consumption of photographic technology by the Baidyas and its connection to the new Ayurveda came more than a decade and a half before Goswami's passing. Hiralal Sen, a Baidya from east Bengal, became in the 1890s the first South Asian filmmaker. Together with his brother, Motilal Sen, he opened a photographic studio in Calcutta and began to simultaneously make films as well. In 1903 or 1905, he put the

2a. Kaviraj Surendranath Goswami in modern attire. Courtesy: Srijib Goswami.

2b. Kaviraj Surendranath Goswami in traditional attire. Courtesy: Srijib Goswami.

new medium of moving films to a new use. He began to make ad films. All three of these pioneering ad films were for medical products, but the best-known among them was for none other than Chandrakishore Sen's flagship product, *Jabakusum Hair Oil.*[108] Once again, ironically, the very product the empty bottle of which started me on the path of this book.

Therapeutic Cliques

Having thus far insisted on the homogeneity and coherence of the social base of new Ayurveda, I would conclude by acknowledging and contextualizing the internal dissonances within this overwhelmingly homogenous group. Charles Leslie, though speaking of a slightly later period, has helpfully alluded to the debate between the proponents of *Suddha* (Pure) and *Misra* (Mixed) Ayurveda.[109] The supporters of the former, he points out, sought to keep out "Western" medical influence on Ayurveda, while the latter advocated a syncretic fusion of the "East" and "West." These debates were not absent in the period I am looking at.

In fact, these differences emerged particularly sharply in the 1920s when government patronage, or at least its promise, for the first time became consistently available to "indigenous" medicine. Sivaramakrishnan points out that by the 1920s, in the Punjab, *Vaid* publicists had made their way into the level of localities and were demanding and obtaining support from district boards and municipalities. This patronage in turn raised new demands for clearer protocols to distinguish the "authentic traditional" from the "quack." A combination of local-level patronage and print-based publicity thus drove the greater factionalism among Ayurvedists in the 1920s.[110] Recently, Rachel Berger has made a similar argument for the United Provinces. Berger, however, links these developments more closely to the administrative diarchy created by the Montague-Chelmsford Reforms of 1919. The provincial ministries made up of elected Indian politicians now became able to dispense patronage in the realm of grants-in-aid to dispensaries and sought to channel some of this to "indigenous" medical traditions.[111] This potential for government patronage in turn led to a whole slew of new alliances, self-presentations, and schisms.

In Bengal, after M. K. Gandhi forced the Congress Party to abstain from legislative council elections in 1921, the local leader, Deshbandhu C. R. Das, led a breakaway faction of the Congress under the name of Swaraj Party to electoral victory in municipal elections in 1923. Das, however, alienated the Hindu right wing by signing the "Bengal Pact," which agreed to share power with Muslim politicians and reserve government jobs according to demographic proportions. This division was to have strong implications for the emergent shape of modern Ayurveda.

It is not clear exactly when lobbies began to form within the Ayurvedic fold in Bengal, but by the early 1920s it was clear that different groups of Ayurvedists were closer to different political patrons. At the outbreak of the Non-Cooperation Movement in 1919, for instance, C. R. Das approached

Shyamadas Bachaspati to set up a "nationalist" Ayurvedic college. The Astanga Ayurveda College established by Jaminibhusan Ray and Gananath Sen was already in operation for about three years at this time.[112] Das's having approached Bachaspati suggests both his proximity to the latter and his alienation from the former group. What may have alienated Das may have been the strident *jati* politics of Jaminibhusan and Gananath. As a Baidya himself, Das would have been well aware of the *jati* politics promoted by those associated with the Astanga Ayurveda College. As a resolutely secular leader committed to nationalist unity in Bengal, Das would have naturally not wanted to associate with them.

In the meanwhile, two years before the Swaraj Party won the municipal elections, elected legislators from the 1921 elections, Kishori Mohan Chaudhury and Dr. Hassan Suhrawardy, had respectively moved two motions for the establishment of two independent committees to investigate the possibilities of reviving and supporting the Ayurvedic and Unani medical traditions.[113] The Ayurvedic Committee was presided over by Dr. M. N. Banerjea, principal of the Carmicheal Medical College and included six Kobirajes: namely, J. B. Roy, Gananath Sen, Hemchandra Kaviratna, Jogendranath Sen, Jadunath Gupta, and Shyamadas Bachaspati.[114] The committee was clearly dominated by the Astanga Ayurveda lobby. Besides Jaminibhusan and Gananath, who were prominent members of the Baidya Brahmin Somiti, Hemchandra Sen (Kobirotno) was the chief executive (*Sompadok*), Jadunath Gupta was a member of the Somiti's Working Committee, and Jogendranath Sen was the joint president of the Somiti.[115] Thus cornered, Bachaspati resigned soon afterward and no one was appointed in his place.[116] The committee had been formed in August 1921 and Bachaspati resigned in January 1922. In the meanwhile, in November 1921, another member of the Legislative Council, Hem Chandra Nasker, moved a motion urging the government to establish a "State Faculty of Examiners" which would regulate the practice of Ayurveda, Unani, and homeopathy. The government recommended this for consideration to the Ayurvedic and Unani Committees, respectively.[117] The one-sided committee attempted to set up a case to implement its vision of Ayurvedic revival. It enthusiastically received a plan from Dr. Girindranath Mukhopadhyay for the establishment of a "Government Central Ayurvedic College" and Sen himself showed a keen interest in the plan.[118] Though Mukhopadhyay appeared before the committee as an independent witness, he was known to be particularly close to Sir Ashutosh Mookerjee, the powerful vice-chancellor of Calcutta University. Sir Ashutosh was also known to be politically sympathetic to the emergent Hindu revivalist political groups. Most of Mukhopadhyay's books on the history of Ayurveda had been personally supported by Sir Ashutosh and funded by Calcutta University, over which he presided.

By this point, the neighborhood networks were also beginning to come into play. Bachaspati seemed to be consolidating the northern Calcutta Ayurvedists, while Sir Ashutosh Mookerjee and Girindranath Mukhopadhyay were both old and respected residents of Bhabanipur and seemed to be consolidating the southern lobby alongside Gananath.

With the Swaraj Party winning the Calcutta Municipal Corporation (CMC) elections, however, the balance of power shifted. As the first mayor of the CMC, C. R. Das appointed Subhas Bose (later to be known as "Netaji") as the first CEO of the CMC on April 14th, 1924.[119] That very month, Bachaspati applied to the CMC for a capital grant of Rs. 25, 000, a recurring annual grant of Rs. 10, 000, and a plot of rent-free land for his college.[120] The application was forwarded to the Ayurvedic Committee by the new corporators. With the tables turned, the committee now changed its strategy. Instead of deciding on the application directly, it now proposed that all three Calcutta colleges be amalgamated into a single, major Ayurvedic college and hospital. Kobiraj Ramchandra Mullick had established the third college, the Gobindo Sundori Free Ayurvedic College, in 1922. Until this time, this last college had been funded exclusively by the royal family of Kasimbazar.

Mullick himself was opposed to Sen's camp. He had written the foreword to the pamphlet published by Haramohan Majumdar, criticizing the inconsistencies in Sen's version of traditional medicine. But the Kasimbazar royals were, like Sir Ashutosh, one of the early sympathizers of the right-wing Hindu Mahasabha and two of the three representatives sent by the Gobindo Sundori College for the negotiations were members of the royal family rather than physicians. It may have been this that led Sen's camp to invite all three colleges in the first place, in the hope that with support from the Kasimbazar royals, they would be able to outmaneuver Bachaspati.

The Ayurvedic Committee invited representatives of the two other colleges, Bachaspati's and Mullick's, to a meeting to consider the proposal. Leading on from there, they set up a fifteen member Special Committee on Amalgamation. The composition of this new committee suggests that an effort had been made to balance the representation of the different lobbies. The Baidya Brahmin lobby was represented by Gananath, Jaminibhusan, Jogeshchandra Sen, and two loyal lieutenants of Sir Ashutosh—namely, his own son, Ramaprasad, and protégé, Girindranath Mukhopadhyay. On the other hand, there were a number of Bengal Congress heavyweights to back-up the Bachaspati lobby. Dr. Kiran Sankar Roy and Dr. Sundari Mohan Das, both known for their proximity to C. R. Das and Subhash Bose, were appointed to the new committee.[121] Mullick and two members of the Kasimbazar royal family represented the Gobindo Sundori College faction. The Special Committee eventually invited detailed plans for the amalgamation. Jaminibhusan and Gananath submitted a joint proposal while Mullick and Bachaspati each submitted their own proposals. None of the other members submitted a plan. As part of the plan, a new Provisional Committee was set up with three nominated representatives from each college.[122]

In 1925, Gananath tried to outmaneuver Das by cultivating close links with Gandhi with whom neither Das nor Bose always agreed. Gananath invited Gandhi to lay the foundation stone of the new building, the Astanga Ayurveda College. Gandhi, whether because he was keen to build bridges with the hugely popular renegade, Das, or simply because he wanted to speak his mind, attended the ceremony and launched into a scathing attack against the pretentious claims made by Ayurvedists. The event backfired. Gananath was deeply embarrassed for having invited Gandhi and engaged in a lengthy correspondence trying to persuade the politico to issue a rejoinder mollifying the attending Ayurvedists who had felt insulted. Gandhi refused to issue such a rejoinder and stuck to his allegations of humbug against contemporary Ayurvedists.[123] Gananath was left with little option but to gravitate further and further toward the Hindu right wing. In the course of time, such stalwarts of the right wing as Madan Mohan Malviya and Shyamaprasad Mookerjee, Sir Ashutosh's son and a founding father of the right wing Jana Sangh, became his political patrons. Bachaspati, though briefly marginalized from the Congress after the forcing out of Bose by Gandhi, was soon rehabilitated into the mainstream of Congress politics. His son, Bimalananda Tarkatirtha, went on to become the Congress Legislative Party's chief whip in the West Bengal Legislative Assembly in the 1950s.

The careful balancing act in constituting the committee in the 1920s was effective in blocking Sen from taking over the project of transformation of Ayurveda officially, but it also acted as an impediment to any kind of progress. Neither the common curricula nor the amalgamation of the three colleges were achieved till decades after independence. Though a State Faculty of Ayurveda was established by 1937, its impact was marginal. Its licensing protocols were undermined by the lack of any standard examination system and a perusal of the registers maintained by the faculty shows numerous physicians were able to register by obtaining certificates testifying to their having practiced medicine for a certain number of years. In effect this meant that there was no standardization of practice that was achieved at either the educational or the regulatory levels.

Ironically, today this history of bitter factional feuding is often telescoped into the history of the conflict between *Suddha* and *Misra* Ayurveda. Shyamadas's grandson and Bimalananda's son, Brahmananda Gupta, has largely contributed to the constitution of this memory. But to frame the entire conflict simply as a conflict between "purists" and "syncretists" is patently misleading. As I have shown, there were many other dimensions to the conflict that overdetermined the bitterness and barren politicking that went into it. These dimensions included both the development of new forms of political patronage as well as the very idea of *jati*-based political mobilization. In fact, one could even notice a faint glimmer of the older rivalries between preceptorial lineages in the conflict, with most of Gangaprasad Sen's intellectual heirs coalescing under Gananath Sen and most of Gangadhar Ray's heirs gravitating toward Shyamadas Bachaspati.

While the bitter and centuries-old Vaishnava and Shaiva-Tantric rivalries had been abating by the 1920s, the sedimented bitterness of those rivalries still cast their dim shadows on the new configurations. The demise of the older religious divide had, however, partly been replaced by the new religious antagonism to Islam. On the one hand was the Hindu Mahasabha affiliates with their strident Hindu chauvinism, while on the other hand were the followers of Das and Bose with their more inclusive vision of society. These complex and shifting groupings—some already past, but yet casting a shadow, others still emergent but already beginning to acquire heft—overdetermined the factional politics that riddled modern Ayurveda in 1920s Bengal. Straightjacketing that complex conflict into a simple dispute over "purity" and "syncretism" is a caricature.

Moreover, to the extent the dispute was about *Suddha* and *Misra* Ayurveda, the "purists" and the "syncretists," at least in the 1920s, it was largely a matter of theory rather than therapeutics. They all clearly agreed on therapeutics. To portray the "purists" as opposed to any "Western" influence is to gravely misunderstand their position. All the people involved had adopted "Western" inventions in therapeutics. Mullick is a good exemplar of this. Through his preface to Haramohan Majumdar's pamphlet, Mullick had explicitly opposed Gananath's syncretism in print. Yet the same Ramchandra Mullick also authored two medical textbooks that included a number of *daktari* drugs as well as diagnostic techniques such as reading the pulse by the pocket watch. Some perspective of Bimalananda's views on the matter can be gathered from the novel *Arogya Niketan*. The novel had been written by his close friend and fellow Congress legislator, Tarashankar Bandyopadhyay, and Bimalananda's son, Brahmananda, recalled the author's long conversations with Bimalananda about the plot of the novel. Not only were the characters in the novel thus modeled on Bimalananda's family of hereditary Ayurvedists, but he also actively gave his input in its writing.[124] In the novel, the central character, Jibon Moshai, in whom can be seen a fairly clear shadow of Bimalananda himself, voluntarily sought out and learned "Western" medicine. He was revolted by anatomical dissections and never performed them, but he embraced *daktari* therapeutics so far as it was useful to his practice. Moshai's medical repertoire was rightly described by another character in the novel as a "medical tricycle" made up of Sanskrit learning, folksy remedies and some *daktari* methods. This was possibly true of Bimalananda as well. All the factions of the 1920s, despite the shrill, were in effect riding the same "medical tricycle."

Conclusion

In conclusion, what I have delineated in this chapter is the historico-sociological map that underpinned the modernization of Ayurveda. Most importantly, I have etched out this map not by relying on colonial categories of identification, but, following Sahlins, by tracing the identity categories of historical agents themselves in action. The differences that this alternate perspective has revealed at every juncture have been significant. For starters, the Baidya caste identity—as understood by both colonial ethnographies and historians like Bala who depend upon the former—has come undone as a discrete, biologically engendered ethnic referent. Instead, the label "Baidya" has been revealed as a dynamic and evolving identity category tied up with a range of struggles for social power, status, and legibility. Similarly, the idea of Ayurveda as the medical tradition of a discrete community called the "Hindus" has also rung hollow. Instead, what are now seen to be mere sectarian differences within Hinduism have been shown to be much more oppositional, sovereign, and important in an immediate sense for the actors than some larger "Hindu" identity until the 1920s. Finally, the 1920s period has been shown to be a period of stasis engendered by a conflict that was much more multifaceted than merely a clash between "purists" and "syncretists." At a moment of significant changes brought about by alterations in the political and commercial context of modern Ayurveda, a range of old and new oppositions coalesced into inchoately articulated but overdetermined bitterness between powerful factions backed by hefty political patrons. Religious tolerance, epistemic difference, old rivalries between preceptorial lineages, neighborhood antagonisms, political affiliations, and perhaps much else were operationalized in the microsocialities of the 1920s conflict that stalled the modernization project in Bengal.

TWO

The Clockwork Body: The Pocket Watch and Machinic Physiospiritualism

Rangalal *daktar* briefly felt the patient's pulse. Despite her incredible tolerance, she let out a slight sigh of pain. The *snayu*s and *shira*s were all stretched taut to their limit and even a slight touch threatened to tear them apart. Rangalal was thus deprived of the "opportunity to sense the *nadi*" (*nadi onubhaber obokash*). "He simply impressed his thick fingers on the patient's wrist and counted the number of beats. He noted whether it is regular or broken. Nothing more. He did not even try to fathom anything further." The patient's condition was such that Rangalal could not even be certain that he had felt a pulse at all. With brows furrowed in thought, Rangalal turned to Jibon and said, "Have a look, Jibon. Tell me what your *nadipariksha* [Kobiraji pulse diagnosis] tells you."

As Jibon stepped up to the bedside, he recognized the importance of the situation not only for the patient but also for himself. Rangalal *daktar* was a brilliant, self-taught *daktar* and Jibon had tried to learn *daktari* medicine from him. Rangalal had accepted him as a student after much hesitation and sought to make him in his own mold. Jibon, however, had refused to dissect a human corpse. He had resolutely stated that Kobiraji medicine, even without engaging in the inhumanity of cadaveric dissection, had much to offer. An enraged Rangalal had thrown Jibon out, accusing him of not having the courage to break with ossified traditions and face the future. This was the first time Rangalal had spoken to Jibon since. It was a test—a test not only for Jibon, but for that hallowed Kobiraji tradition whose moral universe he sought to uphold. Rangalal was an upright man, a man of strong principles and he was giving his beloved ex-student one last chance to prove to him the truth of his assertions.

Jibon did not pick the patient's hand up from the bed. It lay there just as it was. "He gently put his fingers on the wrist. As he closed his eyes, the curtains came down on the world around him. A barren tree that had lost all its leaves stood in front of him. A single leaf remained. It flickered gently in the slightest of breezes. He had to be able to feel that slightest tremor of the leaf. A moment's inattentive pressure might break the leaf of its stalk. He sat, as though in meditation, seeking to evoke the keenest sensibilities in him." Jibon's father, who taught him Kobiraji, used to say, "It is in the nature of force (*shokti*) to progressively become subtle with use. Sensitivity (*onubhuti*) is a force. The more you use it, the subtler it will be. If you make it gross, it will be a mace."

Impatiently, Rangalal asked him what he felt. To Jibon, Rangalal's voice came from afar. With the slightest nod of the head, Jibon signaled to Rangalal, taking care that the nod not move his fingers even the slightest. By delving deeper and deeper into his trance-like state, he "instigated the flickering light of his knowledge and sensitivity to burn at their brightest" and in that light he sought out that single leaf that trembled in the breeze. Time passed Jibon by. He did not know how long he had been in the trance-like state. When he emerged from it, he confidently told Rangalal that the case was not one of tetanus. It was instead a complicated case where chronic *ojirno* (a Kobiraji pathology arising from faulty digestion) had been aggravated by a bad fall. It was a difficult case, but not incurable as Rangalal had feared.[1]

This dramatic scene is adapted from Tarashankar Bandyopadhyay's critically acclaimed novel, *Arogyo Niketon*. Set in the late nineteenth century, the novel tells the story of a family of hereditary Kobirajes grappling with the onrush of modernity and the then-emergent authority of biomedicine. In this key scene, as elsewhere, Bandyopadhyay, deploys *nadipariksha* as the iconic and exemplary instance of Kobiraji knowledge which marks both its distance and its superiority from "Western" medicine. The rich tapestry of metaphors and allusions through which this alterity between Kobiraji and "Western" medicine is contrasted is in need of unpacking.

Daktari pulse diagnosis is shown to be superficial, quantitative, and incidental. By contrast, Kobiraji *nadipariksha* is mystical, subtle, qualitative, and so on. Its truths are resolutely metaphorical, frequently conveyed through rich visual metaphors such as that of the flickering leaf on the barren tree, in sharp contrast to the numerical truths of *daktari* pulse. Such contrasts mapped on only too well onto the older, orientalist register of binarisms that presented the "Spiritual East" and its "mystic knowledges" in stark opposition to the "Scientific West" and its "precise knowledges." Though Bandopadhyay had been extremely careful to research his topic thoroughly and wrote *Arogya Niketan* in close consultation with his good friend Bimalananda Tarkatirtha, one of the stalwarts of refigured Ayurvedic medicine as well as a scion of one of the most respected families of hereditary Kobirajes, neither Bandopadhyay nor Tarkatirtha could escape their own historical moment. By the time when the novel was written in the 1950s, certain aspects of orientalist myth making, particularly contrasts such as the "Spiritual East" versus the "Scientific West" or "mystic subtlety" versus "numerical precision," had become deeply embedded in the self-presentation of refigured Ayurveda. Rachel Berger's insightful recent study has partly described this process in her discussion of the "invention of biomorality."[2] Behind this work of "invention" that Berger uncovers lies the lurking shadow of a more complex historical reality: a reality that cannot be resolved into easy binarisms; a reality where "systems" are porous, opposites interpolated, and identities scrambled.

It is worth noting here that, well into the second half of the nineteenth century, many "Western" physicians remained loyal to the Galenic tradition of qualitative pulse diagnosis. Stanley Joel Reiser points out that "all of [Galen's] interest [in the pulse] . . . was limited to a qualitative account . . . he did not investigate the value of its rate of beatings."[3] Though attempts to numeralize the pulse started way back with Herophilus in the fourth century BC, these attempts remained marginal till the nineteenth century.[4] Even in the nineteenth century, when the trend acquired new vigour, it did not fully stamp out the Galenic interest in qualitative pulses. A survey of the case reports of British doctors published in the *Transactions of the Medical and Physical Society of Calcutta* in 1835 demonstrates how the new quantitative ideas about pulse and older Galenic qualitative pulses coexisted within the "Western" medical repertoire of the time. Dr. T. Mouat, MD, writing of the "epidemic diseases which occurred in Bangalore in 1833," for instance, wrote of patients with "slow, feeble, or oppressed pulse"; in another instance, he spoke of the "feeble pulse," while elsewhere in the same lengthy essay he described the pulse as "generally quick often hard." It was only in the last instance that he then added that "yet in several cases, and severe ones too, the pulse was little affected or did not rise above 80 or 90."[5] Duncan Stewart, MD, writing on a type of fever prevailing at Howrah in 1834 similarly spoke of one patient having a "quick, hard pulse." On the very next day, he wrote of the same patient, "pulse 110." On the third day, still writing about the same patient, he wrote, "pulse 120 and hard." On the fourth occasion, he put down "pulse rapid and unsteady."[6] Clearly, while numeralization had definitely entered the "Western" medical repertoire, it had not replaced the Galenic qualitative pulses, much less become iconic of "Western" pulse lore. The clear-cut binarism represented in later sources, such as the *Arogyo Niketan*, which were close to the self-representation of modern Ayurvedic physicians, was clearly misleading.

The greatest difficulty in writing the history of modern Ayurveda frequently lies in unpicking the emic self-presentation of modern Kobirajes and the etic realities of their practice. Particularly in the absence of clinical records and relying almost exclusively on the more public writings of these Kobirajes, it is difficult to tease apart their self-presentation and their therapeutics. The historian's natural inclination toward the bigger issues and the ideological and theoretical differences too has contributed toward confusing emic self-presentation with etic therapeutic realities. Nowhere is this difficulty more acute than with regard to *nadipariksha*. Despite being a relatively recent addition to canonical Ayurvedic writing (not predating the thirteenth century under any condition), it has come to acquire such iconic status that any public discussion of it is frequently overlaid with politically redolent self-fashioning and myth making of the sort that can be seen in the contrasting descriptions of Bandyopadhyay.[7] Yet the few surviving nineteenth- and early twentieth-century textbooks, which give a better glimpse into the therapeutic realities of everyday Kobiraji practice, reveal a much more complex and entangled reality.

Perceptive ethnographers point out that "pulse therapy, though peripheral to many textual constructions of the field, remains the sine qua non of the astute clinician who is said to be able to diagnose complex and previously untreatable disorders with a single reading of the afflicted person's pulse."[8] Unfortunately, the synchronic gaze of ethnography fails to notice that the thing called "pulse" has not always been what it is today. Since colonial times it has been a site of intense boundary marking, yet it has also been transformed through those battles. In the end, what appears to "remain the sine qua non of the astute clinician" is not a remnant at all, but an entity born by braiding the old and the new, the indigenous and the exotic, the remnant and the recent.

And the spindle with which the braiding was possible was the pocket watch. It was the humble pocket watch around which the beats of the pulse and the subtle movements of the *nadi* wrapped themselves around each other.

Modern Times

The role of the pocket watch as a medical technology has largely been ignored. Moreover, the history of time-keeping devices in India is also mired in neglect. Though it is recognized that the Mughals had largely rejected mechanical clocks, little beyond that is known.[9] While much has been written about the accommodation and conflict between precolonial and colonial macro-temporalities, much less historically based attention has been paid to the micro-temporal changes ushered in by colonial modernity. As a result, there is a wealth of studies on rival notions of historical or nonhistorical macro-temporalities in British India, but practically nothing on the actual histories of watches and clocks.[10] This is particularly striking given the fascinating work on the adaptation of clocks in other non-Western cultures.[11]

A popular children's rhyme in nineteenth-century Bengal catalogued the remarkable achievements of the age thus:

Ghani Miyar Ghori
Nilamborer Bori
Gokul Munshir Gonphe Ta
Golpo shunbi to Mrityunjoy Munshir kachhe ja.[12]
[Ghani Mian's watch, Nilambar's tablets, Gokul Munshi's mustache,
to hear a tale go to Mrityunjoy Munshi.]

Another proverb from the same time also had a very similar ring to it: "*Nilamborer bori o Ghani Miyar ghori kokhono bhul kore na*" (Nilambar's tablets and Ghani Miya's watch never miss their mark).[13] The Nilambar of these popular utterances was Nilambar Sen, one of the founding fathers of both modern Ayurveda and an eminent dynasty of reformist Kobirajes.[14] Ghani Mian most likely was the Nawab of Dhaka, Sir Khwaja Abdul Ghani Mian (1813–96). The association had possibly resulted from the fact that Nilambar Sen had originally hailed from Dhaka. But it is significant that popular memory had serendipitously connected the watch with the modernizing of Kobiraji medicine.

Though the pocket watch is seldom recognized, even in the "West," as a medical instrument, it was this portable time-keeping device that fundamentally altered one of the most basic, not to mention iconic, diagnostic tools of Kobiraji medicine, *nadipariksha*. From a complex qualitative appraisal of the pulsation at the wrist, the diagnosis moved toward a quantitative measurement of the number of beats per unit of time. Naturally, such a shift entailed a shift in the underlying body metaphor as well, yet it was practically never explicitly remarked upon.

Nadipariksha was conspicuously absent in the classical textual corpus of Sanskrit Ayurveda. It first emerged in the thirteenth century in the *Sarangadhar Samhita*. Since then it gradually grew in complexity and esteem, and in its alleged capabilities.[15] By the nineteenth century, it almost verged on divination and included among other things an extensive death-pulse lore that aimed to predict the exact time of death.[16] *Nadipariksha* had, however, significant differences from what is today recognized as pulse diagnosis. First, the Kobiraj was said to be able to detect the three different para-humors at three different points on the patient's wrist using three distinct fingers. Second, individual pulses were identified with reference to a large number of animal gaits rather than numbers. Third, *nadipariksha* was underwritten by an elaborate anatomical imagination centered on *nadi*s that could stand for both a flow and a channel through which something flowed.[17]

The European medical tradition, to the extent that it can be seen to constitute a single coherent tradition, had long evinced an interest in the temporality of the pulse. The Alexandrian physician, Herophilus, a student of Praxagoras, is said to have built the first water clock to measure the pulse. He called it clepsydra. But it was Galileo Galilei (1564–1642) who took the next important step in this direction. Having timed his own pulse with the vibrations of a pendulum, Galileo built a new device known as the Pulsilogia. Sanctorius (1561–1636), the Paduan, provided illustrations of this device and incorporated it into his commentary on the *Quanoon of Ibn Sina* (Canon of Avicenna). These classical anecdotes do not, however, lead directly to modern European efforts to quantify the pulse. Silas Weir Mitchell, a famous late nineteenth-century Philadelphia physician, looking back at the history of "pulse numbers" reminisced that

> Harvey but once or twice mentions the number of the pulse. . . . In the case descriptions of the time and of Sydenham's it is rare to find it. . . . In Whytte's work, circa 1745, he not rarely mentions the pulse number. . . . The force and other characters of the pulse receive, however, immense attention, and are on the whole more valuable aids than mere numeration.

The modern efforts are generally held to commence with Sir John Floyer (1649–1734), who wrote a highly influential book entitled *The Physicians' Pulse Watch*. The book standardized pulse measurements using a watch or clock. Floyer took a single minute as the standard temporal unit of measurement and provided normal ranges defined by age, gender, exercise, and so on.[18] Somewhat ironically, what inspired and helped the acceptance of Floyer's work was his successful espousal of the alleged similarities with the Chinese sphygmological tradition.[19] What Floyer drew upon and exploited was the interest and knowledge of Chinese sphygmology that had appeared in Europe since the 1670s by authors such as Michael Boym, Andreas Cleyer, Philippe Couplet, and others connected to the early Jesuit mission to China.[20] Floyer's success, however, did not ensure immediate and smooth universal uptake of pulse measurement throughout the "West." Weir Mitchell somewhat perplexedly wrote that,

> about 1710, an English physician, Sir John Floyer, wrote an able and now half-forgotten book, quaintly called the "Pulse Watch. ." . . . It is as true as strange that this convenient method was lost of habitual use in medicine for quite a hundred years. It reappeared in the writings of the time of the great teachers who arose in France and Germany about 1825.[21]

Clearly, the clinical uptake of these technologies had been sketchy and marginal. Writing of the eighteenth century, Malcolm Nicolson explicitly points out that "neither the pulse-watch nor the thermometer came into general use.[22] While the nineteenth century did indeed prove more accepting of such efforts, as Reiser points out, by the end of the nineteenth century much of the energy toward mechanization and numeralization of the pulse had moved on again. From the second half of the century the efforts focused mainly upon building machines that produced graphical outputs rather than chrono-numeral ones.[23]

A reliable material history of clocks and watches in India is still wanting, but this much is certain that Sumit Sarkar was inaccurate in suggesting that watch manufacture in India did not begin till after independence.[24] There is enough anecdotal evidence against it. In the late eighteenth century, Claude Martin, the Frenchman in English East India Company employ who rose to be Resident to the Nawabs of Awadh, spent most of his spare time making watches.[25] Later, one learns of the noted Calcutta teacher, David Hare, having originally been a watchmaker.[26] Besides these well-known examples, there are also lesser-known anecdotes that demonstrate the growing social uptake of watches. Around 1792, for instance, one Alexander Haswall was granted permission by the East India Company for traveling to India with a view to set up a business "as a maker of chronometers, watches, and clocks."[27] In a more sinister vein, in 1800, one hears that a Bengali watchmaker, Brajamohan Dutt, was hanged for having stolen a watch from someone else's house.[28] Similarly, an obituary notice published in 1840 in a local paper in Calcuta mentioned that the departed had been the wife of a watchmaker, one Mr. H. Peters.[29] Mr. Peters, however, was clearly not the only watchmaker in Calcutta at the time, for a list of Calcutta residents published just two years later in 1842 mentioned one G. Grant, a "watch-maker," living at what was then-called Tank Square.[30] Advertisements throughout the nineteenth century offered glimpses into the market for watches (see figs. 3a and 3b).

In 1861, for instance, Hamilton & Co. of Old Court House Street, Calcutta, advertised a wide variety of watches ranging in price from Rs. 40 to Rs. 850. Geared toward the high-end niche market, the company offers to set their watches with jewels, enamel them, or offer them in gold and silver boxes.[31] As larger and larger swathes of Bengali society became exposed to new temporal regimes through railway timetables, fixed office hours and routines in educational institutions, and the loud sirens regulating the workday for factory and dockworkers, it is highly likely that the demand for watches diversified and expanded beyond those who wanted their timepieces encrusted with jewels.[32] By the mid-1880s, the Salaries Commission which investigated the adequacy of salaries for government employees in Bengal stated that living expenses of the average Bengali gentlemen had gone up because they had adopted several elements of European dressing. Among these was the widespread use of gold watches and chains.[33] The expansion in the market for watches is reflected in the numbers employed in watchmaking and repairing. In a census taken in 1891, it was found that 1,626 men and 864 women in the city of Calcutta alone were employed in watchmaking or watch repairing.[34]

What seems to have led Sarkar astray is Ahsan Jan Quaisar's work on the period between 1498 and 1707. Quaisar points out that, despite a long history of Indians being exposed to mechanical clocks and watches through foreign travelers, they did not seem to have adopted them. Unlike in places like Japan or the Ottoman lands, no attempt seems to have been made to adapt the mechanical clocks to extant temporal cultures. Instead, Quaisar points out that "the use of the [older] water-clocks by Indians had such traditional force that even Europeans in India were compelled to adopt it."[35] Quaisar's conclusion, however, relied largely on Mughal perspectives. Beyond the Mughal dominions there seems to have been a fairly sophisticated understanding and interest in—if not actual manufacture of—mechanical time-keeping devices. An intriguing episode documented in Dutch sources testify, for instance, to the appreciation of clocks and watches in non-Mughal courts. In December 1678—at the height of Mughal power—beyond their borders, the Nayaka king of Madurai, Chokkanatha Pillai, had it officially communicated to the Dutch that he did not like the watch presented to him by the local Dutch ambassador, Mukappa Nayaka. Evincing a fairly nuanced understanding of different types of watches and also bearing witness to inter-elite rivalries for their ownership, the king explicitly asked for a watch like the one owned by "Ekoji [Bhonsle, the ruler of Tanjavur], hanging on a small chain around his neck." Few years after this in August 1685, another Nayaka lord, Muttu Virappa Nayaka III once again asked the Dutch at

HAMILTON & CO.,

JEWELLERS AND SILVERSMITHS

IN ORDINARY

TO THE VICEROY OF INDIA,

GOLDSMITHS, SEAL ENGRAVERS, WATCHMAKERS, OPTICIANS, ELECTRO-PLATERS, &C.

7, OLD COURT HOUSE STREET, CALCUTTA.

ESTABLISHED 1808.

PATRONIZED BY THE SUCCESSIVE GOVERNORS GENERAL OF INDIA, GOVERNORS OF PROVINCES, MEMBERS OF COUNCIL, AND NATIVE PRINCES; BY REGIMENTAL MESSES, AND BY VARIOUS CLUBS THROUGHOUT INDIA.

Hamilton and Co. respectfully bring to the attention of the public, their present extensive Stock of Fashionable and Elegant Jewellery, Watches, Clocks, Silver and Electro-Plated Articles, &c., which are replenished continually from England, France, Switzerland, Bohemia, and from China.

H. & Co.'s working departments of Jewellery, Silver, and Electro-Plate are conducted with great efficiency by European artists.

Their Watch and Chronometer making and repairing sections of business are attended to by a thoroughly practical and experienced English workman.

JEWELLERY.

Rings, Brooches, Lockets, Neckchains, Earrings; Guard, Albert, and Safety Chains, Chatelaines, &c.; Seals, Keys, Charms, Bracelets, Scarf, Hair, and Shawl Brooches and Pins; Head Ornaments, Buckles, Tiaras, Combs, Collars, Studs, Sleeve, Vest and Collar Buttons, Crosses, Coat Links, Whistles, Fusee Boxes, Gold and Silver Thimbles, Masonic Jewels, Medals and Decorations including the recently issued "China Medal," and all Novelties that are imported.

CLOCKS,

For travelling, house and office uses, ordinary and chiming quarters, and some to shew the Moon's age. Prices from 15 Rs. to 1,000 Rupees.

WATCHES.

English and Geneva Patent Watches, winding at the handle and requiring no key, in gold and silver cases: English Hunting, Half-Hunting, and Open-faced by McCabe, Hamilton and Co., and others: Geneva Hunting, Half-Hunting, and Open-faced, set with diamonds, enamelled, or in gold and silver cases. Prices varying from 40 Rs. to 850 Rupees.

PENCIL CASES AND PENHOLDERS.

Ladies' and Gentlemen's in great variety, of gold and silver; Leads for patent points, of usual and of extra thicknesses; gold Everlasting Pens. Gold and silver Pocket Letter Weighers, &c.

VINAIGRETTES.

Gold and silver Book, Locket, and Box patterns, and with rings and chains for suspending from the finger, &c., Scent and Toilette Bottles.

SPECTACLES.

With Pebbles and Glasses of every power and variety in steel, tortoiseshell, silver, and gold frames. Large Reading Magnifiers.

EYE PROTECTORS.

With Crape and Steel Gauze, and coloured Glass Fronts, in gold, silver, and steel frames.

TELESCOPES, &c.

For astronomical and field purposes, Binocular, Field and Opera Glasses, Pilot's Glasses, Microscopes, &c.

CHILDREN'S PRESENTS.

Corals and Bells, sets of Knife, Fork, and Spoon, and of Fork and Spoon only in Morocco cases, and ditto with Bowls and Mugs. Variety of silver Mugs, &c.

CUTLERY.

Complete Dinner sets and parts of sets, with silver, M. O'Pearl, Patent and ordinary Ivory handles and with silver plated blades. Dessert ditto, with silver and Plated Blades, M. O'Pearl, Pebble, and Ivory handles.

3a. Advertisements for clocks and watches, 1861.

BRWARE of IMITATIONS!

FREE! A small sized Liver Pocket Watch FREE!

ONLY FOR X'MAS SEASON.

New Model Swiss, only 500 pieces left.

Purchaser of our W. B. Fountainpen, British made, gets a reward of Silvered Dial Pocket Lever Watch Free.

Its beauty of workmanship, careful adjustment and excellent finish will be seen at once, its accuracy of time-keeping and good wearing qualities, its strength and reliability are excellent. Guaranteed for 10 years.

Sale Rs 3-15-0 a Pen & Watch.

DECCAN WATCH CO., 11th Beat, Sowcarpet MADRAS.

— GIVEN AWAY A HOME OR OFFICE PRINTING —

FREE! POCKET PRESS (GR.)

TO THE PURCHASER OF OUR

22-Ct. Golden Wristlet Watch
Swiss Prettiest Model.

Gent's High Grade Perfectly and accurately machined small and stylish, very durable 5 Years guarantee.

Free Pocket Press and a Watch, Reduced Price For This month only Rs. 6-15.

The Pocket Press—Very useful for Printing Name and Address in Letters etc, with solid 150 Rubber Types, Holder, Tongs, Ink and Pad complete FREE.

LORD WELLINGDON WATCH CO. Mount Road, MADRAS.

In writing to advertisers, please mention the Journal of Ayurveda.

3b. Watch advertisements in the *Journal of Ayurveda*, 1927.

Tuticorin for a number of objects including watches.[36] Though these instances testify to the demand for watches, there is no any parallel evidence about efforts to manufacture watches in India. But by the eighteenth century, the picture had changed. With the demise of the powerful Mughals and the rise of a new cosmopolitan culture at courts such as Awadh, Hyderabad, and Murshidabad, the appetite for mechanical clocks seems to have grown. Alongside this more robust demand, local manufacturers started to emerge. Men like Claude Martin, Brajamohan Dutt, Alexander Haswall, David Hare, H. Peters, and G. Grant all testify to the growth of local manufacture. This trend toward local manufacture almost certainly grew exponentially through the nineteenth century and it is this growth that is robustly attested to by the census statistics showing nearly one and a half thousand people working as clockmakers and repairers in 1891 in the city of Calcutta alone.[37]

The sketchy details about watchmaking and watch repairing can be further fleshed out through art and literature. A tragic short story by Nirmalshib Bandopadhyay published in 1913, for instance, furnished a glimpse into the

4. Hindu lady with a pocket watch. Courtesy: Wellcome Library, London.

desperately poverty-scarred lives of peripatetic European watch repairers who went from door to door peddling their wares. Set in the Swadeshi era of a decade or so earlier, the narrative also showed pocket watches as a common possession among middle- and upper-class high school students.[38] In *Arogyo Niketon*, Rangalal Daktar, a late nineteenth-century self-taught *daktar* practicing in a remote Bengali village, is described as wearing a pocket watch prominently like a pendant slung from his neck by a black chord.[39] Similarly, a late nineteenth-century chromolithograph published by the Chitrashala Press of Pune, held in the Wellcome Collection in London, showed a Brahmin woman of means at home with a pocket watch on her lap (see fig. 4).[40] Such representations demonstrated the extent of the social penetration of pocket watches by the end of the nineteenth century when schoolboys, *daktars* in remote villages, and conservative ladies in urban household all began to own and use them.

Pulsating Times

It is a matter of little surprise, then, that by 1871, Kobiraji textbooks began to include watch-based *nadipariksha* in them. Gopalchandra Sengupta published the *Ayurveda Sangraha* in that year with two express purposes. First, he wished to defend Kobiraji medicine against the dismissive comments made by the supporters of "European mode of treatment." Second, he felt an accessible Bengali book summarizing the Sanskrit works would go a long way in battling the plague of "quackery" that was affecting Kobiraji medicine. His efforts had been inspired and sponsored by a group of Bengali gentlemen living around Sookea Street in Calcutta—an area known for its many resident Kobirajes. It was thus a book supported by orthodox physicians, claiming to be based upon classical works and also invested in delineating itself from "Western" medicine. Yet its substantial discussion on *nadipariksha* was admittedly influenced by watch-based pulse diagnosis.

Gopalchandra's efforts at homologizing *nadipariksha* and pulse diagnosis were visibly inchoate. Far from blending the two regimes of examination and the body image they operationalized, Gopalchandra quite literally divided the pages into two horizontal halves that segregated the two descriptions. In the upper half of the page, he described a body imaginary that was organized around 35 million *nadi*s. These included both "gross" and "subtle" (viz. intangible) *nadi*s. They originated from a tortoise-shaped *nadi* at the navel.[41] Among other characteristic features of this body image was a soul-like entity conceptualized as a small, intangible being known as *jib*. *Jib* was said to perambulate throughout the body, riding through the *nadis* astride various *prans* (life breaths), "just as a spider travels along its web."[42] Even more enigmatically, this body constituted by *nadis* and home to *jib* was sometimes said to encompass the universe. Thus the sun and the moon were said to reside in the two spinal *nadis*, *ida* and *pingala*, and the world (*bishwo*) was located in the heart.[43] David Gordon White, commenting on this kind of a body imaginary, says that it does not fit the usual "macrocosm"/"microcosm" mold. Rather, he dubs this mode of the body encompassing the world a "macranthropic" way of imagining the body and the world.[44] In macranthropy, as well as a number of other features, the body image seen in these descriptions of Gopalchandra shared much with numerous much older body-world configurations available in Bengal.

Significantly, some of the more enigmatic features of Gopalchandra's description are more closely reminiscent of heterodox religious traditions available in Bengal rather than the Sanskrit Ayurvedic canon. The *jib*, for instance, is absent in the classical texts now thought to constitute the classical canon. The macranthropic body image, more generally, is less reminiscent of the classical Sanskrit-Ayurvedic canon than the locally available heterodox traditions. These are available, for instance, in difficult-to-date Nathist lore such as the *Goroksho Bijoy*.[45] They are equally available in seventeenth-century hatha-yogic texts such as the *Siva Samhita*.[46] Such body images were also available in the songs of peasant protest religions usually designated as *Baul gan* from the eighteenth and nineteenth centuries.[47] Among these latter, these ideas about the body constituted an esoteric branch of embodied theology known as *dehotottwo* (body knowledge).[48] This body imaginary was resolutely transmaterial—that is, its embodiment went beyond the simply material realm. Starting with the materiality of the tangible pulsation at the wrist, it transcended such tangibility and its narrow material dimensions. Its implication within Baul *dehotottwo*, hatha yoga, and other embodied ritual praxes clearly accented the transmateriality of the body by emphasizing its sacrality and its potential for religious deployment.

The body imaginary engendered by the quantitative, chrono-numeral pulse was in stark contrast to this transmaterial body imaginary. This latter body imaginary was resolutely materialist. It was neither inscribed upon a macranthropic anatomy nor did it posit the existence of an intangible *jib*. In Gopalchandra's writings, this material body imaginary was drawn mainly from the writings of William Augustus Guy (bap. 1810–1885) and to a lesser extent from Robert-James Graves (1796–1853). Guy's influence is interesting for two reasons. First, he was a pioneer of medical statistics. He served as the secretary of the Statistical Society from 1848 to 1863, edited their journal from 1852–56, was vice-president (1869–72), and eventually became president of the society (1873–75). His work on the pulse, therefore, was predictably one of the most numeralized. Second, he is not usually remembered today for his work on the pulse. Instead his work on occupational health and sanitation are seen to be more important.[49] He was thus not the most obvious choice from whom to derive European ideas of the pulse. Why Gopalchandra chose his writings remains a bit of a mystery. His writings on the pulse were published mainly in the *Guy's Hospital Reports* and circulated through their extraction and publication in journals such as the *Eclectic Journal of Medicine*.[50] Some of his work on the pulse also entered into the *Hooper's Vade Mecum*, which Guy enlarged and published several editions of from 1869 onward.[51] It is difficult to tell whether Gopalchandra encountered Guy's writings in a journal or in the Vade Mecum and why he was attracted to them. Graves, on the other hand, was much better known for his work on the pulse. Graves had demonstrated that variations in pulse rates corresponded to clinical changes. He published this work in the 1830s in the *Dublin Hospital Reports*.[52] Once again, however, Gopalchandra's source for Graves's views remains unknown. In fact, he refers to the Irish doctor as "Dr Grave" rather than "Dr Graves."[53]

Whatever the source, Gopalchandra discussed Guy's, and to a lesser degree Graves's, views at length. In effect, this meant giving lengthy numerical lists of mean pulse per minute for people of different gender and age, in different physical postures, etc.[54] Despite the lengthy exposition, Gopalchandra made absolutely no attempt to paper over the incompatibility of the two systems of diagnosis. He briefly described that, unlike in *nadipariksha* where the spatial placement of the *nadi*s within the body is understood to vary according to gender thereby requiring the physician to feel the *nadi* on the right wrist for men and on the left one for women, Guy's system made no distinction between genders.[55] Though made in passing, this single distinction was a fairly important one. As Martha Ann Selby has recently pointed out, Sanskrit Ayurvedic classics have a highly gendered anatomical imagination and the idea that different genders could have the same general anatomy would have been a revolutionary idea.[56] In the case of pulse/*nadipariksha*, the implications were both practical and theoretical. Yet Gopalchandra passed over this as if it were a fairly minor difference. He acknowledged, however, Guy's constitutionalism that held that mean pulse rates varied with reference to whether one had a bilious, nervous, or lymphatic temperament.[57] What remained unstated, but central to the whole process, was the need of a pocket watch.

Using the pocket watch, a new form of data about the pulse had begun to emerge. But Gopalchandra seemed unsure about how this related to the old forms of data. An anatomical diagram showing the nervous system was thus left uncoded—that is, there were no labels next to the arrows. The materialist body image was thus a mere supplement to the transmaterial body image. The two body images were simply co-located on the same page, they were neither supplanted, nor homologized (see fig. 6). Nor indeed did they remain innocent of contact. They came together and sat side-by-side, cheek and jowl, but did not initially cross-fertilize or displace each other.

By 1894, there was a distinct shift. Jashodanandan Sarkar's Bengali translation of the *Susruta Samhita*, published in that year, included a two-page prefatory note. This note clarified those aspects of Kobiraji medicine that, though important to its practice, were not mentioned in the classical works. Apart from discussions of the thermometer, the microscope, etc., it also included a brief exposition of *nadipariksha*. In the latter, however, there was no discussion of the transmaterial body image that was seen in Gopalchandra. Instead, it is simply stated that the different types of "para-humors" could be detected simply by the temperature of the wrist and the "thinness" or otherwise of the flow detected by the touch. A numerical range of normal temperatures then, once again, supplemented this description.[58] While the numerical description was marked off as "Western opinion" (*daktari mot*), the fact that it was present on the same page and that all the incompatible bits of *nadipariksha* were now excluded meant that the two body imaginaries and their different regimes of truth were now much better integrated than in Gopalchandra.

In the same year, 1894, Nagendranath Sengupta published one of the most successful textbooks of Kobiraji medicine. Entitled *Kobiraji Sikshya*, it had gone into seventeen editions by 1930. This immensely influential book too followed the broad pattern noticed in Sarkar's work. Though Nagendranath retained much more of the precolonial *nadipariksha* than Sarkar, he too elided most of the transmaterial dimensions. What he retained over and above Sarkar were the descriptions of different individual *nadi* movements for individual pathologies. These were frequently described as the gait of some animal, bird, or insect. He also retained some of the death pulses that allowed the physician to foretell death. Significantly, though, nowhere was the enigmatic macranthropic body imaginary upon which *nadipariksha* had previously been inscribed articulated. Neither was there any mention of *jib* and his perambulations. Yet, unlike Sarkar, there are several Kobiraji elements that Nagendranath does retain. Most prominently, he insists that the subjective and non-numeralized description of individual *nadi* movements expressed through the metaphor of animal gaits be retained. He also insists that men and women have diametrically reversed bodies and therefore their *nadis* need to be felt respectively on the right and left wrists. Through such insistences, Nagendranath retains a distinctive regime of Kobiraji *nadi-pariksha*. Unlike Sarkar, it does not become identical to the "Western pulse-diagnosis." Yet, neither does it retain the radical incommensurability it evinced in Gopalchandra's writings. For Nagendranath, the Kobiraji *nadi-pariksha* is a subtler and therefore more advanced form of the same truth that is revealed by pulse diagnosis, not an incompatible alternate truth. He writes, "It is extremely difficult to sense the pulsations of the *nadi*, discern its subtleties, or diagnose by it. . . . that is why western physicians have discovered such an easy method of doing it by calibrating it to the minutes of a watch."[59]

What can be heard in all the three texts of Gopalchandra, Sarkar, and Nagendranath is the unmistakable ticking of the pocket watch. But, while in the first, it is a confused and parallel sound that hardly detracts from the esoteric sojourn of the *jib* in all his enigmatic perambulations, in the second, its din has virtually stamped out all other sounds. But neither of these texts is as successful as Nagendranath's. In this latter, the ticking of the pocket watch has come to be partially reconciled with the charisma of *nadipariksha*. The minutes' hand of the watch is now the baseline of truth. Like all baselines, it is humble and lowly and admits higher truths. But those higher truths still need to be built on this baseline. They must be shorn of their radically alternative transmaterial dimensions and can only be retained to the extent that one can still perceive of them by subtler, but still tangible, means.[60]

What can be discerned in each of these attempts is a process of braiding. Distinct strands from within heterogeneous cultural traditions in the "East" and the "West" were singled out and braided together. In some instances, such as Gopalchandra's, this braiding was rather loose and the distinct strands were clearly discernible. On the other hand, in Nagendranath's much more successful book, the strands were braided much more tightly and appeared almost fused together.

Other Times

As the braids were tightened, the distinct identities of the constitutive strands began to further lose their distinctiveness. The chrono-numeral model of the pulse that derived from the ticking of the pocket watch came to gradually be redeployed within other, less conspicuously exotic, temporal cultures. I will discuss two of these attempts to redeploy the quantitative pulse into alternate temporal cultures.

The first of these was Haralal Gupta's *Nadi-gyan Shiksha* (Manual of *Nadi*-Knowledge). Haralal's text was in its ninth edition by 1913 and had possibly first appeared in the 1880s when his other texts appeared. In this immensely successful text, Haralal had retained the numeralized temporal model for *nadipariksha*, but sought to replace the "Western" temporal measure of "minutes" by the indigenous "*pol*." S. V. Gupta has cited contemporary Indian astrological practice that takes sixty *pol*s or one *ghori* to be equal to twenty-four minutes. A single *pol*, according to Gupta, is equal to six *asho*. An *asho* in turn is equal to ten *vipul*, and a single *vipul* is about 0.4 seconds. A *pol* therefore can either be said to be roughly 0.4 minutes or twenty-four seconds.[61] Haralal's statement that the normal pulse for a new-born infant is fifty-six times a *pol*, therefore translates, according to Gupta's conversion, to mean 140 beats per minute.[62] Abul Fazl's Mughal classic, the *Ain-i-Akbari*, provides an older definition of the Indic temporal units that closely approximates Gupta's description. "Hindu philosophers," said Fazl, "divide the day and night into four parts each of which they call *pahr*. Throughout the greater part of the country, the *pahr* never exceeds nine *ghoris* nor is less than six. The *ghori* is the sixtieth part of the nychthemeron, and is divided into sixty parts, each of which is called a *pal* (*pol*) which is again subdivided into sixty *bipal* (*vipal* = *vipul* ?)."[63]

The *Ain* goes on to further clarify how the *pol* and its family of temporal units were tied up with radically different technologies of time measurement. Fazl explained, "In order to ascertain or indicate the time a vessel of copper or other metal is made of a hundred *tanks* weight. In Persian it is called *pingan*."[64] He went on to describe this device at length. The metal bowl was twelve fingers in height and breadth and narrower at the lower end. At the bottom a five-finger-long golden tube weighing one *masha* was inserted through a hole made for it. The instrument was then placed in a basin of water in a place that was undisturbed by wind. Col. H. S. Jarrett, Fazl's Victorian translator, in his footnotes to Fazl pointed out two slightly variant modes for constructing the device, culled respectively from the *Surya Siddhanta* and the *Vishnu Puran*. In the former, the device was called a *kapala yantra*.[65] J. H. Fleet pointed to a far greater diversity of methods of construction, measures, and names for the water clock in 1915. He even showed that not all ancient Indian devices were of the type described by Fazl which took in water and therefore sank in the basin of liquid at a steady rate. At least a few, especially earlier ones, followed the Greco-Roman method of losing water at a steady rate. Fleet's essay also helps us understand the divergence between Fazl's description and the contemporary account of Gupta. Fleet points out that astrological (*jyotish*) works on time measurement tended toward greater accuracy, whereas others writing on water clocks mainly for uses in royal households tended toward practicality.[66]

Remarkably, such devices in British India were not a thing of the past. Fleet himself attested to the fact that even imperial British institutions such as the colonial police continued to use it regularly at far-flung outposts.[67] In fact, historian Ahsan Jan Quaiser points out that "water-clocks (the sinking-bowl variety) continued to be used in India, even in the villages, down to the early decades of the twentieth century."[68] The fact that these alternative devices of time reckoning, along with their distinctive units of temporal measurement were still extant at the time of Haralal's writing, gives his comments an underappreciated cogency. What he seems to be doing then is not simply build a hypothetical indigenous alternative to the pocket watch–based quantitative pulse, but rather reattaching the chrono-numeral pulse to an alternate "temporal culture."

Avner Wishnitzer, in a fascinating study of time in the late Ottoman world, defines "temporal culture" as "the huge ensemble of practices, behaviours, and concepts that concern the social organization of time and fill it with meaning."[69] In British Bengal, there were clearly multiple "temporal cultures." Each was a complex ensemble of practices, behaviors, devices, and concepts. Haralal was attempting to embrace the chrono-numeral model of pulse diagnosis, but not the "temporal culture" which was usually associated with it.

These alternate "temporal cultures" remain largely neglected by historians. Yet, available evidence suggests that the older temporal cultures that had resisted mechanical clocks under Mughal rule were not simply limping on to a slow but inevitable death in the nineteenth century. There were in fact some traces of a new energy and development of the older cultures in the period and Haralal's efforts must be located within such efforts.

A history of Nadia—a minor principality within Bengal known for its intellectual achievements—written in the late nineteenth century reported that one Gokulananda Bidyamani, who possibly thrived sometime in the very late eighteenth- or early nineteenth-centuries, had built an advanced time-keeping device. This device, known as a *ghoti-yantra,* was operated by heated mercury rather than water and had four copper fishes rather than a bowl at its core. Though the essential technique of its operation was similar to earlier water clocks, Gokulananda's device was special in that it was able to directly tell the smaller units of time such as *pol, bipol,* and *onupol.*[70] In the earlier devices, these smaller measures had to be calculated rather than directly read off the device.

The second attempt to implicate a quantitative pulse within an alternative temporal culture was undertaken by some of the eminent physicians belonging to the familial and intellectual lineage of Chandrakishore Sen. Their strategy was to develop an alternate "temporal culture" that did not depend on any external time-keeping device at all. The strategy was first espoused by Chandrakishore's famous nephew, Binodlal Sen. Later, Chandrakishore's two equally famous sons, Debendranath and Upendranath Sengupta, also advocated this mode of measurement. Like Haralal, these heirs of Chandrakishore gave the normal ranges in *pol*s. Thus, a newborn infant's normal pulse rate per *pol* was said to be 56 beats, just as for a youth it was 36 beats per *pol* and for a middle-aged man 29 beats per *pol*. Where they differed was, instead of implicitly relying on indigenous time-keeping devices, they explicitly suggested an alternate, device-independent way for measuring these *pol*s. Binodlal, writing in the first volume of his pioneering work, *Ayurveda Bigyan*, stated that "the time it takes to pronounce the 60 syllables [*guru borno*], that is in one *pol*, the pulse of a just-born infant is 56 beats."[71]

Binodlal's younger cousins followed exactly the same technique and even used the same examples in explaining the technique in their popular, four-volume textbook, *Ayurveda Samgraha*, which remains in print till this day. Fleshing out the technique a little further, all three of them also added a more detailed account of the temporal units connected to the *pol* accompanied by an alternative derivation of the *pol* (though still reliant on the same basic technique). Debendranath and Upendranath Sengupta stated, for instance, that "the time taken to pronounce a single long syllable is called a *matra* or a *nimesh*. Ten *matra*s make one *pran*, six *pran*s make a *pol* and sixty *pol*s make a *danda*."[72] Both versions of the technique—either pronouncing the 60 long syllables or just enough syllables to make up a *matra* or a *pran* (and then adjusting the rates)—was not unlike the various utterance-based, time-measurement techniques learned by American school-children—for example, one-Mississippi, two-Mississippi, etc. These methods do not depend upon external devices for the determination of time, but rather turn the person of the observer into a kind of impromptu time-reckoning device. By freeing the person from reliance on time-keeping devices, however, the technique was in a certain sense more culture-dependent. It required the knowledge of both the Sanskrit alphabet and certain standard pronunciation. In so doing, it functions once again within a distinctive temporal culture.

Such observer-oriented, personalized temporal cultures too had precolonial precedents. Once again, the *Ain* is educative. Fazl wrote, "The Hindu philosophers account 360 breathings of a man in good health as a *ghori* of time, and each is formed of six inspirations and respirations, of which 21,600 are drawn in the course of a nychthemeron."[73] The Sengupta's technique possibly built upon such earlier attempts to use the observer's body and its own internal processes as a time-keeping device, but it also broke with the earlier tradition in making the utterance of Sanskrit letters central to this temporal culture.

Both Haralal and Binodlal attempted to appropriate a chrono-numeral, quantitative pulse into their diagnostic repertoire. Such efforts followed in the wake of earlier efforts by Gopalchandra and others to assimilate the ticking of the pocket watch into the refigured Ayurvedic repertoire. What is significant in these efforts, however, are the attempts to align this new quantitative Ayurvedic pulse with alternate temporal cultures. In fact, insofar as all of these attempts are about embodied time—that is, pulse time—modifying Wishnitzer's more general concept, they can be thought of as alternate physio-temporal cultures.

The plurality of such physio-temporal cultures was most likely also linked to the prices of time-keeping devices. Watches, despite their steadily dropping prices, remained expensive instruments even into the twentieth century, when upper- and middle-class boys used them in Nirmalshib Bandyopadhyay's short story. It might well be that some of these alternative temporal cultures thrived at the lower end of the social ladder where poorer physicians operated. After all, as Rabindranath Tagore reminded us in his *Shesher Kobita*: "The time taken shouldn't be the same for everybody. There's no such thing as uniform time, the pace of the pocket watch depends on the pocket. That's Einstein's view."[74]

Poly-Normal *Prokritis*

Precolonial *nadipariksha* accepted a number of variations to the pulse. Eschewing chrono-numeral pulses, it held that the qualitative pulse varied with season, time or day, diet, etc. It is worth pointing out that these temporal rhythms of seasonality, diurnal phases, etc., were all nested cyclical temporalities—one fitting within the other, rather than a single linear one.[75] In fact one of the most significant areas of growth within the *nadipariksha* tradition was the correlation of pulse variations to a variety of diets. For instance, Pitambar Sen's *Nadiprokash*, possibly the very first published text on the subject, after having described the typical pulses for various fevers, then went on to describe a wide range of variations such as the fevered patient's pulse upon consuming curd (*dodhi*), too much sour food, oily stuff, jaggery, etc.[76] Some texts, such as Amritalal Gupta's *Nadigyan Rohosya*, even organized the dietary variations under a separate chapter heading entitled *Rosogyan* (Knowledge of tastes).[77] Even a general medical textbook, Annadacharan Barman's *Sarkoumudi*, included a small section on dietary variations of basic pulses.[78] These variations related to a dynamic, transformative body, not a static, quintessentially unchanging one.

The only area where *nadipariksha* recognized a stable physiological difference was between men and women. It posited that the entire complex circuit of *nadi*s arising from the navel of the subject was inverted in men and women. As a result, for women *nadipariksha* was performed at the left wrist, whereas for men it was at the right wrist. But here too, the embodied difference was not as stable as it appeared at first sight.[79] Most texts explicitly included one or more verses clarifying that in the case of a person who did not fit into either gender (*nopungsok*), the physician had to decide which wrist to check by first determining whether the patient's *prokriti* (constitution/temperament) was "male" or "female."[80]

By contrast, late nineteenth-century biomedicine was already creating a plethora of numeralized normal scales within which to imagine patient's bodies. Especially through advances in biochemistry beginning in the second half of the nineteenth century, clinical biomedicine came to rely more heavily on pathological laboratories doing routine blood or urine tests.[81] Christopher Crenner has drawn attention to how the routine laboratory work led to the development of numerous handbooks and textbooks that included "normal values" for various measurable aspects of the human body. Crenner argues that "as medicine in the nineteenth century gradually came to embrace measurement as a key means to assess health, physicians reasserted [a] two-part structure with new force, allowing the normal to serve as both optimal and average, mixing norms and descriptions."[82] One of the more respectable of the kind of middlebrow clinical textbooks Crenner writes of was one entitled *A Textbook of Physiology for Medical Students and Physicians.* Published simultaneously from London and Philadelphia in 1909, the book was authored by William H. Howell, professor of physiology at Johns Hopkins University. The book is therefore typical of how cutting-edge biomedicine conceptualized "normality" in pulse diagnosis. Howell wrote, "When we speak of the normal pulse rate we mean the rate in an adult when in a condition of mental and bodily repose." Yet, even under these conditions, Howell pointed out there were significant variations. "The average normal rate for man may be estimated at 70 beats per minute; for woman, 78 to 80 beats; but the normal rate for some individuals may be much lower (50) or much higher (90)." Howell briefly described three major factors for variation—namely, gender, physical size, and age.[83]

It is noteworthy that when modern Ayurveda incorporated the chrono-numeral pulse into its diagnostic repertoire, along with it it also imported the much more stable sense of bodily difference in the process. Gopalchandra Sengupta had already affirmed that pulse rates varied by age and gender in 1871.[84] All later authors followed suit. Haralal Gupta in his enormously popular *Nadi-gyan Shiksha*, which went into at least nine editions, did exactly the same.[85] Nagendranath Sengupta's more comprehensive and arguably even more successful work, the *Kobiraji Siksha*, agreed.[86] Writing in 1907, Basantakumar Pramanik also gave tabular lists of normative pulse rates that varied by age and gender.[87] All this was perfectly in line with prevailing biomedical opinion which too held that the normal values for adult male pulse rates had to be adjusted for age, gender, etc.[88] Pandit Salimuddin Ahmed Bidyabinod, who wrote a *nadipariksha* text entirely in verse, expressed the watch-based, poly-normal pulse rates thus:

Western Pandits by a watch
Have found a way to tell the pulse
In a minute an infant's pulse, if it be healthy
Beats hundred and forty, upto one thirty
From seventy to eighty a minute for a youth
In an old man fifty to sixty five times

Women's pulse compared to men's
Ten to fifteen beats more per minute.[89]

Naturally, this broad consensus did not rule out minor points of contention. The precise normal values stated by each textbook writer varied, the number of age brackets included varied, and a minor disagreement over whether gender differences were congenital or opened up at a particular age, especially after puberty, remained unsettled. Gopalchandra Sengupta held that gender differences opened up after eighteen years of age.[90] The brothers Debendranath and Upendranath Sengupta asserted that the gender differences did not open up until middle age.[91] By contrast, their cousin Nagendranath held that gender differences were congenital and perceptible at birth.[92] Basantakumar Pramanik agreed with Nagendranath.[93] One striking difference from the contemporary biomedical perspective was that no attempt seems to have been made to adjust pulse rate to body size.

Modern Ayurveda's move toward relatively stable bodily differences was also accompanied by an effort to create a distinctive poly-normal framework. At the core of this distinctive poly-normality was the introduction of a new typology of "constitutions" or "temperaments." As Dominik Wujastyk points out, the idea of distinctive constitutions was only "hinted at" in the *Charaka Samhita*. And, notwithstanding the fact that it did indeed figure relatively more prominently in other parts of Ayurvedic literature, the subject has only now emerged as "one of the most prominent parts of New Age Ayurveda."[94]

The brief statement in the *Charaka Samhita* merely stated that while certain people are born with their three *dosh*es mutually equalized, others are born respectively with *vayu*-dominant, *pitta*-dominant, or *sleshma* (*kapha*)-dominant constitutions (*prokriti*). Those whose *dosh*es are mutually balanced at birth are seldom ill, whereas those possessed of a *dosh*-dominant constitution are frequently ill. The *Charaka Samhita* advises the latter to always consume food that instigates the opposite *dosh* from the one they are constitutionally dominated by.[95] There is nothing to suggest that the discussion on *deho-prokriti*s (lit. "bodily natures") had advanced any further than this in any of the *nadipariksha* texts. None of the several *nadipariksha* texts I have seen from the eighteenth and nineteenth centuries dwell upon it at all. Such a disjuncture is hardly remarkable since *nadipariksha* itself is a very late development and is absent in the canonical works of classical Ayurveda.[96]

Yet Gopalchandra Sengupta incorporated *deho-prokriti*s into his discussion of the chrono-numeral pulse. Gopalchandra allowed for variations of normal pulse according to different types of *dhatu* ("tissues") that constituted a person. He posited that those respectively with bilious, sanguine, and nervous constitutions would have different normal rates of the pulse.[97] The word *dhatu* is a polysemic word with numerous connotations in the medical context, but one of the most prominent ones pertaining to seven basic tissues of the body. The Sanskrit Ayurvedic canon speaks of these seven tissues: namely, *ros, rokto, mangso, med, osthi, mojja*, and *shukro* as *dhatu*s. These substances are the material basis of the body and constituted through progressive stages of the "cooking" of the food in the bowels.[98] Hence the mapping of the term *dhatu* onto a notion of "constitutions" or "temperaments" is not only somewhat stretched, but also heavily loaded.

The choice of the three constitutions—bilious (*pitta-prodhan dhatu*), sanguine (*rokto-prodhan dhatu*), and nervous (*snayu-prodhan dhatu*)—was interesting in that they clearly did not derive from the humoral translations of Kobiraji *dosh*es. With the exception of the first, bilious, the other two constitutions, especially the "nervous," have no conceivable antecedents in the Kobiraji discussions of *dosh*es. They seem instead to be related to the notion of "temperaments" available in the "Western" tradition. Originally, the term *temperaments* had designated "the specific mixture, in a particular subject, of the four [Greek] humours, in both quantity and quality."[99] The original model of four temperaments based on the four Greek humors had been canonized by Galen around the second century CE.[100] This original set, however, had comprised of sanguine, phlegmatic, choleric, and melancholic temperaments. It was in the eighteenth century that a "nervous temperament" began to emerge.[101] Yet in the early decades of the nineteenth century, the four temperaments were usually still the Galenic foursome.[102] By the 1870s though, a new set of four temperaments had emerged. John Murray, an MD, writing in the *Anthropological Review* in January 1870, sought to relate the four temperaments to racial types. The four temperaments he outlined were the nervous, the bilious, the lymphatic, and the sanguine. Each of these, he added, was characterized by "a certain shape of the head, the face, the body, different colour of hair, of skin, of eye."[103] Gopalchandra's model clearly seems indebted to models such as Murray's, the only exception being the absence of the lymphatic temperament. Arguably, the most visible outcome of this absence is the superficial symmetry between the three different constitutions mentioned by Gopalchandra and the three distinctive *dosh*es of Kobiraji texts.

Significantly, despite having access to racialized poly-normalities, none of the major Ayurvedic authors seemed to have advocated a strictly racialized set of pulse rates. This is particularly interesting since, as Gyan Prakash points out, Bengali *daktar*s did in fact occasionally advocate such racialized pulse rates in the early twentieth century.[104] It is perhaps cogent here to recall Shruti Kapila's observation that "race" in colonial India was a "remarkably mutable" category. In the vernacular print sphere at least, "race" functioned as part of a larger ensemble of what Kapila, following Michel Foucault, calls "insurrectionary knowledges" which, while amenable to the reconstitution of Indian selfhood, did not necessarily become a tool for perpetuating racial hierarchies.[105]

Even while the racialization did not catch on, the language of "temperaments" and "constitutions" caught on. Pandit K. P. Mukherjee, writing the preface to Rasiklal Gupta's translation of a traditional *nadipariksha* text, justified the publication on the grounds that the contemporary "Western" medical tradition has not undertaken a "complete and correct examination of all the various modifications, which the natural pulse assumes in different periods of life, in either sex, in different physical constitutions."[106] Elaborating further on this, Rasiklal included a fairly detailed description of "temperaments" in his longer and original work on *Hindu Anatomy, Physiology and*

Therapeutics. The section collated disparate fragments from the Ayurvedic corpus and fitted them into neat "temperamental" silos.

A closer inspection, however, reveals that Rasiklal's entire section on "temperaments" was copied, verbatim, from a similarly titled section in Thomas Alexander Wise's *Commentary on Hindu Medicine* (1860).[107] Wise's writings were extremely popular among the modernizing Ayurvedists (Binodlal Sen, Udoychand Dutt and Jashodanandan Sarkar, for instance, all explicitly cited Wise), and Rasiklal's appropriation of it comes as no surprise. It demonstrates that the history of poly-normality—something seen to be almost characteristic of Ayurveda in its New Age renditions—was thickly braided with Europeans and European knowledges.

Wise's writings preceded Gopalchandra's by little over a decade. Clearly, this was the common source from which had begun the braiding of the European discourse on "temperaments" and the dispersed observations of embodied difference in Ayurveda. Though Wise had presented his discussion explicitly as the "opinion of the Hindus," most of his material in the section on "temperaments" was drawn from a single chapter in the *Susruta Samhita.*[108] Discussions such as the one about *deho-prokriti*s in the *Charaka Samhita* were absent in Wise. According to Wise/Rasiklal, there were a total of seven "temperaments" depending on whether any single, dual or triple *doshe*s predominated at the moment when the parental semen combined with the parental female blood. Wise's "temperaments" were a motley list of characteristics. Following from this, Wise outlined three "temperaments" in detail and stated that the remaining four were proportionate mixtures of these. The first "temperament," arising out of an excess of *vayu,* was marked by such characteristics as

> the person is not inclined to sleep or become warm. His disposition is bad and he becomes a thief; is proud, and has no honour; is always singing and dancing; his hands and feet split, his hair and nails are dry; and he is always angry and boisterous; he speaks untruths and is always grinding his teeth and biting his nails; he is always impatient, he's not a firm friend, is changeable and forgets good actions. His body is slender and dry, he always walks fast, is always in motion, and his eyes are always rolling. He dreams that he is flying about the air, friends are few and his riches of little value. . . . [He has] the disposition of the goat, jackal, hare, camel, dog, vulture, crow and ass.[109]

The list grew even more chaotic as Wise moved to more "temperamental" types. Wise, however, was hardly alone in his confusion. The British discourse on the subject was characterized by similarly confusing traits being lumped together. Thus, in John Murray's description of the four temperaments from 1870, for instance, the "nervous temperament" is described thus:

> Body being small and almost feminine, with narrow sloping shoulders. The bones are small, muscles soft and fine, capable of sudden and great effort, but soon fatigued. Step short, quick, elastic; hands soft, small and delicate. . . . Countenance intelligent . . . large dark blue sparkling eyes, situated near the nose with much white . . . frequent winking or twitching of one eyelid . . . those who are in easy circumstances dress smartly, button up their coats, and are usually seen wearing their hat on the side of their head . . . clever, sensitive and aspiring, will give and take much flattery, have strong likes and dislikes, make good soldiers, novel writers, public speakers, professional men, merchants or teachers.[110]

Though almost all of the material is drawn from the fourth chapter of the book of *Shorir-sthan* in the *Susruta Samhita*, the end result is equally beholden to the metropolitan British discourse of the time. The already chaotic combination of characteristics ended up further confused by their being braided together. Wise made a largely unsuccessful but valiant effort to synthesize this seething diversity of forms by writing that "the predominance of one or more of the humours, and the qualities of goodness, passion and inertness, explained the peculiarities of habit, and character, and, as more shades were observed, in these the simple excess of one or more of these agents, were explained by their peculiar mixtures which produced the dispositions of Gods, sages, demons, and the lower animals."[111] Unfortunately, the contents of the chapter do not bear out his attempt to reduce the last and most diverse set to merely combinations of the former sets.

One principal source of confusion arises out of Wise/Rasiklal's attempt to combine two very different typesets: one based on *dosh*es and another on *guna*s. While *dosh*es are the parahumors, *guna*s are more ambiguous. Dagmar Wujastyk points out, "though *Charaka* and *Susruta* have somewhat different interpretations of the meaning and functions of the *guna*s, they share the fundamental notion that the *guna*s determine how a person thinks, feels and acts. Unfortunately they don't explain how exactly a person comes to have his or her personality type . . . whether they are "genetically" preordained constitutions . . . determined at conception or birth . . . or something a person acquires."[112] Since there is indeterminacy about whether *guna*s are congenital or acquired, they clearly cannot be equated with the *dosh*es that are constitutive of the very fabric of the body.

It is striking that both Wise, who was a fairly conscientious scholar and did not take too many interpretative liberties, and practicing Kobirajes, like Rasiklal after him, overlooked such serious conceptual muddles in their bid to construct a poly-normal framework.

By 1894, Jashodanandan had inserted the poly-normal framework developed by Wise into the very source text upon which the original formulation was supposedly based: the *Susruta Samhita.* Not only did Jashodanandan introduce the poly-normal framework in the prefatory note preceding the actual text, but even in the section from which Wise had drawn his material, Jashodanandan repeatedly inserted Wise's English renditions in parentheses. Such textual practices clearly sought to guide the reader in their interpretation of the classical text and did so along the lines laid down by Wise.

Not content to simply reproduce Wise's poly-normal pulse rates like Gopalchandra and others before him, Jashodanandan introduced further innovations. Drawing upon the then emergent paradigms of Tropical Medicine, Jashodanandan wrote that "in summer-dominated countries [like India], people's bodies are *vayu*-dominated, since otherwise they will not be able to tolerate the excess heat." Similarly, in cold countries, like England, people were *pitta*-dominated, in excessively cold countries, like Kabul, people were dually dominated by *pitta-kapha,* while in excessively warm climes, like Abyssinia, people were dually dominated by *vayu* and *kapha.*[113]

A geo-climatic imagination by itself was not entirely new to Ayurveda. As Francis Zimmermann's pioneering classic points out, "an ecological theme" functioned as the basic classificatory rubric in classical Ayurveda. Yet, there is much that is novel in Jashodanandan. The classical Ayurvedic classificatory scheme Zimmermann outlined was based on only two basic geo-climatic zones: *jongla* (dry lands) and *onup* (marshy land).[114] There was no sense of political territories or countries like England, "Kabul" (Afghanistan), "Abyssinia" (Ethiopia), etc., attached to these geo-climatic zones. Moreover, the ecological theme was used to categorize everything, including foodstuff and medicines, not "temperaments." Finally, and most importantly, these differences were not calibrated to differential numerical pulse rates.

Such geo-climatic alignment of "temperaments" though novel in Ayurveda was common in British authors like Murray. Of the "nervous temperament," Murray wrote, for instance, that "this temperament was bequeathed to us by the Celts, or ancient Britons, and are seen more or less pure in Wales, Cornwall and the Highlands of Scotland, and in Ireland, where, however, the bilious temperament was infused into the south and the west by the Phoenicians or Darkmen."[115]

It is cogent to point out that, such "temperamental" poly-normalities were not racialized as such. As is clear from Murray's comments, different temperaments could remain submerged within a single race. In fact, the distinction that was made between "race" and "temperament" among European medical men in the colonies is clarified by Mark Harrison's discussion of "seasoning." Harrison points out that many colonial medical men, such as the Anglo-Indian doctor Charles Curtis, explicitly stated that not all European men—though they belonged to the same racial group—could acclimatize in India with equal ease. It was men with "sanguine temperaments" who were best suited to the region and "seasoned" most smoothly.[116] Jashodanandan's typology adheres closely to this rubric and is therefore not clearly racialized.

Poly-normalized pulse rates emerged around the pocket watches' ability to numeralize not only health but also differential susceptibility to disease. Modern Ayurveda acquired this poly-normal framework by braiding together an European discourse on "temperaments" with a very stretched, partial, and occasionally mutually contradictory reading of Indic sources. Yet, the braid that emerged proved remarkably durable and, happily, not explicitly racialized.

Bedside Watches

Therapeutics, as Charles Rosenberg has repeatedly reminded us, is not just about the materiality of the curative actions. It is also about the emotional, social, and cultural encounters at the patient's bedside between physician, patient, and often several others. The transformations occasioned by the pocket watch were therefore not limited merely to conceptual reworking. It included a shift in bedside manners and precipitated new emotional responses.

Yet it is virtually impossible to obtain any firsthand account of exactly how watches were used at the bedside and how this affected patients. Unlike other small medical technologies such as the thermometer or the microscope, somewhat expectedly there is practically nothing in contemporary medical writing explaining exactly how to use a watch to check the pulse. Silas Weir Mitchell's reminiscences are exceptional in talking directly about the watch as such. Intrigued by Sir John Floyer's book, the *Pulse Watch*, Mitchell wrote, "I am pretty sure he was the first to put a minute hand on the watch to enable him to time the pulse beat, but nowhere in any English collection have I been able to find one of his watches."[117] Mitchell's comments draw attention to the absolute materiality of timing the pulse. How did a doctor use the watch to time the pulse? Mitchell's comment about the minute hand is also interesting, but perplexing. Today, those of us who still use watches with hands, use the second hand rather than the minute hand to time the pulse. Using the minute hand, would entail observing the dial of the watch very closely. It can be surmised that this in turn would interrupt eye contact with the patient and make a patient used to a traditional physician's bedside manners seem aloof.

But there is some confusion about what exactly was being referred to as the minute hand. Jimena Canales describes the French scientist, Francois Arago in 1853, as "marvelled [by] a rare instrument [then] 'in [his] possession, made in Vienna' in which the hand, instead of making a complete turn in a minute, 'made a complete turn in a second.'"[118] This suggests that most of the usual watches by the 1850s in France had hands that we today, in our modern vocabulary, would call "seconds hands"—hands that complete one circle of the dial in the span of a single minute. A nursing manual published from London in 1876 also confirms the foregoing linguistic usage. It advises nurses that "the minute hand of the watch should make at least one complete while the pulse is being counted, or, in other words, it should never be counted for less than a whole minute."[119]

The appearance of these instructions in a nursing manual also demonstrates the new potential for delegating pulse checking to semi-skilled subordinates, or perhaps even the patient's family members, so long as they knew how to use a watch. In the Ayurvedic context, this was an enormous shift. Guy Attewell's study of Unani medicine points out that by the nineteenth century, "pulse" in the traditional medical setting had acquired an almost quasi-divine aura. Pulse checking was an occasion upon which the physician displayed his virtuosity, often verging almost on the mystical.[120] Yet, by the end of the nineteenth century, it is clear that in some cases pulse checking was delegated to nonphysicians. In Debendranath and Upendranath Sengupta's four-volume textbook, for instance, in describing how to exactly hold the patient's wrist while taking the pulse, they describe two unprecedented postures. The first of these is for female pulse takers, while the second is for anyone trying to check his own pulse.[121] As I have mentioned above, precolonial *nadipariksha* texts mentioned that a man's *nadi* was to be checked on his right wrist, a woman's on her left wrist, and for a person of indeterminate gender (*nopungsok*), the wrist was to be decided by reference to the *prokriti* (nature) of the patient. Nowhere in these descriptions was anything mentioned about the gender of the physician. Nor was there any reference to people checking their own *nadi*s.

The use of technology to delegate certain aspects of the physician's work has been observed in the "Western" context by Margarete Sandelowski and Christopher Hamlin.[122] They have also commented on the gendered nature of this delegation where the male physician usually delegated the work to female nurses. In the Bengali context, however, these delegations would have taken on a very different social and cultural resonance. Ayurvedic hospitals, until the very end of my period of study, were marginal. Most patients would not have wanted to go to the hospital.[123] Simultaneously, the nursing profession also progressed hesitantly in India. Mridula Ramanna has recently shown how, despite repeated efforts, social and cultural anxieties about the nature of the job impeded the growth of the profession, and at the time of independence from British rule the numbers of nurses were still abysmally low in the country.[124] In this overall context, it was the immediate family members, including both men and women, who undertook much of the long-term care for patient. If there were any delegation of medical duty therefore, it would likely have been delegated to the patient's family members rather than trained nurses at hospital or home.

Another possible scenario where the new directions given by Debendranath and Upendranath Sengupta might have been useful is in situations where the patient was an ultra-conservative woman. One report from 1885 that lauded the work of the newly founded Association of Medical Women in India explained the reason behind the founding of the association thus: "The association owed its origin to the fact that a great proportion of the women inhabitants of India are forced by custom to lead a life of seclusion, so complete that they are not allowed to see a male practitioner, however urgent may be their need for medical aid."[125] Despite the condescending language and the imputed lack of volition or agency to the women, the report clearly describes the problems male physicians had to face in accessing some of their women patients. By delegating the task to a woman or even the patient herself, such objections could be overcome.

These delegations did not happen in a vacuum. There is clear evidence to suggest that by the time Debendranath and Upendernath Sengupta published their book, the *Ayurveda Samgraha*, in 1892, some resentment had already begun to crystallize around the physician's watch. A virtually unknown Bengali poet published an experimental prose-poem entitled *Ghori* ("Watch") exactly a decade before the *Ayurveda Samgraha*, in around 1882. Unfortunately, the poem is no longer traceable, but an extract of it survives in a book review of the entire collection done by the foremost Bengali litterateur of the day, Bankimchandra Chattopadhyay. Therein the forgotten poet satirizes in the voice of the watch:

> Various communities have various gods. I am the god of all communities. My powers are endless and my voice never ceases. No one in the universe can equal my powers. Worship me every day. I am the physician's capital. As long as I am there, the physicians are invincible. If I fail, their luck runs out. So, remember Doctor, to feed me some sweet oil every six months. Remember too that I am in peak health when my stomach is clean.[126]

The physician's increasingly rapacious commercialism and the gradual depersonalization of the clinical encounter was both metaphorically and literally inscribed onto the figure of the watch. The watch had become the "physician's capital" rather than an instrument of cure, and it demanded care and attention like a jealous god. In this regard, it is worth noting that Brahmananda Gupta recounts how one of the first moves by modernizing Ayurvedists was to introduce standard consultation fees that "equalled or surpassed the fees of British physicians."[127] Gupta, as an heir to this hallowed legacy, naturally views these developments positively as steps in the right direction, but contemporary patients may have resented the flagrant and rapid commercialization of the doctor-patient encounter. That such sentiments were expressed in 1882 strongly suggests that the efforts by the Sengupta brothers a decade on to redistribute authority around the taking of the chrono-numeral pulse might well have been a response to pressures from below.

Yet the strength of the sentiments seemed to have been too strong to die down. About thirty years later, in the early 1920s, a much more prominent literary figure, Rajsekhar Basu, who wrote under the pseudonym Parasuram, produced another stinging satire; this time in the form of a short story with a very similar motif. In this story titled *Chikitsa Sonkot* ("The Treatment Crisis"), a good-natured, aristocratic hypochondriac is preyed upon by a series of physicians belonging to different medical systems ranging from the allopath to the Ayurvedist and the Unani Hakim to the homeopath. Each tradition's characteristic diagnostic skills are satirized in the story, but much of the action and humor turns around the pulse taking scenes. A particularly stark contrast is drawn between the traditional Ayurvedist's *nadipariksha* and the modern allopath's pulse taking. While the former scene is played out in an overly informal setting where the Ayurvedist pretends to divine aspects both medical and nonmedical about the patient by holding his pulse in between keeping an eye on the family cat trying to steal food from the kitchen, the latter scenario plays out in an exactly opposite setting. The allopath's chamber is a highly formalized and daunting space regulated by formal appointments, secretaries, etc. Most interestingly, however, the allopath does not even touch the patient to take his pulse. Instead, he fits an elaborate contraption that includes a car's spark plug onto the patient's wrist and reads the results at a distance.[128] Parasuram was equally critical of both the traditionalists and the modernists, but the specific points of his satirical critique are illuminating. The traditionalist was faulted for trying to over-read the physical signs, whereas the modernist was depicted as distant and enamored by machines. In their own ways, both silence the patient and exude a clear sense of authority.

E. Valentine Daniel, in his fascinating early study of pulse reading in Siddha medicine, argued that at the moment of contact when the physician held the patient's wrist, the individual subjectivities of the physician and the patient briefly merged into a "consubjectivity."[129] Though Daniel's study was limited to Siddha medicine and he described how the Siddha tradition sought to develop specific techniques for temporarily fusing the patient's and the physician's subjectivities, the general argument was later extended to Chinese pulse taking by Elisabeth Hsu.[130] Even if one does not go as far as Daniel and Hsu in saying that traditional Kobiraji *nadipariksha* seamlessly fused the patient's subjectivity with that of the physician, this much is clear that it was often the only moment of actual physical contact between the two. This physical contact, irrespective of whether it produced consubjectivity or not, was definitely a metaphor for sympathy. Parasuram's satirical contrast pits this sympathy against a certain posture of distance enabled and underwritten by the reliance of modernists on mechanical devices.

Yet what is equally significant in Parasuram's story is that neither the traditionalist nor the modernist—which in this case also doubles up as a contrast between Kobiraji and *daktari* medicine—are shown to place much importance on the patient's narrative. The shift to modern biomedicine is frequently depicted in the extant literature as a silencing of the patient, with either "traditional" or "alternate" medicine perceived as more open to hearing the patient's voice.[131] In this regard, it is important to note that some key contemporaries of the change did not see it in this light. They found both traditional Kobiraji and modern "Western" medicine to be equally objectifying toward the patient. At one point in the story, for instance, when the traditional Kobiraj, upon reading the patient's pulse, asks if he has been vomiting in the mornings, the patient answers in the negative, but the Kobiraj insists that it must have happened and the patient did not know it. Though the encounter is clearly a humorous episode, it still underscores the degree to which the Kobiraj exalts claims that his *nadipariksha* is more reliable than the patient's testimony.

The issue of the patient's silencing also resonates with the allied question of the physician's authority. Once again, the tendency is to contrast the highly skewed distribution of authority in a biomedical context with the allegedly more egalitarian physician-patient relationship in Ayurveda.[132] Yet once again, encounters, such as the one just recounted where the physician alleges that the patient had vomited without knowing it, show—albeit through a humorous incident—the extent of the authority enjoyed by the Kobiraj, which authorized him to dismiss the patient's own body knowledge with such élan.

The larger point that these issues of authority and objectification raise is that Parasuram's story identified these traits with "tradition" rather than "modernity." According to contemporary observers like Parasuram therefore, these elements of the bedside encounter were not novel. Taking this contemporary perception seriously, I argue that the pocket watch was able to slide so easily into the bedside encounter precisely because the logic of medical practice in Kobiraji medicine was already conducive to the trends the watch promoted. The watch and its data further enhanced the physician's authority in a very obvious way: the physician held the watch and looked at it, while neither the patient nor others could, unless they unceremoniously peeked over the physician's shoulder. Similarly, it produced a form of data about the state of the patient's health that, while significant and intelligible to the physician, did not require the latter's voice to be a part of it.

Such a view is partially confirmed by other novelistic accounts from the late nineteenth century when the watch-based pulse diagnosis is described in pretty much the same way as the older qualitative pulse was described. In Taranath Ganguli's *Swornolota*, for instance, the doctor's examination is described thus: "On his arrival, that doctor at once gave the patient some stimulant. Sitting down by the bedside of Sorola, he then proceeded to acquaint himself with the particulars of her illness. This done, he opened his watch to examine the pulse. He next set about the examination of her chest and sides."[133] The description was not much different from the way contemporary authors described the taking of the qualitative pulse. It was neither more nor less disempowering for the patient. It is doubtful that contemporaries used to the authority and status of the traditional physicians even thought of the changes in terms of a loss of patient power. The only difference was possibly that the watch-based encounters were shorter and the doctor was less focused on the patient as such, since he had to follow the hands of the watch. This would fit well with the mechanized distance, formality, etc., caricatured by the forgotten poet of *Ghori* in the 1880s, and again by Parasuram in the 1920s.

What was new in contemporary accounts and was resented was the new air of formality introduced through the mechanical mediation of pulse taking as well as the purely commercial character that the doctor-patient relationship was taking on. The lament was not organized as a loss of voice or authority by the patient, but rather the simple fact of the loss of informality, empathy and physical intimacy. In this regard, it is worth noting that sociometric studies done in the late 1970s found that among hospital patients in particular and among people in professional/functional situations in general, touching produced positive affective, behavioral, and psychological reactions for women. For men, the response was more ambiguous and sometimes even negative.[134] These studies and their conclusions of course are culturally and historically limited. Similar studies, had they existed during my period of study, might well have produced very different conclusions. Yet, what is clear is that touching hands in specific contexts does have affective consequences. Moreover, the critiques of commercialization might well fold into this larger laments over depersonalization and lack of intimacy and empathy.

The pocket watch, to be fair, did not eliminate physical contact altogether. In fact, I have shown there were some attempts to actually use the watch to redistribute clinical authority, and Parasuram's satire after all had to deploy a fictional instrument rather than a watch to make his point. But the watch, as noted in the scene from *Arogyo Niketan* recounted at the beginning of this chapter, did attenuate that moment of contact. It made the physician's touch shorter, more perfunctory, and less reflective. This in turn precipitated new anxieties on the patient's part—not about the loss of control, voice or authority, but over the loss of intimacy, informality, and empathy.

The Clockwork Body

Nadipariksha was iconic of Kobiraji therapeutics, and the insertion of the pocket watch into it fundamentally transformed the way that therapeutics accessed the body, diagnosed illness, and conceptualized normality. By the second decade of the twentieth century, so great was the transformation and so central the role of the watch in it that the watch itself became a metaphor through which to apprehend the body imaginaire.

The body imaginary, to the extent it existed in the consensual therapeutic space of precolonial Bengal, was a body of receptacles connected by a variety of specialized conduits and channels. Dominik Wujastyk points out that contrary to earlier scholarly opinion, it is indeed possible to find fairly specific references to particular internal structures that are resonant with, though not identical to, biomedical ideas about organs. Though illustrative traditions were weakly developed, that precolonial Ayurvedists did not see the body as a single undifferentiated space united by the flow of *dosh*es can also be discerned from the way in which *dosh*es themselves were described in great detail as collecting at or moving from specific locations within the body.[135] Overall it was seen to be a universe of sorts, or a huge cavern frequently referred to as *deho-bhando* or the body container within which were specific receptacles and channels. In fact, it was frequently asserted in both medical and lay contexts that "that which is not in the body, is nowhere [in the universe]."[136] In both the image of the universe and the one made up of receptacles and channels, there is a largely spatialized body image that is internally differentiated and heterogeneous, not unlike the real social space of a city or village. Most importantly, particularly in the absence of any robust illustrative tradition, its overall unity did not rule out a certain level of incoherence or incompatibility at the practical level. At the level of daily practice, it was far more important to be able to relate particular symptoms to the vitiation of *dosh*es in particular receptacles within the body, than to think of the overall body image as a whole. Thus, in a case of common fever, for instance, the symptoms had to be recognized and related to the corruption of *dosh*es in the part of the stomach known as the *amashoy*. The treatment had to redress the accumulation of corrupted *dosh*es in the *amashoy* as such and not worry about the achievement of overall balance, at least not as the preliminary goal.

In sharp contrast to this, Nagendra Nath Sen Gupta of Barisal (not the Nagendranath Sengupta who authored multiple textbooks) writing in the journal *Ayurveda Bikash* in 1915–16, explained the working of the body thus: "Imagine that having applied all your mind to its fullest for long you manage to build a machine (*yantra*). The moment you wind up this machine it starts turning round and round and saluting you and me . . . such a machine created by science is no different from those made by the Goddess Nature (*prokriti debi*) we see all around us."[137] Human beings and automata were thus seamlessly compared. Both were supposed to run on a clockwork mechanism. Nagendra developed this analogy over two extremely long essays. In it, he cited Sir Issac Newton and Sir Jagdish Chandra Bose among others. In seeking to establish his metaphor, he manfully reworked passages from the canonical *Susruta Samhita* (a work incidentally replete with organic metaphors of trees and creepers). Using the term *jib-yantra* (creature-machine), he argued the need to rethink the boundary between the animate and the inanimate. There were, Nagendra posited, two types of sentience. The first was a sentience present in all matter. This was a part of the universal, undifferentiated ultimate consciousness (*poromatma*). The other was a sentience marked by the capacity to desire/attract (*iccha*) and hate/repel (*dwesh*). This latter was the individual consciousness (*jibatma*) and comparable to the English use of the word *soul*.[138] Demarcating and distinguishing these two forms of sentience allowed Nagendra to propose that the clockwork body was perfectly in line with the canonical works of Kobiraji therapeutics.

Nagendra, however, was not a major figure in the intellectual milieu of modern Ayurvedic therapeutics. He was based in a small village in south Bengal well away from the intellectual ferment of Calcutta, Dhaka, and Benares. His formulation, though published in a major journal and at great length, need not have had much impact. But that the new body imaginaire was not merely the quirky formulation of a rustic outlier became clear in the very next year. Gananath Sen, the founder of two Ayurvedic colleges in Calcutta, the first dean of the Ayurvedic Faculty at the Benares Hindu University, and the doyen of modern Ayurveda, wrote in his journal *Ayurveda* that "just as in order to repair a watch we must first know how the watch works, how many wheels there are within it, which wheel is adjacent to which other wheel, which wheel turns which way, why the watch runs faster, etc.—know all such subtle matters, similarly to treat the body we must first be aware of the subtle matters about the internal workings of the body."[139] Continuing further, he claimed that just as knowledge about the various parts of the watch allowed one to find out exactly what was wrong, similarly knowing the anatomy of the body allowed one to know what was wrong.

The popularity of the clock metaphor is not surprising. Recent work on science in the Arab world has found the widespread familiarity and adoption of the argument that the body, or even the world, was in fact a well-made clock thus implying the existence of an intelligent Creator. Marwa Elshakry calls it "one of the ironies of the introduction of Darwin into Arabic," that so many Arab intellectuals came to embrace the natural theology engendered in the watchmaker argument.[140]

Gananath was a major proponent for the introduction of anatomical education into the teaching of Kobiraji medicine. He also claimed that modern anatomy was already anticipated in classical Ayurvedic writings. Many contemporary Kobirajes, however, challenged both claims. Haramohan Majumdar, for instance, a proponent of unmixed *Suddha* Ayurveda, published a pamphlet challenging Gananath's version of history, stating that there was no single unified anatomical tradition that could be worked out since even some of the key texts were on occasion diametrically opposed to each other.[141] This debate over anatomy and dissection, however, was a red herring. For both camps, whether they supported dissections and anatomy or not, eventually came to adopt the clockwork body as the body imaginaire. The *Ayurveda Bikash*, where Nagendra had published the year before, after all was a journal associated with the camp opposed to Sen.

In fact, there was nothing in the clockwork body that intrinsically promoted anatomy or dissections. One could easily hold a mechanical model of the body imagined through the model of a clock and yet not be led to opening the clock up to look within. In fact, Descartes, one of the founders of mechanical philosophy that held the clockwork mechanism as the key metaphor in the "West," in his methodological writings explicitly deployed the metaphor to promote hypothetical thinking rather than explicit empirical examination. He wrote that

> just as an industrious watch-maker may make two watches which keep time equally well and without any difference in their external appearance, yet without any similarity in the composition of their wheels, so it is certain that God works in an infinity of diverse ways. . . . And I believe I shall have done enough if the causes that I have listed are such that the effects they may produce are similar to those we see in the world, without being informed whether there are other ways in which they are produced.[142]

Clearly, Descartes is deploying the clock metaphor here but also at the same time saying that it is unimportant, or impossible, to find out the exact mechanism underlying it and asserting the sufficiency of making plausible hypothesis. As Laurens Laudan sums it up for Descartes, "we can never get inside nature's clock to see whether nature's mechanisms are what we think them to be." Yet, the clock metaphor remains central to Descartes' thinking.[143] Gananath makes exactly the opposite case in arguing that the body must be opened up to see how exactly the clock works.

For Gananath, the clock metaphor naturally leads to the next step of dissection and anatomy. He repeatedly deployed the metaphor, which he had first published in 1916. It appears in his *Ayurved Samhita* in 1924 once more in almost identical form, and finally again in 1938 in his Sanskrit anatomical textbook entitled *Pratyaksha Shariram*. These prominent deployments of the clockwork body by Gananath should not distract us. His particular agenda had many opponents among the proponents of refigured Ayurveda, but the metaphor was shared by all of them and thus underwrote a new, shared body imaginaire. Thus Pandit Shiv Sharma, a militant champion of "pure" or *Suddha* Ayurveda who clashed with Sen on his championing of syncretic medical education in Ayurvedic colleges,[144] still borrowed from Roy and Sen extensively in describing the "human machine."[145]

Given the similarity of the clock metaphor to the iatromechanical philosophies of the seventeenth-century Europe, it is useful to look back at them in order to decode the cultural consequences of its adoption by refigured Ayurveda. Rivka Feldhay points to three important consequences. First, the metaphor "related observable phenomenon (clock hands) to invisible mechanisms (clock weights and wheels)" through "explanations," "causes," or "hypotheses." Second, by equating the natural objects to machines, the metaphor underlined regularity and therefore their potential for obedience of "mathematical principles." Finally, while potentially preserving and even consolidating the role of God as creator/watchmaker, it made His intervention in the regular running of the universe redundant.[146] Each of these three broad consequences can be observed in the case of refigured Ayurveda as well. Hypotheses and explanations relating the seen to the unseen mechanisms behind it proliferated. Bodies and bodily processes were invested with a new capacity for numerical regularity that could be numeralized. Finally, while the theistic aspects of therapeutics were repeatedly asserted, the scope for direct divine intervention through either a transmaterial presence in the body (such as *jib's* travels in the body) or immediate invocation through prayer was attenuated.

But the obvious similarities should not blind us to the equally obvious differences. Let us not forget that Sen Gupta called the clockwork body a *jib-yantra*, an "animate machine," and then proceeded at length to posit his idea of different forms of sentience. Reading him closely it is clear that he does not have a purely mechanistic model of the body in mind. Here it is useful to think with John Tresch's recent discussion of "romantic machines." Tresch's exploration of "machines" in nineteenth-century France distinguishes between two types of idealized mechanisms. On the one hand were the "classical machines" exemplified by the clock, the lever, and the balance and embodying the "primary qualities of mass, position and velocity." Classical machines were seen as a "passive transmitter of external forces, as a symbol of balance and external order. It implied a stable determinist universe"; it implied a "lifeless agglomeration of points and forces." In stark contrast to these stood the "romantic machines." These latter "were powered by steam, electricity and other subtle forces, [they] could be seen to have their motive force within them, they were presented as ambiguously alive." If the classical machines were exemplified by the likes of clocks, the romantic machines were materialized in "steam engines, electromagnetic apparatuses, geophysical instruments, daguerreotypes and industrial presses."[147] In many ways, the clockwork body described by Sen Gupta and Sen was a romantic, rather than classical, machine.

Just as Sen Gupta fell back on what Ashis Nandy has called J. C. Bose's "alternative science," Sen drew upon the subtle forces described by the then-emergent science of endocrinology to power his machine. These subtle and internal sources of motion clearly rendered the clockwork body akin to a romantic machine, though they also partook of the rule-bound and regular character of the classical machines. Moreover, other more easily recognizable metaphors of romantic machines in the form of electromagnetic apparatuses and steam engines were often found in close proximity to the clock metaphors. Hence, despite the iconic association of the clock with idealized classical machines, in the case of the clockwork body of refigured Ayurvedic therapeutics, the lines between the classical and the romantic machines blurred. The elaborate image of hidden, internal gears and wheels acquired the subtlety of electromagnetic and endocrinal apparatuses. What emerged through this reworking was an amorphous notion of *machinic physiospiritualism*.

Machinic physiospiritualism was a loose understanding of the body as a machine characterized by regularity, inter-connectedness, and complex internal mechanisms, etc., yet also by life, internal power, and substantial constitution of substances that repeatedly brought the physical and the spiritual in contact. It was an entity constituted by the braiding not only of particular strands of "Eastern" and "Western" sciences, but also of materiality and spirituality, of mathematical regularity and mystic vitalism.

THREE

The *Snayubik* Man: Reticulate Physiospiritualism and the Thermometer

Sita, the nurse who resented the idealist young doctor, Prodyot, for calling the aged Kobiraj, Jibon, a quack and a charlatan, quietly handed the thermometer to Prodyot and returned to the sickroom. Despite the silence, the look of proud vindication on her face could not avoid Jibon Moshai's seasoned eyes. "It seems to be hundred and four all right," murmured Prodyot now slightly unsure of himself. Hundred and four was exactly what Jibon, relying on his *nadipariksha*, had predicted the instrument would say.[1]

We had met Jibon Moshai, the central character in Tarashankar Bandyopadhyay's novel *Arogyo Niketon*, in the last chapter when he was still a young man in the shadow of Rangalal *daktar*. At the time, his qualitative *nadipariksha* was forced to engage with Rangalal's quantitative, pocket-watch-reliant sphygmology. By the time we find him interacting with Prodyot and Sita, he is an old man. Rangalal is long dead, and Prodyot is the new, state-employed, idealistic, young biomedical physician. Though the dates in the novel are sketchy, it would seem as though about half a century had elapsed since the day Jibon and Rangalal compared their skills at *nadipariksha*/pulse reading. By that time, pulse was no longer an important diagnostic tool for biomedical physicians like Prodyot. The thermometer had replaced the iconic bedside ubiquity of sphygmology. The bodily truths revealed by the thermometer, however, seem, in the brief encounter cited above, to be wholly compatible to those revealed by *nadipariksha*.

There is much to unpack here. To begin with, the chronology, though not explicitly incorrect, is somewhat misleading. The clinical thermometer, though undoubtedly younger than the pocket watch, was not as late as the novel's plot suggests. In fact, James Long's catalogue of Bengali printed books listed an unfortunately now-lost book by one S. C. Karmakar published in 1853 that discussed the use of thermometers in medicine.[2] From the point of view of Ayurvedic therapeutics, both the pocket watch and the thermometer therefore arrive around the same time. Their chronological distance in the "Western" medical tradition is, in this case, of little putative import for Ayurvedic therapeutics. Yet it is significant that in Bandyopadhyay's dramatization this chronological gap is implicitly reopened.

Despite the misleading chronological gap, it is surprising that the truths of *nadipariksha* and the truths of the thermometer are seen to be so easily compatible. Whereas in the case of the pocket watch, the data it revealed and the "Western" diagnostic technique of sphygmology shared at least a family resemblance with *nadipariksha*, the latter had very little obvious overlap with the thermometric measurement of body heat. They were two completely distinctive diagnostic techniques operationalized on different parts of the patient's body—the thermometer usually inserted in one's mouth or armpit, whereas *nadipariksha* was typically conducted at the wrist—and seeking different kinds of data about the body. Yet Bandyopadhyay and others held the truths revealed by these diagnostic techniques to be entirely transposable. This transposability required a further reimagination of the body metaphor that undergirded modern Ayurvedic therapeutics.

From Thermoscopes to Thermometers

Though today thought of primarily as a medical instrument, the thermometer first came to India as a meteorological device. Its medical life emerged only gradually. As a result, Bengali medical authors continued to call it the *Tapman-yantra* (climatic temperature instrument). Like the pocket watch, Bengali Kobirajes gradually adopted it in the last decades of the nineteenth century. Like the watch too, Galileo is credited with the development of the first, crude temperature-measuring devices, and Sanctorius (1561–1636) is said to have developed the first workable mouth thermometer. The instrument was still far from perfect and underwent several more modifications and improvements in the centuries following Sanctorius. It was through the influence of the redoubtable Herman Boerhaave (1668–1738) and that of his students that the instrument's medical uses began to expand and it came to be used at the bedside.[3] Boerhaave's acceptance was largely a consequence of improvements in the technology devised by instrument maker Gabriel Daniel Fahrenheit. But it did not translate into universal acceptance within the medical profession. Moreover, by the dawn of the nineteenth century, there was a reversal in the trend and diagnostic advances that localized diseases led physicians to ignore the lead of doctors like Boerhaave. It was only from the 1840s that medical thermometry gradually grew among the mainstream of "Western" physicians. It did not become a mainstream diagnostic tool among Anglo-American physicians until the 1860s.[4]

Partly what delayed the uptake of the thermometer in medical circles was a combination of crucial conceptual and technological problems in thermometry. There were two crucial issues: first, what fluid indicator to use in the tube; and second, against what standard values heat could be calibrated. With regard to the first question, a number of alternatives ranging from linseed oil and "spirit of wine" to mercury were tried. With regard to the second question, suggestions varied from using the boiling point of water to using the temperature of a certain English cave.[5] Some of the early devices, including Fahrenheit's instrument, actually used "blood heat" or the human body temperature as a fixed point. Obviously, this was misleading since human body temperature, we now know, does in fact fluctuate within a small range even if we are not ill. Though these early instruments produced a kind of numerical data, Hasok Chang refuses to call them "thermometers" since the numbers calibrated on them were in a sense arbitrary. What they in effect told the physician was still a kind of qualitative data about the relative hotness or coldness of two bodies. Chang calls these instruments "thermoscopes" as opposed to the thermometers proper where the numerical data is defined with reference to fixed points and therefore bears an internal mathematical relationship.[6] Heinz Otto Sibum's work on James Prescott Joule has pointed out that the standardization of heat measurement around the middle of the nineteenth century owed much to changes in the brewing industry. These changes allowed and encouraged brewers and malsters to convert their embodied "gestural knowledge" of heat into precision instruments.[7] Building on these earlier developments in thermometry in general, in the 1860s Carl Wunderlich and Thomas Clifford Allbutt independently determined a standardized measurement of human temperature as well as modernized the design of the clinical thermometer.[8]

The first thermometers that appeared for regular sale in London were designed by William Aitken of the Netley Hospital using Wunderlich's studies for inspiration. They were available in London shops in 1863 although they were actually manufactured in Leipzig. This early design had a curved end and required twenty to twenty five minutes in axilla and had to be read in situ.[9] This unwieldy design was eventually overtaken by the Albutt thermometer that largely remains the prototype for most contemporary thermometers. Writing of the first thermometer he designed, Albutt wrote:

> I wanted a thermometer which should live habitually in my pocket, and be as constantly with me, or more constantly, than a stethoscope. Such a one

> Messrs. Harvey and Reynolds, of 3, Briggate, Leeds, have had made for me. It is scarcely six inches in length, and being slipped into a strong case, not much thicker than a stout pencil, is carried in the pocket easily and safely. It is made as I proposed—namely, by slightly widening the thread above the bulb, such as to allow the mercury to expand for 20 degrees of heat without rising much in the thread. The graduation begins a little above this widened portion at 80°, and runs upto 115°.[10]

Albutt had originally approached a London instrument maker, Casella, but eventually worked with Harvey & Reynolds of Leeds. The first instrument the latter produced was calibrated according to the Fahrenheit scale, but Albutt insisted on changing it immediately to the Celsius scale. The instrument cost 7s 6d.[11]

The uptake of the Albutt's clinical thermometer was rapid. He wrote the above lines in 1867, and by 1872 Dr. T. Edmonston Charles of the Calcutta Medical College had already conducted a series of important studies on dengue using the clinical thermometer. His lectures on the subject to his Bengali students at the college might well have been one of the early conduits for the dissemination of clinical thermometry among Bengali physicians.[12] Another study, published in the *Indian Medical Gazette* published the very next year, was undertaken by Dr. Edward Lawrie upon European and Indian prisoners at Calcutta to examine the impact of hard labor on body temperature.[13] Despite these early studies, there is no evidence to suggest that any attempt was made to manufacture the instrument locally.

Clinical Thermometry and Ayurveda

As clinical thermometry progressed in the second half of the nineteenth century, it increasingly came to be a part of the everyday operation of hospital medicine. Expectedly, its immediate consequence was to standardize the human body temperature.[14] As Christopher Hamlin, commenting on the promise of the thermometer in nineteenth-century "Western" medicine, puts it, "numbers clarified, personhood confused," thus "data" from instruments like the thermometer increasingly displaced "embodiment."[15] Yet its long-term impact on the clinical encounter remains ambiguous. Some have argued that far from disempowering the patient by objectifying and standardizing her body, the thermometer gave the patient a shared vocabulary through which to engage the doctor.[16] Others have pointed to the role of thermometers in the development of more mediated forms of medical care by the doctor delegating regular monitoring functions to the emergent profession of nurses.[17] Emphasizing the intimate connection between clinical thermometry and the nursing profession, Margarete Sandelowski writes that "the thermometer was rather quickly incorporated into routine nursing practice and almost as quickly became associated with, and even to represent, nursing."[18] We should, however, be careful about generalizing these arguments. In the context of modern Ayurveda in Bengal, hospitals and nurses both played a limited role until much later in the twentieth century. Instruments such as the thermometer therefore acquired very different social and cultural roles than it did in the Euro-American biomedical milieu. Thermometers after all are a form of everyday (medical) technology, and like all other "everyday technologies," when they are re-embedded in new social and cultural contexts, they "undergo adaptation, assimilation and compromise through which they acquire new meanings and usages."[19] Thus, Charles Rosenberg, for instance, points out that in the mid-1860s in New York, the average practitioner was loath to use thermometers despite enthusiastic pleas in their favor.[20]

Unfortunately there is limited data on the manufacture, distribution, and general uptake of the thermometer in British India. By the late 1860s, the *Indian Medical Gazette* does occasionally discuss thermometers, but more often than not it is in relation to meteorology rather than clinical examinations.[21] The thermometer or the *Tapman-yantra* begins to appear in Kobiraji books from the 1890s.

Jashodanandan Sarkar's interpolations into the *Susruta Samhita* describe the instrument in a single short paragraph. In it, Sarkar only mentioned that the instrument was to be placed in the armpit, that the normal temperature was 98.4 degrees, and that this was actually the temperature of the blood in the patient's body. Sarkar also stated that, though any temperature higher than 98.4 signaled fever, the readings had to be compared with the pulse to arrive at a firm decision and that it was not advisable to rely on the *Tapman-yantra* by itself.[22] This hesitant reliance on the thermometer was not unique to the nineteenth-century refigured Ayurveda. Hamlin points out that while thermometric heat did gradually emerge as the central feature of fevers in "Western" medicine, pulse diagnosis was not easily displaced. Referring to the experimental work of James Curie (1756–1805) that helped promote the importance of thermometry, Hamlin states that, "if heat seemed to be the cause of the symptoms, pulse still defined the disease."[23]

What is also important to keep in mind in this context is Robert Aronowitz's argument that the very concept of "symptom" was changing in the late nineteenth century. Aronowitz argues that a single notion of "symptom" at this point began to bifurcate into an objectively traceable "sign" and a more functionalist and subjective new notion of "symptom." Instruments like clinical thermometers emerged within this broader quest for objective "signs" as distinct from the still inchoately objective "symptoms."[24]

In contrast to the pocket watch, which was itself almost absent from medical discussions that presumed its use, the thermometer in most cases was described at length. Nagendranath, for instance, included a lengthy exposition of how to use the instrument together with an accompanying illustration (see fig. 5). Deploying his familiar trope, Nagendranath contended that Western physicians had invented the thermometer because diagnosing by *nadipariksha* was exceedingly difficult for them. He then proceeded to give detailed instructions on how to use the *Tapman-yantra.* He gave detailed instructions on which parts of the body one could place the instrument in and for how long. Patients, Sengupta advised, ought to be made to lie on their sides and asked to keep any kind of physical movement to a minimum for at least an hour before the application of the thermometer. Once applied, he said, sometimes the physician might even need to keep the thermometer in place for up to ten minutes. He even pointed out that each of the markings on the tube represented a "degree." Expectedly, he explained that the normal temperature was 98.4 Fahrenheit, but then also proceeded to describe normal temperatures for groups of diseases. For a certain group of complaints, he said, the temperature would likely be between 106 and 107 degrees. For another group he set the limit at 104–5, but then added that the temperature did not usually hold at this level. Going further, in certain cases he also laid down death-temperatures—that is, certain temperatures for certain diseases that foretold certain death. For *Bisuchika* (often translated as cholera), for example, he stated that if the temperature fell to between 77 and 79, the patient was going to die. For certain other febrile diseases, he stated that temperatures of 109 and 110 foretold certain death.[25] These "death temperatures" were clearly reminiscent of "death pulses" that were well-known in *nadipariksha* texts and were seeking to replace them.

In another later work entitled *Chikitsaratna*, Dwarkanath Bidyaratna, followed exactly the same model as Nagendranath. The only variations occur in the level of detail. On the one hand, Bidyaratna added more details describing the actual use of the instrument such as an elaborate explanation of the decimal markings in between degrees on the thermometer; while, on the other hand, he reduced the level of details about individual clusters of diseases and their normal ranges by merely stating the absolute temperatures that could be fatal for all diseases.[26] Overall, it was a general simplification of Nagendranath's model without any fundamental alterations.

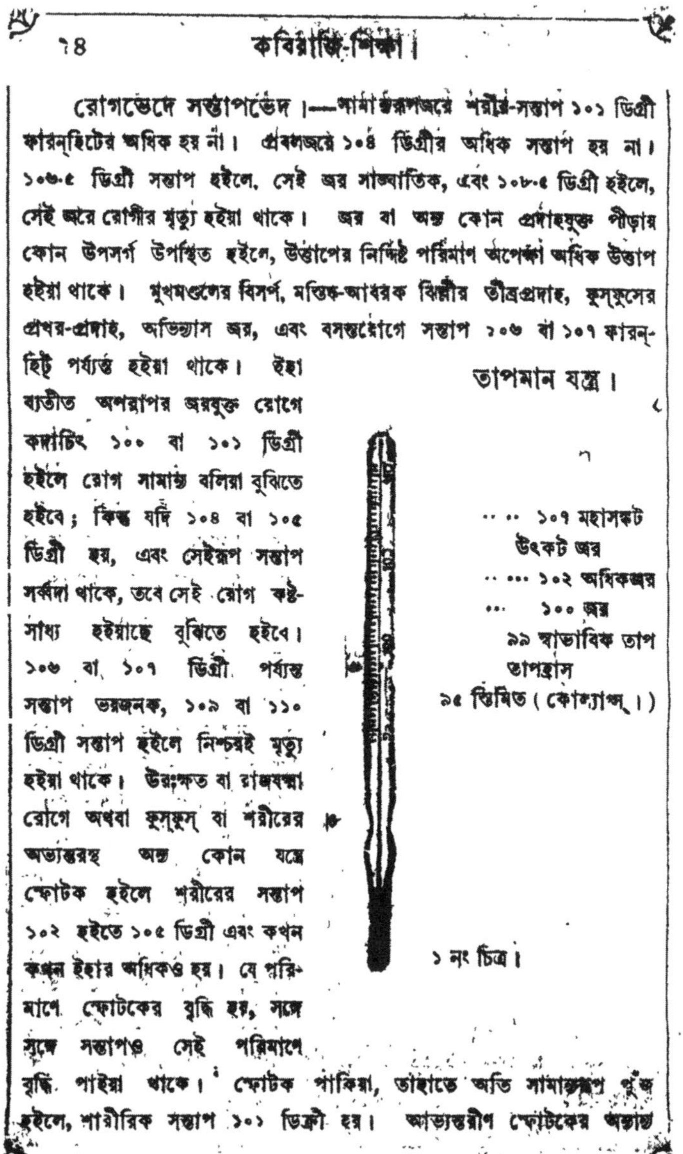

রোগভেদে সন্তাপভেদ।—সামান্যরূপজ্বরে শরীর-সন্তাপ ১০১ ডিগ্রী ফারন্‌হিটের অধিক হয় না। প্রবলজ্বরে ১০৪ ডিগ্রীর অধিক সন্তাপ হয় না। ১০৬·৫ ডিগ্রী সন্তাপ হইলে, সেই জ্বর সাঙ্ঘাতিক, এবং ১০৮·৫ ডিগ্রী হইলে, সেই জ্বরে রোগীর মৃত্যু হইয়া থাকে। জ্বর বা অন্য কোন প্রদাহযুক্ত পীড়ায় কোন উপসর্গ উপস্থিত হইলে, উত্তাপের নির্দ্দিষ্ট পরিমাণ অপেক্ষা অধিক উত্তাপ হইয়া থাকে। মুখমণ্ডলের বিসর্প, মস্তিষ্ক-আবরক ঝিল্লীর তীব্রপ্রদাহ, ফুস্‌ফুসের প্রথর-প্রদাহ, অভিন্যাস জ্বর, এবং বসন্তরোগে সন্তাপ ১০৬ বা ১০৭ ফারন্‌হিট্ পর্য্যন্ত হইয়া থাকে। ইহা ব্যতীত অপরাপর জ্বরযুক্ত রোগে কদাচিৎ ১০০ বা ১০১ ডিগ্রী হইলে রোগ সামান্য বলিয়া বুঝিতে হইবে; কিন্তু যদি ১০৪ বা ১০৫ ডিগ্রী হয়, এবং সেইরূপ সন্তাপ সর্ব্বদা থাকে, তবে সেই রোগ কষ্ট-সাধ্য হইয়াছে বুঝিতে হইবে। ১০৬ বা ১০৭ ডিগ্রী পর্য্যন্ত সন্তাপ ভয়জনক, ১০৯ বা ১১০ ডিগ্রী সন্তাপ হইলে নিশ্চয়ই মৃত্যু হইয়া থাকে। উরঃক্ষত বা রাজযক্ষ্মা রোগে অথবা ফুস্‌ফুস্ বা শরীরের অভ্যন্তরস্থ অন্য কোন যন্ত্রে স্ফোটক হইলে শরীরের সন্তাপ ১০২ হইতে ১০৫ ডিগ্রী এবং কখন কখন ইহার অধিকও হয়। যে পরিমাণে স্ফোটকের বৃদ্ধি হয়, সঙ্গে সঙ্গে সন্তাপও সেই পরিমাণে বৃদ্ধি পাইয়া থাকে। স্ফোটক পাকিয়া, তাহাতে অতি সামান্যরূপ পূঁজ হইলে, শারীরিক সন্তাপ ১০১ ডিগ্রী হয়। আভ্যন্তরীণ স্ফোটকের অন্যান্য

তাপমান যন্ত্র।

১০৭ মহাসঙ্কট
উৎকট জ্বর
১০২ অধিকজ্বর
১০০ জ্বর
৯৯ স্বাভাবিক তাপ
তাপহ্রাস
৯৫ স্তিমিত (কোল্যাপ্স্।)

১ নং চিত্র।

5. *Tapman-yantra.* The Thermometer in Vernacular.

While the details and the illustrations immediately catch our eyes and seem to be a glaring contrast to the almost surreptitious ways in which the use of the pocket watch was introduced into modern Ayurveda, there are many aspects of Kobiraji thermometry which are much more difficult to discern. Most conspicuously, whereas the superficial equivalence set-up between *nadipariksha* and pulse diagnosis immediately opens up the possibility of comparing the two regimes of embodied truths, the thermometer, by appearing to be a wholly novel diagnostic tool, makes it more difficult to detect what exactly changed in Kobiraji medicine in the course of its adoption. In order to locate this difference, we must look deeper and over a more varied field.

At its heart, what the thermometer measures is of course body heat, and it is this that needs to be focused on. Thinking in terms of "Western" medicine, the most obvious place to look for discussions of body heat or *sontap* would be in discussions of fever. The *Oxford English Dictionary* mentions that the word *fever*, since around 1,000 CE, has meant a morbid condition "characterized by undue elevation of temperature."[27] Hamlin's more rigorous investigation of fevers in the "Western" medical tradition, however, categorically denies any such long and uniform connection between body heat and fever. Instead, Hamlin argues that "the coming centrality of heat, its objectification in thermometry and the foundation of both in systematic experiments, were achievements of that magnificent knowledge-generating engine, the nineteenth-century German university system."[28] This "coming centrality," however, remained dogged by multiple figurations of "heat" itself in the "Western" tradition of the long nineteenth century. Chemical heat did not always correspond to physical heat, and the emergent but often imprecise and polysemic language of "energy" further complicated matters.[29] Reiser quotes a late eighteenth-century French physician rejecting thermometric investigations on the grounds that the quantification of body heat told the physician nothing about the more important qualitative aspects of that heat.[30] It was only toward the end of the nineteenth century that a quantifiable, monolithic notion of body heat as a measurable entity came to chiefly define fevers in the "Western" tradition.

Kobiraji medicine's alignment between body heat (*sontap*) and fever (*jwor*) emerged at the same time. But both the German university system and its experimental foundation were distant and mediated realities for the Bengali Kobiraj. Instead, the latter accessed the thermometer at the intersection of imperial techno-commodity culture and the emergent technoscientific orientation of colonial medicine. Predictably, this divergent route produced an alignment between *sontap* and *jwor* that, while still privileging thermometric heat, remained distinctive from the emergent alignment in "Western" medicine.

This distinctiveness arose from modern Ayurveda's braiding of emergent "Western" thermometry with strands of older, premodern Kobiraji praxis. Madhavakar's *Rogavinischaya,* an eighth-century medical work popularly referred to as the *Nidan,* is a good window into precolonial medical opinion on fevers in the region, since the *Nidan* was undoubtedly the most widely read and authoritative medical text in Bengal on the eve of colonialism.[31] In it, while *jwor* was given the pride of place among all afflictions, *sontap* did not always accompany it.

The general description of *jwor* began simply by stating that it originally arose from Shiva's angered sigh upon being scorned by King Daksha. It then proceeded to provide a brief classification of various types of fevers. Madhavakara then provided two slightly distinct descriptions of fever. The first of these is close to the canonical *Charaka Samhita,* but not identical. According to it, "By improper regimen of diet and conduct the [para-humors], residing in the receptacle of undigested food, after having thrown outside the visceral fire, will give rise to fever, when they follow the nutrient fluid in its course."[32] The phrase "having thrown outside the visceral fire" might here connote heat, but it is significant that even if it is so, it is presented not simply as heat but the throwing out of the "visceral fire," which later is a specific concept in the classical Sanskrit Ayurvedic tradition. This is crucial because Vijay Rakshita's influential *Madhukosa* commentary on the *Nidan* had clarified that the "the visceral fire is mentioned in order to exclude the fire of the elements (of the body) etc."[33] Clearly, there is more than one type of "heat" within the body and these cannot be taken to imply thermometric heat in any simple sense. What constitutes true body heat, then, is far from self-evident, and its relationship to fever remains obscure.

In the next definition, borrowing from the *Susruta Samhita,* Madhavakara stated that "the disease in which obstruction (of the flow) of perspiration, a general sense of glowing heat and also a seizing (pain) in the whole body occurs simultaneously, is designated fever."[34] This is a clear and important definition. "Heat" here is clearly said to be a "general sense"—that is, something that the patient feels. It may or may not be felt by others. In Bengali renditions of the *Nidan,* two different words are often found deployed in these discussions. The first is *sontap* or temperature more generally, and the next is *doho* or burning that seems to suggest a more patient-subjective sensation.[35] Even without the linguistic difference in Bengali, serious doubt as to whether the phrase "general sense of glowing heat" implied anything akin to an objectively perceptible heat or not is introduced by Rakshita. In the *Madhukosa*, Rakshita, in glossing this verse wrote that "a deviated state of the senses should be known as a characteristic of a general sense of glowing heat," before further adding that "a deviated sense of the mind, disinclination and languor are the characteristics of a general glowing heat of the *manos* [very roughly translates as 'mind']."[36] In fact, a subjective, as opposed to an objective, conception of body heat could generate dramatically different conclusions in cases of fever since, as Nicholson points out, "in fever patients might shiver and complain of feeling cold, although the thermometer recorded an elevated temperature."[37]

Further evidence that fever does not necessarily entail any objectively perceptible rise in temperature can be seen in the descriptions of the individual types of fevers. In the fevers caused by the wind (*vayu*), for instance, Madhavakara, drawing again on Susruta, said that "shivering, incongruous paroxysms, complete desiccation of throat and lips, insomnia, inhibition of sneezing and certainly dryness of the limbs, pain in head, cardiac region and limbs, an altered taste in the mouth, tightly packed faeces, piercing pain, inflatedness and yawning [occur] in a fever arising from Wind."[38] There is no necessity in this, one of the commonest of fevers, of any outwardly perceptible rise in temperature.

Three things thus stand out. First, fever is not necessarily accompanied by an outwardly perceptible rise in temperature. Second, even when there is a rise in temperature, whether it is a state of mind, a subjective sensation of the patient, or something more objectively perceptible, remains doubtful. Third, to emphasize the obvious, "heat" was never thought of as a measurable entity. What is clear then is that there is emphatically no notion of a calibrated, measureable entity called body heat. *Sontap*, the term that is later made to carry the connotation of a measurable body heat by authors like Nagendranath, is more ambiguous in its precolonial deployments, overlapping with terms like *doho* that (in these contexts) have a stronger orientation toward subjective sensations. None of this is unique to Kobiraji medicine. In the "Western" medical tradition there are almost identical but parallel processes by which externalized, thermometric body heat gradually displaces more subjective indicators of fevers. This "gradual recognition of the discrepancy between feeling and facts," Hamlin argues, "would ultimately become paradigmatic of the superiority of instrument truths in clinical medicine."[39]

In Kobiraji medicine, however, there are discussions of *sontap* outside febrile contexts as well. Heat is particularly prominently discussed in the context of smallpox and the cult of the pox goddess, Sitala. Frederique Apffel-Marglin in her fascinating study of smallpox writes that

> suffering from the disease can be spoken of in terms of an excess of heat, the pustules being the visible signs of an overboiling blood erupting through the skin and fever being an immediate indication of an excess of heat escaping from the body. It can also and simultaneously be spoken of as the anger of the goddess . . . material substances transform and refine themselves (*songskar*) into feelings, thoughts and consciousness. Anger is an excess of heat; heat is fire; heat is bile; heat is the sun. Heat out of congruence is disease, which is the anger of the goddess. The distinction between the literal and the metaphorical is blurred to the point of being dissolved.[40]

Another context in which "heat" is discussed in Ayurvedic circles is that of *rasayana* or rejuvenation therapy. As Lawrence Cohen following McKim Marriott points out, alongside other polarities such as windy/still and wet/dry, hot/cold constitutes one of the most ubiquitous frameworks for structuring experience. This structuring of experience is as much about the physiological body of the old person as it is about the position of the old person in the social body. As a result, Cohen pithily states that "heat, particularly in the context of life cycle, can be read as the externalization of power."[41]

Moreover, the *Nidan* itself went further and blurred the very lines between the human and divine, the historical and mythic. It located the origin of fever in the anger of the god Rudra/Shiva. This was the story of Jworasur: the fever demon. Often misleadingly spoken of as a "folk deity" with no immediate connection to the classical Sanskrit Ayurvedic textual corpus, Jworasur in fact is eminently present in texts such as the *Nidan*. There are reports from the nineteenth century that tell of eminent teachers of Kobiraji medicine who insisted upon beginning the lessons on fever by recounting the myths of Jworasur. They insisted it was more than just a story. It was integral to understanding the Kobiraji conception of fever.[42] There, febrile heat was never simply an objective and measurable entity. It was instead always a polyvalent entity fluctuating between the mythic, intangible realms of a god's (i.e., Rudra's) anger, a patient's subjective sensations, a physician's objective determinations and always already as externalized social power in one form or the other. It was impossible to reduce this polyvalence to a measurable physical reality.

In stark contrast to this polyvalent and frequently transmaterial understanding of *sontap*, Kobiraj Ramchandra Mullick, founder of the Gobindo Sundori Free Ayurvedic College, advanced an alternate definition[43]:

> During the process of digestion the foodstuff that we consume undergoes chemical change in the stomach and intestines (*pakostholi o ontrer modhye*), it

> is then that heat (*tap*) is produced. Besides this, the "tissues" in our body are continually being destroyed. This process of destruction also raises the temperature. These two stated reasons are the fundamental causes of heat. Using a *Tapman-yantra* is extremely helpful in diagnosis, prognosis and treatment.[44]

Not satisfied with the radical redefinition of "bodily heat," Mullick went further and illustrated with examples how the precise determination of the temperature in numeric terms was crucial to clinical practice. The thermometer, for Mullick, was an indispensible diagnostic tool. Particularly in cases of contagious diseases such as "scarlatina" and smallpox (*bosonto*), he stated that "many people of a suspicious disposition (*sondigdhochitto lok*)" began to suspect the disease as soon as they encountered one or two uncertain symptoms. This in turn introduced doubt into the physician's mind. A *Tapman-yantra* could provide certainty and assurance in this atmosphere of doubt and clearly establish the presence or absence of the dreaded diseases. Moreover, Mullick continued, "all types of fever (*jwor*), consumption (*joksha*) and internal bleeding (*rokto-srab*) are made particularly discernible by the *Tapman-yantra*."[45] For prognosis too, the instrument was invaluable. Often, Mullick pointed out, an increase of body temperature from 104 to 105 would not be discernible to the Kobiraj merely by touch. Were he to proceed on simply his sense of touch, he might have well mistaken the rising fever for a static and slight one. This in turn could have led to grave and fatal mistreatment.[46] Unlike in the case of the pocket watch therefore, at least some Kobirajes like Mullick were willing to concede a seemingly superior diagnostic power to the *thermometer* than had previously been available.

Patient Voices, Measured Tones

Not only were the thermometer's diagnostic powers superior to simple tactile impressions, but, according to Mullick, it also allowed for more regular monitoring of fevers. He recommended that the temperature of any patient be checked regularly every morning and evening. In more severe cases, he thought the temperature needed to be checked between five to seven times per day. Naturally, such regular and frequent monitoring was not possible if one depended upon diagnostic methods like *nadipariksha* that required the presence of a highly skilled physician. With the coming of the thermometer, greater regularity of monitoring became possible because the act of monitoring no longer required rare and specialized skills. The act of monitoring a fever could therefore be delegated to nonphysicians attending to the patient.

It was this act of delegation that sped up the dissemination of the radically new idiom for speaking about fevers that the thermometer inaugurated. In other words, the rapidity of the lay uptake of the thermometric idiom was aided by the fact that the thermometer itself helped reorganize the nursing of fever patients. Temperature monitoring now became a technical aspect of doctoring that could be separated from the person of the physician and undertaken by the nurse.[47] Since close relatives often nursed fever patients, they quickly acquired the technical and linguistic competence of using the thermometer thereby leading to a rapid popular dissemination of the thermometric idiom of speaking about fever. By the end of the nineteenth century, nonmedical, lay interlocutors came to speak among themselves referring to febrile conditions through the thermometric language of degrees of body heat. But there are also hints to suggest that this new idiom, far from universal, worked to further accentuate social divisions.

Letters provide an excellent, even if somewhat self-selecting, source for lay conversations about illness. Clearly restricted to the literate sections of society, letters, particularly for the early or precolonial period, tended to overwhelmingly be written by relatively well-off men. Women seldom featured as letter writers. Literacy also somewhat overlapped with caste and hence there was a preponderance of upper-caste and particularly Brahmin letter writers. Despite these exclusions, letters provide an unparalleled and candid window into how people spoke about illness among themselves. Luckily for us, illnesses were actually one of the most common topics about which people needed to write, and many such letters have been preserved at the Visva Bharati University's archives in Santiniketan.

In striking contrast to later nineteenth-century letters about illness, early letters seldom mentioned a specific disease by name. Most letters, written either to inform a kinsman or woman living away, or to solicit some material help for the illness, referred to the illness through generic words such as "*byamo*" (disease) or "*pira*" (pain/suffering). There was no clear notion of disease specificity. On rare occasions, one or two additional characteristics of the disease were mentioned, but most frequently it was simply left at the generic level with no attempt to further specify.[48] Comparing these to medical books of the day it can be assumed that a significant majority, if not the vast majority, of these afflictions would have been identified as some form of "fever" or the other. Yet most of these letters neither say so nor mention any temperature.

Only a small handful of letters mention specific illnesses by name, and then, predictably, they refer most often to *jwor* ("fever"). One letter, written by a Brahmin named Krishnadas Sharma to his uncle, Shibchandra Nayalankar, stated, for instance, that "upon arrival I found the diseases (*byamo*) of the gentlemen [at home] had become specific (*bishesh*) [.] *Jwor* (fever) and *ham* (measles) have abated, the *pora-gha* (burn) has become specific (*bishesh*) but has not fully healed."[49] Another letter, addressed to a Brahmin named Kamalapati Bhattacharya by an unnamed older relative, stated that "*Jwor* (fever) and thereafter *pala-jwor* (relapsing fever/malaria) along with cough has caused me great suffering (*boro pirito*) [.] Most recently for a few days [I have not] sensed fever (*jworanubhab hoyni*) [but the] arrival of [the] cough has made me adopt a convalescent diet (*pothyo joge achhi*) [.] Do not worry."[50]

In marked contrast to these late eighteenth- and early nineteenth-century letters was the correspondence of Rabindranath Tagore at the end of the nineteenth century. Writing to his wife, Mrinalini, about his grandson Nitu's illness from Calcutta in November 1900, Tagore stated that "now his *jwor* is 99°, cough is simple, breathing is easier and the *nadi* is strong."[51] Again during the same ill-fated illness a little over a month later in mid-December, Tagore wrote, "[His] *jwor* is close to 100° today."[52] In another letter, perhaps written a few years earlier in the mid-1890s by Rathindranath, Rabindranath's son, to his mother informing her of the illness of another family member, stated that "last night the *jwor* left Nidda after he broke into a sweat. This morning it is 100. On other days it is usually 101."[53] These references are revealing in many ways. Not only did the men of the house use the idiom of thermometric measurements, but it was obviously also intelligible to Mrinalini Debi to whom all the letters were addressed. While recognizing that the Tagore family was in many ways atypical in their cultural and educational accomplishments, it is important to note that these were intimate conversations about and among close family members and therefore revealing of the extent to which the thermometric idiom for describing fever had become naturalized—at least in elite households.

While the Tagore family's accomplishment and status have led to the preservation of their correspondence and made it a natural source to turn to, that they were not wholly atypical can be seen from a few stray letters by lesser-known authors who somehow survived. One contemporary letter, reprinted in a book on dreams, showed one brother writing to another in another town about their sister's illness and mentioning that she "had a 105° *jwor*."[54] The author of the book and the recipient of the letter was an obscure figure. His book is equally obscure and little is known about him or his family other than what he mentioned in the book. They seemed to have been a very ordinary middle-class Brahmin family of professionals with the author the only one in the family to have obtained a university degree. Yet their letter is hardly distinguishable from those written by the illustrious Tagores.

While Brahmin authors who were reasonably affluent wrote all the letters compared here, there is an unmistakable and obvious shift in the way fever was perceived and discussed. The letters of Krishnadas Sharma and Kamalapati Bhattacharya stand in stark contrast to those of the Tagores as well as the obscure Kishorimohan Chattopadhyay. While the former at best mentioned "sensing fever," remaining entirely silent about temperature, let alone its exact measure, for the latter it was almost impossible to speak of fever without specifying its thermometric value. The contrast is enhanced by the fact that none of these were public letters as such. They were all intimate letters written to close family members describing the illnesses of other kinsfolk. This was a register that was both intimate and informal. It is also noteworthy that none of the interlocutors had any medical training. Their deployment of the thermometric idiom signaled therefore a truly epochal alteration of the way lay people—at least of a certain class and caste—imagined fever.

There is, however, some hint that not everyone shared this idiom. *Daktarbabu* was a play staged on Calcutta's hugely popular theater circuit in 1875. In 1886, another popular play called *Thengapathik Bhuinphod Daktar* was staged, and four years later in 1890, a third play once again titled *Daktarbabu* followed. Each of these plays was about medical corruption, and their plots entirely revolved around medical matters. In all three, the central character was a *daktar*, rather than a Kobiraj, and much effort was put into caricaturing their "modern" ways. Yet not one of the three plays at any point referred to fevers through the thermometric idiom. Given the ubiquity of "fevers" as the preeminent pathological designation, this absence is telling. While naturally wary of overreading this absence, it surely permits us to hold an impression that until the 1890s not all classes were equally familiar with the thermometric understanding of body heat.

This uneven shift in the idiom for conceptualizing what was frequently referred to as a "king of all diseases"—namely, *jwor*—demonstrates that even though we can speak of a broadly shared body image within refigured Ayurvedic therapeutics, that modern image is no longer one shared universally with the laypeople throughout society. This, however, is not unique to refigured Ayurveda. Rosenberg has pointed to a similar breach progressively distinguishing professional and lay understandings in the "West."[55]

What distinguishes Rosenberg's American case from my Bengali Ayurvedic example is that instead of the breach mapping wholly onto a divide between "professional" and "lay" understandings, in my case it resonates more fully with a class divide. Contemporary scholarship on the middle classes in India has pointed out that "class" cannot be understood simply as either a purely economic station or in terms of a position in the system of production. Instead, much of class identity is tied up in consumption practices. The consumption of both material and educational goods associated with modernity, a combination of university degrees and such material objects as the thermometer, together created a class of *bhadralok* ("genteel folk") patients and physicians who stood apart from those who could not or did not consume such goods.

That class limited the use of thermometers, and therefore the thermometric idiom, can easily be surmised by returning to Ramchandra Mullick's writings. He had specifically recommended the thermometers of three companies as reliable instruments. These were respectively Casella, Hicks, and Mason.[56] Unlike the use of watches, which, as I have shown were often locally manufactured, for thermometers, one clearly and explicitly still relied exclusively upon foreign manufactures. Besides the higher prices of imported instruments, Mullick's recommendations also pointed to the emergence of branded consumption within a medical milieu.

As is clear from Mullick's recommendations, the reliability of the diagnostic conclusions was made contingent upon the reliability of the instrument, and the latter in turn equated with the consumption of specific brands. The skill and expertise of the instrument user went unremarked, while the brand name became the guarantor of accuracy. This was in perfect contrast to *nadipariksha* where the skill and expertise of the physician were paramount. The patient's reliance on a specific physician was replaced by the patient's reliance on a particular brand. However, the consumption of branded commodities also serves to consolidate a sense of group or class identity. Whereas price barriers are implicit boundaries that enclose fellow consumers together, brand names are iconic markers of that commonality.[57] Moreover, as Douglas Haynes has recently pointed out, branding is a form of disciplining of consumption that helped redefine middle class identities in strictly coded normative ways.[58] Hence, the uneven social distribution of the thermometric idiom, far from accidental, was a part of the complex politics of consumption that engendered the techno-modernity through which both middle class identities and modern Ayurveda were coproduced.

Electromagnetic Vitalism

For physicians like Mullick to be able to hold the high opinion he did of the powers of the thermometer, body heat obviously had to be refigured as a purely material entity, shorn of all transmateriality. The process through which this concretization was achieved was one of metaphorization of the transmaterial. Thus the mythological narratives about Jworasur were reread as metaphors of something else. What this "something else" exactly was, however, kept changing. In the late nineteenth century when "body electricity" and "animal magnetism" were influential, Jworasur and other transmaterial elements of the fever lore were read to be "in reality" metaphors for "electricity," "magnetism," or something akin to these. But the exact nature of these entities—namely, "electricity" and "magnetism"—remained contested. Even while accepting that *sontap* or body heat was in reality either electricity, or magnetism, or something very similar to these entities, considerable debate existed over the nature of such entities. Some thought them to be a "substance" (*drobyo*), and others thought them to be "forces" (*shokti*), while a third group imagined them as forms of "motion" or *goti*. Each of these figurations resonated on multiple registers. Such understandings kept competing and inflecting each other until the 1920s when they were eventually replaced almost entirely by the refiguration of body heat as "hormones."

The indeterminacy was not unique to the Bengali or Ayurvedic context. When the physical sciences borrowed the word *energy* around 1850 from earlier literary and philosophical lexicons, those older significations did not disappear completely or immediately. Writing of the complex polysemy of "energy" and "heat" within Victorian thermodynamics, Bruce Clarke says that, "then as now, the authors of scientific discourses reached into the archives of mythology and literature, not just for the rhetorical embellishments, but also for the terms that named new conceptions of natural objects and processes. Untidy and promiscuous significations enter the mansion of science through the back alley, like Mr Hyde, and then step out through the front door, like Dr. Jekyll, in fresh lab coats."[59]

One of the earliest reimaginations of *sontap* appeared in Binodlal Sen's highly influential two-volume work, *Ayurveda Bigyan*. As a scion of the highly influential Chandrakishore Sen family and an influential early modernizer, Binodlal's views have a special place in the history of modern Ayurveda. Not only was Binodlal one of the first to forestage *sontap* ("heat") as the key element in fevers, but he was also one of the first to try to understand *sontap* as "motion." According to him, "the speedy circulation of blood within the body, various bodily pulsations, inhaled air's tryst with blood, the amount of food consumed as well as the latter's various dis-aggregations and aggregations within the body, all of these cause body heat to arise. It is one of the prime symptoms of life."[60] Sen's model, which was clearly one that saw heat as engendered through mechanical actions of material components of the body, did not explicitly say what "heat" was, but merely how it arose.

Binodbihari Ray developed a more comprehensive new configuration of *sontap*. Ray was one of the few modern Ayurvedists to not be based in Calcutta. Though possibly originally trained in "Western" medicine at one of the medical schools of Calcutta, he was mostly based in Rajshahi and published from there. Ray, writing two years after Binodlal in 1889, was one of the first to explicitly reimagine body heat along electric lines. In the very first issue of his short-lived Ayurvedic periodical, *Chikitsak*, Ray wrote that "Ayurveda tells us that fever arose from the enraged breath of Shiva when he was insulted by King Daksha, but a truly inquisitive person (*jothartho onusondhitsu byakti*) cannot be satisfied by this." He then proceeded to state that "normally everybody's body is slightly warm, when that heat increases beyond the normal it becomes fever."[61] Like Sen, Ray began by explaining the process by which body heat arose. He was explicit in stating that dissecting a corpse did not illuminate the issue since the body heat could no longer be seen. Instead, Ray pointed out that body heat arose at two points: first, when blood entered the lungs and came into contact with fresh air; and second, when the purified blood from the lungs came into contact with impure blood. At both these points of contact, heat was generated and then disseminated throughout the body by the constantly mobile blood in our veins, he explained.[62] In a later issue of the same journal, Ray explained what body heat really was. In the course of reimagining the three *dosh*es or parahumors, Ray stated that *pitta*—often translated as "bile"—is also known as "Usmo" (lit: heat, warmth, rage, excitement, etc.) or "Agni" (lit: fire). Classical Ayurvedic definitions of the *dosh*es, however, are plural. Each *dosh* is defined in multiple, overlapping ways. There were substantive as well as deified and agential definitions. Hence Ray quickly added that the Lord Brahma was an embodiment of fire and it was Brahma himself who, in the form of *pitta*, resided in the human body. He also clarified that it was because of the bodily residence of Brahma as *pitta* that the human body was bequeathed with "extraordinary beauty, grace and memory."[63] Later Ray added that "even western physicians admit that *pitta* is one of the factors for the maintenance of body-heat." Through an elaborate analogy with another para-humor, *vayu*, Ray explained that while *vayu* was electricity, body heat was something akin to electricity. Thus, while "nerves" (*snayu*) were made of electricity-conducting material to transport the bodily electricity, *pitta* was that part in the blood which was capable of carrying or conducting body heat.[64] But Ray also wrote that

> just as it is only when the lame ride on the back of the blind can they perform all functions, similarly it is only when a force [*shokti*] seeks refuge in a

> container that they become functional. Without *shokti* no work is done clearly then the *snayus* must perform their work by the strength of some *shokti* . . . this *shokti* is nothing but electricity.[65]

Clearly, Ray equated body heat with *pitta* and saw *pitta* to be akin to electricity in one way or another. But exactly what electricity is seems unclear. At times he clearly imagined it as a force or *shokti*, while at other times he wrote of it as a "part of the blood." Ray developed his ideas further in a book he wrote almost two decades later. Written in the versified *punthi* form, he made his substantive understanding of *sontap* explicit. In this account, Lord Shiva taught his disciple Shantiram about body heat thus:

> The Rail Engine as long as it is cold
> Remains like a corpse in the same place
> Water and fire once added it becomes warm
> Notice how the machine comes alive then
>
> Similar is the human-body so long as it is warm
> So long as it is alive
> In the midst of the blood I have placed a substance [*drobyo*]
> In it heat is always stored
> Flowing with the blood it spreads through the body
> By that the heat [*usnota*] spreads evenly through the body
> The substance that resides in the blood and stores heat [*tap*]
> According to Ayurveda, Shanti, that is called *Pitta*.[66]

Ray's confusing figurations of electricity as both a *drobyo* and a *shokti* were not unique to him. This conundrum repeatedly surfaced among authors following Ray. Another eminent Ayurvedist, Surendranath Goswami, of whom we will hear more in the subsequent chapters, writing in 1916, asserted that *vayu* was "motion" and *pitta* was "heat," but then went on to emphatically disagree with their translation as "humors."[67] Instead, he argued that *vayu* and *pitta* were respectively the "correlative" and "sustentative functions." Drawing extensively on the writings of Herbert Spencer, Goswami argued that these two "functions" were the key to adjusting the relationship between the body and its environment. He went on to argue that *vayu* was an entity "like electricity, but more sublime (*sukshmo*)."[68] Through a series of complicated equivalences, Goswami ended up arguing that *vayu* was in effect akin to the alpha, beta, and gamma rays found in radium and that it resided around a core of "positive electricity."[69] Goswami was more equivocal than Ray, but his equation of types of atomic radiation with "functions" demonstrates a certain amount of confusion about the nature of *pitta*. Was it a substance or a type of action/function? Goswami approvingly cited Lord Kelvin to say, "That the primitive fluid is the only true matter, yet that which we call matter is not the primitive fluid itself, but a mode of motion of that primitive fluid."[70] An editorial published around the same time in a short-lived medical journal published from Dhaka also articulated a similar confusion. On the one hand, the editorial posited that *sontap* was generated by the "friction" caused by "molecular vibrations," while on the other it said *pitta* was "the name of a heat-producing substance [*podartho*] present in the body."[71] The anonymous author went further to identify both "red blood corpuscles" and bile as forms of *pitta*, but then stated that in the heart, this very substance took the form of the "heart's electric force" (*hridoyer boidyutik shokti*).[72]

The most influential physician to write on the issue in the nineteen teens, however, was neither the anonymous Dhaka author nor Surendranath Goswami, but another eminent Calcutta-based physician, Amritalal Gupta. Goswami acknowledged that much of his thinking had been shaped by hearing Gupta speak at a meeting, and it was Gupta who sought to find a way out of the dual ascriptions of substantive and force-like characteristics to *sontap*. Writing in 1915–16, he argued that body heat was essentially a concentrated form of "solar energy" called *tej*, but that in the body it took the substantiated form of *pitta*.[73] Its functions within the body, mirrored the sun's functions vis-à-vis the earth, and it was the main agency through which the body was able to communicate with the surroundings. Despite agreeing with those before him in translating body heat as *pitta*, in describing the actual action of thermometer, Gupta curiously argued that "disease has many motions and it is *vayu* that gives rise to these movements. One of *vayu*'s motions, for instance, gives rise to fever and makes the mercury in the thermometer rise above the 98 degree mark, while another motion cools the body and makes the mercury move beneath 98 degrees."[74] Gupta therefore spliced off the two contradictory tendencies in refiguring *sontap*, rendering them thus independent of each other. While body heat itself was figured as a substance in the form of *pitta*, the interactions of body heat and the thermometer, on the other hand, were figured as a motion.

Even as electromagnetism formed the leitmotif against which these novel figurations of *sontap* were developed, there remained a strong current of vitalism that was forever being braided into these figures. Sen had explained that "increase or decrease in body-heat, and thus disease, occurs when nature's protective force [*prokritirokshini shokti*] is forced to repeatedly counter-attack."[75]

Binodbihari Ray versified a position almost identical to Sen's. Through him, one finds the god Shiva saying:

> There resides a great force [*mahashokti*] in the body
> Keeps it working smoothly
> Should any disease enter the body
> [That] Force trounces it with ease without your knowing
> Only when [the] disease is stronger than the force
> Then it requires support.[76]

Though expressed in the guise of indigenous ideas, the actual source of such vitalism is obscure. Mid-nineteenth-century "Western" medicine too had a powerfully resurgent vitalism at its core. Charles Rosenberg points out that by the 1830s it had become a cliché in erudite medical circles to question the powers of traditional therapeutics. People spoke of "self-limiting diseases" and there was a new "respect for the healing powers of nature."[77] By the 1850s, this skeptical line of argument became much more powerful and led to significant therapeutic changes. The belief in the body's "vital powers," Rosenberg points out, while not encouraging the extreme "therapeutic nihilism," did produce significant moderation in the level of intervention (including a gradual decline in dosage) a physician undertook and much more interest in the overall management of the patient's diet, regimen, etc.[78] Accenting a somewhat different causal directionality, John Harley Warner has argued that the peaking of interest in the "natural powers of healing"—namely, the *vix medicatrix naturae*—from the 1850s was a consequence of the therapeutic and intellectual changes that had already begun to assail traditional pathological models based on "inflammation" (*phlogistics*) and "strength" (*sthenos*).[79] Whereas for Rosenberg, the mid-century interest in "vital powers" was part of a gradual renegotiation of the doctor-patient relationship, for Harley Warner, the chronology cannot be explained without noticing a sense of crisis among the physicians.

Though the idea of the *vix medicatrix naturae* draws upon a long and hoary intellectual genealogy, as the works of Rosenberg and Harley Warner demonstrate in the American context, its modern uptake has a very specific history. In Victorian Bengal, both the intellectual and social histories of this uptake were more complicated. To begin with, the "Western" medical beliefs overlapped fairly closely with the Unani views that drew upon shared Galenic roots.[80] Moreover, Bengali *daktars* trained in "Western" medicine had also appropriated and developed their own vernacularized version of the *vix medicatrix naturea.*[81] Given the multiple sources and versions in which the idea was available to modernizing Kobirajes such as Binodlal Sen and Binodbihari Ray, it is difficult to firmly delineate the exact source of the vitalism in their writings. Its social history, however, for that very reason is easier to explain. The uptake of the idea would have seemed relatively more plausible when so many others in a heterogeneous medical milieu already espoused similar views. The concept of "mimetic legitimation" developed by Michael Taussig and then used by Jean Langford in her study of contemporary Ayurveda is useful in explaining the ease and smoothness of the uptake.

Yet, interestingly, this vitalism in both Sen and Ray was figured as a "force." By contrast, in Amritalal Gupta's writings, the same vitalism figured as a sentient being, rather than as a blind force. Gupta asserted that "*jibatma* [life-soul] is happy at 98 degrees, any movement [of the mercury] upwards of it caused unhappiness. *Jibatma* is he who experiences happiness and sorrow."[82] Gupta's use of the compound word "*jib-atma*" at once aligned it with the figure of the traveling, anthropomorphic life creature simply called "*jib*" or life that was available in traditional *nadipariksha* texts. Gupta's vitalism was, however, not simply limited to this alignment. It was a much more expansive vitalism that spilled over the limits of the body itself. He ended up explaining the movements of the mercury in the thermometer itself in vitalist terms. Gupta explained that "the glass tube of the thermometer encompasses the mercury in exactly the same way that the three [principle] *nadi*s [viz.] Ida, Pingala and Sushumna, enclose the subtle-body [*sukshmodeho*]."[83] "Life," Gupta continued, resided within these three *nadi*s and it was the sorrows and happiness of that Life or Soul that was detected by the "transformations" (*bikriti*) of the mercury in the thermometer.[84] He had already declared that the Soul experienced emotions but remained unmoved or untransformed (*nirbikar*). Punning on the notion of "transformation," he argued that it was the Soul's emotional state that was expressed through the "transformation" of both bodily constituents like the *dosh*es as well as the mercury in the thermometer.[85] Gupta's vitalism, using the notion of the fundamental unity of the constitutive elements of the body and the universe, was thus able to inject the very instrument through which the new vitalism in the body was to be calibrated with its own vital powers.

Though vitalism, for much of the Victorian period, had operated "within the mainstream of chemistry and biology as a countertheory to materialist mechanism," by the late 1880s an "energetic vitalism" had begun to hesitantly emerge. This latter drew upon the notion of energy from the physical sciences and its thermodynamic laws to produce a new vitalism. "The cultural history of vitalistic ideas," Clarke writes, "is a good example of the way critical scientisms can be a form of ideological defense, a resistance to the violence of partisan scientific dogmatizing itself."[86] Gupta's critical "energetic vitalism" not only resisted the tendency toward objectivization of the patient's body, but infused the instruments of objectivity itself with a palpable vitality.

To varying extents, vitalism thus came to be intricately braided into a new figuration of body heat. This new figuration reimagined body heat within modern Ayurveda in the dim and sometimes contradictory image of electromagnetic forces, but then also added tinges of vitalism to that figure. The thermometer was at the heart of such new figurations: it instigated the new figure, supplied the data for it, and even came to be fundamentally re-identified in the process. This figure was to remain strong and well-etched until the 1920s when a new figure based on hormones overcame the shadows of electromagnetism.

Balancing *Sontap*

One of the most curious aspects of the newly braided figure of *sontap* that emerged in modern Ayurveda was its concern with balance. Serious scholars of Ayurveda have challenged the frequent assertions in popular writing that Ayurveda conceptualizes healthfulness as a state of balance between the three *doshes*.[87] I will discuss this more general idea of "balance of humors" later in this chapter. The concern for balance that was articulated with regard to *sontap*, however, was not the balancing of "humors" *within* the body as such. It was rather an idea akin to energy conservation that made a complex connection between the body's internal balance of heat with its balanced uptake or output of heat from and to the body's environment.

The earliest and most crisp statement of this position is to be found in Binodlal Sen's writings. He conceptualized *sontap* to be a fixed and finite energy whose increase or decrease were both harmful for the body. To avoid such increase or decrease, above all one had to protect the body from both cold and hot winds. Even the elimination of bodily wastes such as urine, stools, spit, etc., was potentially harmful since every time one ejected such matter, the body lost some heat.[88] Clarifying further, Binodlal wrote, "just as the erosion of body-heat is the cause of affliction, the increase in body-heat gives rise to disease."[89]

Developing this idea of balancing body heat further, Amritalal Gupta argued that the very notion of "disease" (*byadhi*) could be defined as any deviation from the thermometric reading of 98 degrees. Gupta argued that the thermometer gave a reading of 98 degrees only when each of the three *doshes* was true to their nature (*prokriti*). Any fluctuation from it rendered the body in a state of disease.[90] In line with Sen's formulation, Gupta also conceptualized body heat in a relationship with the external environment that constantly threatened to upset the balance at 98 degrees. He insisted that the elements in the body, particularly the three *doshes*, had exact correlates in the external environment. As a result, the *vayu* in the body was essentially the same as the wind outside, and in winter the blasts of cold wind touching the body threatened to lower the temperature of the body. Similarly, the *pitta* in the body was the same as the sunlight, and therefore strong sunlight in the summer could easily raise the body's temperature. Water was similarly connected to *kapha* and when the body came into contact with it, there was a risk of the body temperature falling once again.[91] These potentials for change forever threatened to upset the fragile balance achieved, in Gupta's view, at 90 degrees. He argued that if the body temperature often "came down to 96/97 degrees upon the cessation of certain types of fevers," any deviation above or below 90 degrees was said to be unnatural and owing to an unnatural state of the *pitta*.[92]

These ideas are reminiscent of nineteenth-century European "energetics." Anson Rabinbach, drawing attention to the multifaceted popularity of such ideas, writes that

> this image of the body as the site of energy conservation and conversion . . . helped propel the ambitious state-sponsored reforms of late nineteenth- and early twentieth-century Europe . . . [and] lent credibility to the ideals of socially responsive liberalism, which could be shown to be consistent with the universal laws of energy conservation: expanded productivity and social reform were linked by the same natural laws. The dynamic language of energy was also central to many utopian social and political ideologies of the early twentieth century: Taylorism, bolshevism, and fascism.[93]

This image of the body as the site of energy conservation was most clearly and explicitly stated in Binodlal Sen's writings, but others continued to develop the model. In Amritalal Gupta and his contemporaries can be found a model that is clearly reminiscent of Julius Robert Mayer's ideas that "the organism in its overall measurable relations with the external world—that world serving as source and drain for the organism's energy supply—was an energy conversion device."[94] In fact, Mayer had also held that the sun, in "terms of human conceptions," was "an inexhaustible source of physical energy."[95] Moreover, connecting the dots, Mayer also explicitly wrote that "the stream of this [solar] energy which . . . pours over our earth is the continually expanding spring that provides the motive power for terrestrial activities."[96] Later, in the 1850s and 60s, William Thomson, later Lord Kelvin, further developed these ideas positing the sun to the main source of all terrestrial energy.[97] Gupta's model thus was almost exactly identical to Mayer's and Kelvin's with the exception that in the former, *pitta* played an important circulatory function.

The editors of the journal *Ayurveda Hitoishini* expressed a more deistic version of the same idea of energy balance. While describing in detail how the body could gain or lose heat through a variety of acts and engagements ranging from being in the sun, having sex, fasting, feeling angry, etc., the authors, who were avowedly pious Vaishnavas, also posited that this energy or power was ultimately Vishnu's force or Vishnu's heat (*Bishnutej*). They even went so far as to assert that the body's "normal temperature" was the point at which this form of deistic force was most clearly and robustly expressed. They thus named this "normal temperature" *Satyabishnu Tej* (True-Vishnu Force).[98]

Such moral and deistic figurations of "heat" were by no means unknown in Europe. Clarke points out that a group of eminent Victorian British scientists, mostly based in the north, had "fashioned an influential discourse of thermodynamic scientism that infused physical concepts—energy, dissipation and equilibration—with moral contents—life, sin and death." Clarke calls this "energy theology."[99]

Surendranath Goswami, one of the most persistent braiders of "Indic" and "European" knowledges, articulated one version of "energy theology" in developing his ideas about "balance." Goswami, drawing upon a single verse from the *Sarangadhar Samhita*, suggested that the *vayu*, *pitta*, and *kapha* in the Ayurvedic writings had brought together domains that his European contemporaries would think of as belonging to the distinct domains of physiology, pathology, and morbid anatomy. Having established this correlation, Goswami went on to baldly state that "today Western scientists are saying that it is impossible to determine a clear boundary between Physiology and Pathology. The maintenance of a physiological maximum or minimum must be regarded as pathological. That is, whatever is the "'Normal Standard' [of] nature or its state of balance is health. Any deviation towards its 'Maximum or minimum' is illness."[100] Though Goswami did not himself cite a source, the sentence he quoted appeared in the introduction to a popular textbook on pathology and morbid anatomy by Thomas Henry Green.[101] In Green's work, the comment is made directly in the context of the emergent regimes of urea testing and the importance of "normal range" figures. It is a comment made therefore within the larger context of seeing numerical ranges as the key indicators of health that emerged on the back of late nineteenth-century medical biochemistry. When transported into the Ayurvedic context, on the other hand, the biochemical applications disappeared. Regular and repeated blood or urine work was still extremely rare if not wholly absent from Ayurveda, yet the comment was grafted onto preexisting ontologies of health to create a new sense of conception of health encapsulated in the figure of balancing energy.

One interesting consequence of modern Ayurveda's investment in balance was that, whereas vitalism had operated to blur the boundaries between the human body and the measuring instrument, the thermometer, in one direction, the notion of balance operated to blur that distinction in the opposite direction. In Binodbihari Ray's writings, therefore, there can be found an explicit comparison of the body with a sensitive measuring scale. He wrote:

> The body is a weighing balance, think about it
> Upon the entry of affliction, it lets you know.[102]

Besides blurring the boundaries between the body and the material world around it, the notion of "balance" also positioned the body as a porous site where the subject and its environment could interact and shape each other. This is precisely what Langford notices in her study of contemporary Ayurveda. She dubs this interaction "fluency," and speaks of the "fluent body." Many, if not most, of these conceptions, however, were shaped in the period under study and the concretization of *sontap* or body heat in the image of electromagnetic energy was one of the key modalities through which this "fluency" was constituted.

While the contemporary articulations of this "fluent body" undoubtedly draw upon older intellectual resources and claim for itself a hoary genealogy, how far such an idea of health as an intricate balance between elements moving in and out of a porous body was indeed available, let alone dominant, before the advent of colonial modernity is doubtful. That the acceptance of the idea of balance was not universal in precolonial India is proven by the existence of works such as Viresvara's *Rogarogvada*. Writing in 1669, Viresvara had pointed out the inherent contradictions about the *tridosh* pathology as a whole. Viresvara argued that

> the greatest authorities define disease as identical to an inequality in the humours. And yet, in other places they say that the humours may naturally exist in different quantities, without causing illness, such as when phlegm

> naturally predominates at the start of the day, or after a meal. This is not to say that one is always ill after a meal. And so the central doctrine that humoral inequality is identical with disease must be wrong.[103]

Viresvara's attack was not merely on the idea of "balance," but on the entirety of classical medical theory. As Dominik Wujastyk points out, Viresvara's radical attack on classical medical theory may not have become mainstream, but it was surely known to contemporary scholars, and manuscript copies of his book have been found at leading centers of Sanskrit intellectual life in Western and northern India.[104] Another anonymous Sanskrit work, popular in Bengal and translated into Bengali by Anandachandra Barman in 1867, entirely eschewed the discussion of *dosh*es and argued that all diseases were a result of the perversion of the "digestive fire" (*mondagni*).[105] Nowhere did this book articulate any notion akin to energy conservation or balancing.

But it was not just these heterodox texts that eschewed an idea of "balance." Even the most mainstream of texts and physicians, who wholeheartedly accepted the pathogenic role of the three para-humors, did not subscribe to an idea of "balance," let alone a balancing of body heat or energy. Thus Jashodanandan Sarkar, in his Bengali edition of the *Susruta-samhita*, baldly stated that commentators [*tikakor-ra*] point out that "just because *vayu, pitta* and *kaph* are mutually antagonistic, does not mean that the increase or decrease in any one would lead to a complimentary rise or fall in the others . . . it is possible for all three to rise or fall in tandem."[106] Clearly the position outlined by Sarkar cannot be described as one reliant on the "balancing" of the three *dosh*es. While they are admittedly locked in a complex mutual relationship, that relationship is not thought of in terms of "balancing" as such.

The concept of balance, particularly in the ways that it was discussed together with *sontap*, entailed the imagination of a closed system with a fixed quantity of one or more substances. If the body itself is imagined as this system, then any rise or fall in the proportion of one para-humor must be commensurate with a complimentary fall or rise of one or both of the other para-humors. If, on the other hand, the body-in-milieu is imagined as the closed system, once again a zero-sum game of complimentary rise and fall must be mapped, but this time at the level of the body-in-milieu, rather than the individual body. The systems outlined by Binodlal Sen and others functioned mostly at the level of the individual body. Though Gupta's writings also have hints of a larger system that included the individual body and its milieu, they clearly conceptualized the body and its milieu as a single, closed system.

This is where, I believe, the huge popularity of "energetics" in the nineteenth century came into play to promote a sense of the body inhabiting a closed system from which it drew resources and into which it discharged valuable constituents including heat. The immediate context for the uptake of these ideas about "energy conservation" within closed systems, however, was almost certainly the growing familiarity of urban, middle class physicians with electricity and electric-driven infrastructure.

Authors like Sen, Ray, Goswami, and Gupta, while regularly citing European authorities who explicitly addressed issues of energy conservation, such as Lord Kelvin, also equally frequently made metaphors out of the new energy-driven material culture that surrounded them. Gupta, for instance, compared *sontap* or body heat to heat in "electric lamps and gas lamps."[107] To describe the relationship between solar heat and *sontap* similarly, he fell back upon the metaphor of two "telegraph machines" being "able to communicate without a connecting wire between them."[108]

I will discuss the telegraph metaphors in relation to the nervous system a little later on, but here I want to attend to the fact that telegraph machines, electric lamps, and gas lamps—very different kinds of gadgets by themselves—could also embody a certain common-sense. All three entities, I argue, helped consolidate a sense of closed systems or circuits. In the case of telegraph or electric lamps, the importance of circuits and systems is self-evident. Any break in the line or the circuit would render both telegraphy and electric lamps redundant. In the case of the gas lights, this is not as self-evident, but when read in the context of early twentieth-century Calcutta, it does become self-evident, for not only did the city continue to use mainly gas lighting on the streets well into the twentieth century, but all the gas for lighting was centrally supplied by a single monopolistic private company, the Oriental Gas Company. Writing in 1908, Maurice Graham stated, for instance, that "the public lighting [of Calcutta] is all done on the incandescent gas system; there also being high powered lamps of all kinds. . . . There are only nine electric lamps, which are flame arcs, put up merely as an experiment. Electric lighting has been kept out by the skillful diplomacy of the Gas Manager, and by the intensity of the gas lighting."[109] As a result, the large infrastructure connected to gas lighting made it yet another instance of a closed, circuit-like system. What I am arguing then is that the ease with which ideas of balance that had their basis in theories of energy distribution and conservation slipped into modern Ayurvedic thinking owed at least as much to the everyday familiarity with closed systems and circuits as it did to the writings of Lord Kelvin and others.

While the ideas of both a closed system or circuit and the balance within it survived and became central to modern Ayurveda, the electro-magnetic and energetics that had formed its basis began to wane around the 1920s. The first person to have equated the three *dosh*es to hormones was said to have been an MD, Dr. Hemchandra Sen. Later it was taken up and robustly propagated by Gananath Sen and other regular contributors to his journal, the *Journal of Ayurveda*.[110] But it was around the 1920s that these new idioms began to congeal into a new figuration of *sontap*. Instead of the earlier complicated statements about heat inhering or being conducted by the *pitta*, authors like Saracchandra Sengupta, who wrote a number of essays on the subject in the journal *Ayurveda* in 1922–23, argued that the body temperature was directly proportional to the amount of *pitta*: that is, the higher the temperature, the more the *pitta* in the body and vice versa.[111] Yet what remained common to both the electromagnetic and hormonal figurations was the dual emphasis on closed systems and balances.

Reticulating Electromagnetic Vitalism

The electromagnetic vitalism, along with its dual accents on systems and balances, helped promote and consolidate a new and privileged position for the *tridosh* theory. Most medieval and early modern texts that formed the educational curriculum of erudite precolonial physicians in Bengal paid scant, if any, attention to the *tridosh* theory. While it was undoubtedly present, it was hardly ever elaborated upon and featured in texts alongside several other theories such as the *soptodhatu* theory, the *ponchomohabhut* theory, the *rasa* theory, and so on. By braiding *dosh*es such as *vayu* and *pitta* together with the electromagnetic vitalism and incorporating a sense of balance deriving from energetics, modern Ayurveda gradually promoted the *tridosh* theory to a position of preeminence.

Udoychand Dutt's first published edition of the precolonial classic, Madhavakara's *Rogavinischaya*, provided a telling example of this elevation of *tridosh* theory. Dutt's preface to the first edition published in 1873 clarified that the *Rogavinischaya*, better known as the *Nidan*, had been the "text-book on Medicine for native physicians for a long time past."[112] Other reports confirm that the *Nidan* had been the main textbook that classical Ayurvedic physicians of the precolonial period had followed. Canonical works such as the *Charaka Samhita* and the *Susruta Samhita*, etc., were seldom part of the actual curriculum of medical education; it was instead the *Nidan* that was central to the curriculum.[113] Despite its centrality in medical education, it had very little to say on the so-called *tridosh* theory. For generations of Kobiraji physicians in Bengal, this had obviously not been a problem and they had continued to use the work. Yet in 1873, Dutt found this to be a drawback and added a prefatory note to supply this newly perceived deficiency. Dutt began his note by declaring that "according to Ayurveda *vayu, pitta* and *kapha* support [*dharon*] the body and their corrupted [*dushito*] state leads to the genesis of disease."[114] Yet he confessed that "the description [*biboron*] of these [i.e., *doshes*] have not been compiled anywhere in the *Nidan*." Though he went on to say that everything he had to say about these *doshes* could be gathered either from scattered references in the text or through similar scattered remarks found in commentaries on the main text, it was clear that this was an effort to retrospectively create a single coherent "theory" for the emergent modern Ayurveda. This elevation of *tridosh* and its framing as the centerpiece of Ayurvedic medicine were most clearly presented in Bhagvat Singh Jee's influential and popular work, *A Short History of Aryan Medical Science*, published in 1896. In a chapter titled "Theory of Indian Medicine," he stated that "Indian medical science attributes all morbid phenomenon to the disordered condition of the three principal humours of the body, called DOSHAS , viz. wind, bile and phlegm. These fluids pervade the whole microcosm of man. . . . The three humours fill the whole body which they support."[115]

While this elevation of *tridosh* was undoubtedly also connected to the parallels with Greek medicine that a generation of orientalist scholarship by authors such as H. H. Wilson, T. A. Wise, and others had developed, the particular form this newly articulated *tridosh* theory took, including its overemphasis on balance, owed much to the theory's braiding with electromagnetic ideas about flow. Two aspects of this new version are remarkable in the extent to which they actually break with the precolonial formulations. First, the sheer centrality of the *tridosh* in pathogenesis, whereby all other pathogenic models are rejected and every disease said to follow from this single model, is utterly novel. Classical Ayurveda most certainly did not attribute all afflictions to the three *doshes*. There were several exceptions such as the several "invasive" or *agontuk* ailments like being struck by an arrow or the numerous afflictions caused by demonic possession.[116]

A second and even more telling change was the way in which the three *doshes* were recast as fluid or fluid-like entities coursing freely around an enclosed body space that resembled a closed hydraulic system. This broke with precolonial interpretations on two significant levels. On the one hand, *doshes* had always had a somewhat ambiguous ontological status between willful agencies akin to specific gods and bodily entities. Hence, the rendering of them as simply types of fluids or energies was clearly an act of serious refiguration. On the other hand, *dosh*es in precolonial literature were always understood to be entities in-place. What caused illness was their being out of place. Recently Anthony Cerulli, for instance, in describing the *tridosh* framework, says that "each humor has a natural seat in the body. Wind predominates in the pelvic region, bile in the abdomen and phlegm in the chest."[117] Yet these are constantly intermixed with the body's seven *dhatu*s (a category of fundamental substances) and three types of bodily wastes. Thus, "to describe the breadth of the humoral influence on the body, vata, pitta and kapha are [again] subdivided according to particular somatic location. For example, *pran* is a type of vata-humor located in the head, chest and neck; *pachok* is a pitta-humor located in the duodenum and intestines; *kledok* is a type of kapha-humor located in the chest and stomach, and so on."[118] In a more crisp comment, Dominik Wujastyk similarly points out that "disease aetiology is mainly a matter of misplacement or displacement, rather than imbalance."[119] The potential for displacement and misplacement clearly demonstrates that the internal space of the body was not imagined as a single, homogenous, or unmarked space united by the free flow of the so-called humoral fluids.

Unfortunately, much scholarship, instead of unpicking these recent reformulations, has actually promoted and consolidated these modernist reinventions. Poonam Bala, one of the most prominent authors of a number of works on Ayurveda writes that, "in Ayurveda, body functions are regulated by humours, recognized as air, bile and phlegm which formed the basis of a tri-humoural system." This "theory of body humours," Bala tells us, "occupied a central role in the diagnostics and therapeutics of Ayurveda and Unani." "Practitioners of both," she continues, "viewed proportions and balance of body humours to conditions of health and disease; imbalance or disturbance in these resulted in the impairment of health."[120] Commenting on statements of this nature, Dominik Wujastyk writes illuminatingly: "While the idea of balance is certainly present in Ayurveda, the platitude that one finds repeated numbingly without exception in all secondary sources on ancient Indian medicine—namely, "disease is caused by an imbalance of the humours" is an inadequate characterization of the disease causation as described in the original Ayurvedic texts." Wujastyk also goes on to suggest a possible source for this widespread misconception. "I suspect," he writes, "that the exclusive focus on this statement, usually presented as the cornerstone of Indian medicine, the very essence of the Indian humoral view, is a reading back into Indian medical history of Hippocratic or Galenic thinking, in the Aristotelian interpretation."[121]

Whatever its origins, the privileging of a version of *tridosh* that imagined the three *dosh*es as freely flowing around the body helped to establish a reticulate image of the body. Instead of being internally differentiated and ill-defined around the edges (more about this in chapter 5 in connection to the chakras), the new body that emerged was structurally reticulate. Its boundaries were clearly defined, even if porous, and its internal space was smooth, interconnected and undifferentiated. The mutation of the differentiated inner space of precolonial Ayurveda into the undifferentiated reticulate body space of modern Ayurveda can be clearly glimpsed by comparing the descriptions of fever in two texts. In Madhavakara's *Nidan*, the cause of fevers is described thus: "In people who indulge in false (*mithya*) food, residence, travel etc. (*ahar-biharadi*), the *dosh*es located in their *amashoy* (part of the stomach where mucous collects) throws out the [digestive] fire (*agni*) located in the large bowels and corrupts the chyle (*ros*) thus giving rise to fever."[122] Clearly the *dosh*es are discussed here as entities located in specific physiological places rather than in terms that constantly circulate the entire body.

By contrast, in Nagendranath Sengupta's *Kobiraji Siksha*, one of the early textbooks that pioneered the use of thermometers, slight but telling alterations to this model of pathogenesis can be noted. Sengupta wrote that "irregular (*oniyomito*) food and enjoyment irritated (*kupito*) *vayu* and the other *dosh*es which then entered the *amashoy* and polluted (*dushito*) it, thereafter they brought the heat (*sontap*) out of the bowels (*kosh*) and caused fever. It is because of this externment of body-heat (*sontap*) that skin is warm in all fevers."[123] Though still fairly close to Madhavakara's description, Sengupta's discussion introduced some telling changes. By recoding "false" diet and enjoyments as "irregular" ones, for instance, he clearly introduced a positive notion of regularity that was lacking in the original. A classification of foods that are "false" or "illegitimate" is different from a diet that is "regular." A more significant change for the present discussion, however, was the inversion of the location of the *dosh*es. Whereas Madhavakara had clearly spoken of the *dosh*es as located *in* the *amashoy*, Sengupta spoke of them as entering the *amashoy* after they become corrupted. He also modified the "digestive fire" or *jotharagni* to simply mean body heat or *sontap*. Each of these slight shifts taken together makes it easier to speak of the body as a unified and unmarked whole. Internal physiological places lose their spatial specificity and the body becomes a reticulate body—that is, a body that is internally largely undifferentiated and understood as a single reticulate unit.

The *Snayubik* Man

The reticulation of physiospiritual matter concretized in a physiogram organized around an ill-defined and acephalous network of *snayus*. *Snayus* are frequently glossed in the contemporary literature as "nerves." Yet their location at the interface of matter and spirit invites us to explore these channels more closely before moving onto their map as such.

It was possibly Kobiraj Sitalchandra Chattopadhyay of Magura, Khulna, who first proposed the gloss that transformed *snayus* into "nerves." Chattopadhyay's intention was to update the understanding of the three *doshes* and hence also update the conception of the channels that carried them. His ideas were taken up and further dilated upon by Pulinbihari Sanyal in the popular medical periodical, *Chikitsa Sammilani.* Sanyal drew heavily on electromagnetic theory and defended the conception of *snayus* as "nerves" against a series of protracted and often personalized attacks from Kobiraj Prasannachandra Maitreya of Umarpur, Pabna.[124] Maitreya argued in favor of a more sacral notion of the body space. He insisted that the Brahma, Vishnu, and Shiva—the Hindu trinity—actually (and not simply metaphorically) resided within the body (*jib dehe birajaman*) respectively as *vayu, pitta,* and *kapha.*[125]

In the end, it was Chattopadhyay and Sanyal's formulations that became more popular, and Maitreya's enigmatic and alternate formulations found few supporters. Binodbihari Ray, for instance, joined the debate through his journal, *Chikitsak,* in support of the Chattopadhyay/Sanyal formulation. Sanyal and Ray both agreed that *snayus* were indeed "nerves" and that *doshes* circulated through them.[126] Both of them described a network of *snayus* stretched all over the body as the pathways through which *vayu*-as-electricity traveled all over the body. *Pitta,* the mysterious heat-conducting material submerged in the blood, was also said to similarly circulate through these *snayu*-pathways. "*Snayu* is built of electricity-conducting material," wrote Ray. "Hence it is able to conduct electricity. Blood is similarly a conductor of heat and therefore is able to carry the heat (*tap*) throughout the body," he wrote.[127] Drawing explicitly on Tantric understandings, Ray continued that "the *snayus* are dispersed throughout the body by being attached to the spinal chord (*merudondo*). Within them the principal ones are the *gyanshoktibahini* (intelligence-power-bearers), *icchashoktibahini* (will-power-bearers) and the *kriyashoktibahini* (action-power-bearers) ones."[128] Clearly, while these authors energetically insisted on seeing *snayus* as "nerves," their descriptions exceeded and rearranged the "Western" understanding of "nerves."

To be fair, when Ray wrote in 1889, "nerves" were a fairly enigmatic structure even within the "Western" tradition (to the extent that the tradition can be homogenized) and their relationship with electromagnetic forces well-established. There had in fact been a long tradition since the eighteenth century of relating electricity and the nervous system. It was Luigi Galvani (1737–98) who had "produced the first widely publicized evidence that the nerves and muscles of animals produced their own internal electricity."[129] And though this had been opposed by scientists like Alessandro Volta, Galvani's ideas continued to grow and mature in the nineteenth century through efforts of scientific men like his loyal nephew, Giovanni Aldini. Aldini undertook a grand tour of Europe to defend and disseminate his uncle's views.[130] By the opening decades of the nineteenth century, the electric physiology of the nervous system had become firmly embedded into intellectual, theological and political debates in Britain. Thus embedded, it continued to evolve along multiple lines producing new objects, debates, experiments, etc. Over the next half a century, Michael Faraday's discoveries of electromagnetic induction provided further weight and direction to the development of the relationship between nerves and electricity.

Yet the "nerves" of the "Western" tradition remained fairly distinctive from the descriptions of Sanyal, Ray and others engaged in refiguring Ayurveda who sought to make them carry entities as varied as *dosh*es, *kriyashokti*, etc. Whereas European scientists sought to find in electromagnetism a single force with multiple applications, the Bengali scholars refiguring Ayurveda repeatedly deployed electromagnetism as a mere hint that further and more subtle but powerful forces awaited discovery. "Nerves" thus became conduits not only of electromagnetism, but also of the hitherto undiscovered potential forces that correlated with *pitta* and *kapha*. One near contemporary British author, Thomas Stretch Dowse, writing only five years before Ray, described nerves and the nervous system thus:

> The nervous system pervades every part of the body. It is to man what the solar system is to the earth. It engenders, distributes and regulates nerve force. The engendering and transmissions of nerve force is the especial office of masses grey and nervous matter, called central ganglia. The conduction of nerve force is performed by nerves.[131]

Here the functions are pluralized, but the force that is carried through the nerves remains unitary. By contrast, in the case of Sanyal and Ray, there is an attempt to make them carry at least three distinctive forces or energies that could co-relate to the three *dosh*es.

Stacey Langwick, in her fascinating study of "traditional" medicine's tryst with "modernity" in Tanzania, has qualified Margaret Lock's call to give "materiality its due" by firmly stating that in order to do so, "we must allow it the diversity we allow such concepts as agency, history, body and modernity."[132] In other words, to take materiality seriously we cannot assume that any one epistemic regime has a monopoly in defining it. We must be willing to acknowledge alternate materialities that are materialized in therapeutic situations where "Western" scientific categories do not enjoy a definitional monopoly. Despite the translation of *snayus* as "nerves," what is clear is that the entirety of the materiality of *snayus* cannot be encompassed through our more familiar uses of the term *nerve*. To physicians like Ray, "nerves" clearly meant something more than what it meant for his peers in "Western" medicine.

In order to fully understand the nature of this alternate materiality, however, it is not enough to simply assert a reified non-equivalence between "East" and "West." Exploring its constitution further reveals a fascinatingly braided image that confounds any absolute distinctions between "East" and "West." Iwan Rhys Morus tells an illuminating tale of the shifting relationship between electricity and "nerves" evolving throughout the nineteenth century.[133] By the 1870s, early neurology, in more than one context, had begun to overlap with the rising tide of American spiritualism.[134] Hence the interpretations proposed by Ray and Sanyal were not as alien to the "West" as might appear at first blush. Indeed much of their vocabulary would have been fairly familiar to one "Western" intellectual tradition or another.

Despite this shared three-way connection between electromagnetism, nerves and spiritualism in the "West" and the "East," the chronology seemed hopelessly confounded in the two cases. In the "West," and more specifically in Britain, by the 1880s and 1890s, electricity's relationship to the body had evolved in two distinctive directions. On the one hand, it had been elevated to a respectable philosophical and physical science that had largely successfully sundered its connections with an earlier tradition of bitterly partisan theological disputation and spectacular, if macabre, public displays. By this time, men such as William Thomson, Lord Kelvin, had made electricity something "real, purchaseable [and a] tangible object" by making it measureable through standardized units. Along this trajectory, medical electricity too was increasingly repackaged as a discrete and measurable "energy." A second and only partially overlapping trajectory rendered medical electricity an object of entrepreneurial creativity. A number of ambitious late-Victorian entrepreneurs produced and aggressively marketed commodities dispensing medical electricity and claiming to work upon the nerves.[135]

All this was in fact in stark contrast to the earlier epochs of electricity's tryst with the body and nerves. Ideas about electricity being a mysterious anima mundi, a subtle life force or even an invisible fluid that vitalized all creation, had died out around the 1830s. Later, around mid-century there had emerged another idea wherein both electricity and magnetism were in effect limited aspects of a still more subtle and powerful "force" that ran through the world at large. Authors such as Edward Bulwer-Lytton, the first Baron Lytton, a friend and a contemporary of Charles Dickens and the father of India's Viceroy Lord Lytton, gave wide public airing to ideas about this mysterious and mystic "force" permeating all creation in works such as the hugely popular, *Vril: The Power of the Coming Race.*[136]

These amorphous and mystic inscriptions of electricity's tryst with nerves and the body were, however, clearly dated by the late 1880s. By then the measured respectability of Lord Kelvin and the entrepreneurial posturing of men like Cornelius Harness, Henry Tibbits, and Thomas Stretch Dowse had come to redefine electricity's relationship with bodies and nerves. Mysticism had been replaced by measurability. Universal permeation had been replaced by local applications. And metaphysical meditations on the divine had been replaced by a ribald commodity culture. In the midst of this, Ray and Sanyal's alternate materiality looked back at the earlier eras. It looked back at the anima mundi of the 1820s and 1830s and it looked back at the mystic capacities of *Vril*-like super forces. In short, they looked behind the contemporary trends toward quantification and commoditization and sought to recover a mystic and qualitative past of electricity's dalliance with the human body and its nerves.

In fact, Sanyal explicitly drew on the researches of Victorian physiologists like William Carpenter from the 1850s who had argued for a single "vital force" that would correlate with the forces in inorganic matter, such as electricity, heat, light, and so forth. Even though Carpenter had been seen to be a radical at his time, he held the body to be a "machine" which "mediated between the ego and the external world."[137] It was this machinic physiospiritualism that the Bengali authors sought to revive.

The discussions on body heat went further. They did not simply make vague gestures toward machinic physiospiritualism. They inscribed it onto a fairly concrete physiogram. Unlike the physiogram of the clockwork body I described in the last chapter, this new physiogram was not limited to textual exegesis. It was materialized in a concrete illustration (see fig. 6). Gopalchandra Sengupta was the first to include an unnamed image that I call the *Snayubik* Man. The *Snayubik* Man was an unlabeled human outline with numerous, imprecise channels drawn throughout the body, but lacking a brain. As I have discussed in the last chapter, Gopalchandra had often struggled to reconcile the divergent truths of the body generated by different

ধনু ও ছিদ্র দৃষ্ট হয় না, নিশাতে দুই চন্দ্র অথবা চন্দ্র সূর্য্য উভয়ই দেখিতে পায় । ১৫৬

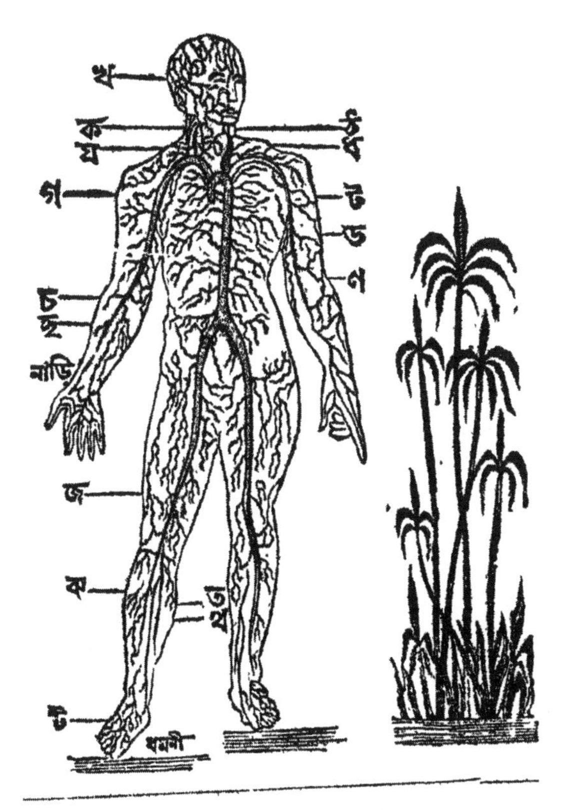

ইংরাজী মতে নাড়ী পরীক্ষা ।

ডাক্তর গাই নামক জনৈক চিকিৎসক পর্য্যাবেক্ষণ করিয়া দেখিয়াছেন যে, সুস্থ যুবার প্রতি মিনিটের নাড়ী স্পন্দনের সংখ্যার মধ্যম পরিমাণ দণ্ডায়মানাবস্থায় ৭৯, ও

6. *Snayubik* Man. Refiguring nerves as *snayus*.

diagnostic practices and usually been satisfied with dividing the printed page into two horizontal halves where each truth was simultaneously but separately presented. In the case of the illustration, these unreconciled contradictions militated against a clear legibility of the body. Hence, though the image was scrupulously provided with numerous arrows, the labels next to the arrows that would allow us to read the image were missing. The empty arrows became a concrete testament to the hesitation and polyvocality of this physiogram, rather than its opacity. Gopalchandra's *Snayubik* Man was acephalous. It had no brain where the European illustrations it was modeled on always had one. It was made up of seemingly endlessly branching, nonspecific, wobbly lines, not unlike the way river systems are depicted on two-dimensional maps. There were innumerable similar European images.[138] But these latter always had a brain and usually also a spinal column. The *Snayubik* Man had neither. The wobbly reticulations were everywhere, including in the cerebral cavity where the brain should have been. Moreover, while there were numerous arrows, none had any labels.[139]

The German biologist, Ernst Haeckel, used an image almost identical to Gopalchandra's, albeit with a brain, only seven years later. Haeckel, like many other European authors of the time, had compared the nervous system to the telegraph system. Both Sanyal and Ray did exactly the same. Sanyal wrote, for instance, that

> that which can be called "nerve force" [English original] or the active power of the nervous machine (*snayu-yantrer kriya-shokti*) can also be called *vayu*. The actions of these nerves are performed by a kind of electric substance (*torinmoy podartho*) called "animal magnetism" [English original]. Hence the nervous threads (*snayu-sutro*) can be said to be like telegraph wires.[140]

Writing in a similar vein, Ray sought to explain the difference between *vayu* as electricity and *snayu*s as the conductors by stating that

> we can see that when a [snapped] telegraph wire is lying around, it does not have the capacity to convey messages. But as it is connected to an electricity-producing machine (*toritotpadok yantra*), news etc. travel by it. Hence we must call the telegraph wires electric-conducting material, rather than electricity itself.[141]

In his later poetic work, Ray again returned to the metaphor of telegraph wires, writing that

O Shantiram, Just as on earth there is the Telegraph
That carries message here and there
Similar electric force is there in the human body
That is called *vayu* in Ayurveda
Like wires there are *snayu*-threads
Electric resides in them, O Shanti . . .[142]

On the one hand, these authors were clearly drawing their metaphors from the material culture of colonial telegraphy.[143] On the other hand, the tremendous popularity of the telegraph metaphor among European authors of texts on the nervous system prevents us from reading too much of the colonial material context into these metaphoric elaborations. Laura Otis points out that "throughout the nineteenth century, scientists' electrophysical understanding of the nervous system closely paralleled technological knowledge that allowed for the construction of telegraph networks."[144]

Among European neurologists such as Haeckle, Katja Guenther points out that the telegraph metaphor at the time came to represent a system whereby multiple independent, autonomous, and conscious elements within the body were coordinated. Individual cells were considered by many authors to have autonomous "cell souls." These independent cells communicated with and collaborated through the nervous system. The brain and the spinal column both, but especially the former, retained a strong organizing role. In fact, in the hands of German neurologists in the wake of the German imperial unification, the brain's functions were often metaphorically represented as being akin to the centralized German monarchy. Guenther observes that the telegraphic system, the state, and the animal, at the time, formed three overlapping "metaphorical realms" for most neurologists.[145]

Not only do we not find a similar overlaying of telegraphic and political metaphors in Ayurvedic authors, but we also notice a conspicuous absence of any privileged or centralizing position for the brain. As already pointed out above, Gopalchandra's *Snayubik* Man lacks both a brain and a spinal column. For that matter, even the heart is absent. There is absolutely no definite center to the nervous system of the *Snayubik* Man. Selfhood is much more conspicuously unified and internally undifferentiated than in the case of the European neurologists of the time. The nervous system, despite its telegraphic metaphors, operated to consolidate a reticulate physiogram defined by undifferentiated flows lacking any pronounced center. The vaguely defined machinic physiospiritualism of the first physiogram—namely, the clockwork body—was overlaid by the more reticulate and seemingly hydraulic, uncentered, and electric physiospiritualism of the second physiogram—namely, the *Snayubik* Man.

By the beginning of the twentieth century, Nagendranath Sengupta sought to add a few more details to this physiogram. He distinguished *snayus* and *shiras*. He argued that *snayus*, of which there were a total of nine hundred in the body divided into four distinct types, were akin to the "diverse bindings that held the wooden planks of a boat together and allowed the boat to carry the weight of human passengers." Similarly, "through numerous *snayu*-bindings humans are able to hold [their own] weight."[146] By contrast, according to Sengupta, it was *shiras* that provided the hydraulic conduits for the circulation of *doshes* throughout the body. In an updated version of the illustration used by Gopalchandra, Nagendranath actually filled in some of the arrows and named the main *shiras*. In the textual description accompanying the illustration, Nagendranath went further. He said there were actually distinctive *shiras* which respectively carried *vayu*, *pitta*, and *kapha*. Moreover, sticking closer to the precolonial descriptions of *nadis*, Nagendranath stated that the *shira* system originated from the navel. He also went on to state that the total number of *shiras* were seven hundred, divided into four groups of one hundred and seventy five *shiras* each, respectively carrying *vayu*, *pitta*, *kapha*, and blood. Despite these significant alterations, the illustration itself remained surprisingly unchanged.[147] Other than the names which now were attached to the previously empty arrows, there is little to distinguish Nagendranath's *shira* system from Gopalchandra's *Snayubik* Man. Essentially, the illustration as well as the rest of Nagendranath's two-volume text, retained the same hydraulic, uncentered, physiospiritual physiogram. Though by the 1920s, hormones came to displace the various electromagnetic and quasi-electromagnetic forces and energies, it all still circulated in a smooth, uncentered, and reticulate space, just as the circulating entity still had both material and spiritual aspects.

FOUR

The Chiaroscuric Man: Visionaries, Demonic Germs, and the Microscope

Two men locked in mortal combat in a dark house. A third enters to help one of the two combatants but, unable to see in the dark, his blows have an equal chance of vanquishing his friend as it does of killing his foe. This is how one author described the clinical encounter in 1875. The two men were "life" and "disease." The third man was the physician, hoping to help "life" but often ending up inadvertently helping "disease" bring about death. The drama played itself out in the "dark house" of the human body.[1]

Incidentally, this conceit appeared in the very first article of the first issue of a newly launched periodical entitled *Onubikshon* (Microscopy). Upon the cover of every issue of the short-lived journal was the iconic image of an *onubikshon-yantra* (microscope) (see fig. 7). The hope, though not explicitly stated in the article, was evidently that the new visibility engendered by the microscope and such technologies would illuminate the darkness within the human body and allow the physician to be more discerning in his blows.

The powerful rhetoric made modernizing Kobirajes keen to embrace the new illumination. They did so in diverse ways and saw novel things with the microscope. But the simple, quasi-Biblical narrative of the microscope suddenly and miraculously enlightening the darkness within the body was a mere rhetorical ploy. Victorian ontologies of vision by the 1870s were a lot more complicated.[2] The new ways of seeing that emerged in modern Ayurveda were much more an intricate product of the selective braiding of "Eastern" and "Western" modes of seeing than a simple displacement of one visual regime by another.

[১ম খণ্ড]　শ্রাবণ ১২৮২ সাল।　[১ম সংখ্যা।]

অণুবীক্ষণ।

স্বাস্থ্যরক্ষা, চিকিৎসাশাস্ত্র ও তৎসহযোগী অন্যান্য শাস্ত্রাদি বিষয়ক

মাসিক পত্রিকা।

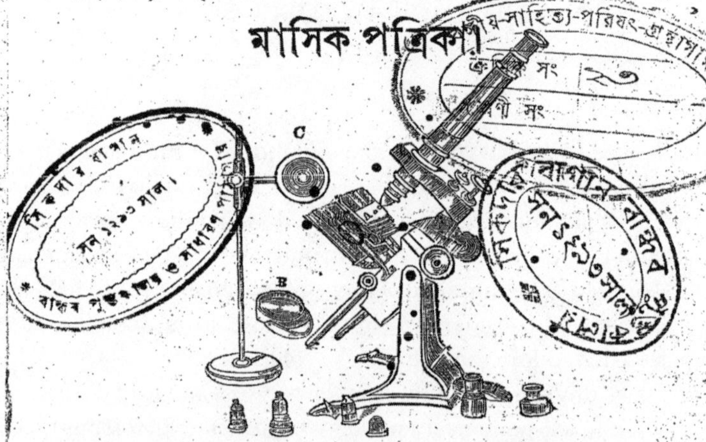

“ দৃশ্যতে ত্বগ্র্যয়া বুদ্ধ্যা সূক্ষ্ময়া সূক্ষ্মদর্শিভিঃ।”
“সূক্ষ্মদর্শী ব্যক্তিগণ একাগ্র সূক্ষ্মবুদ্ধি দ্বারা দৃষ্টি করেন।”

অবতরণিকা।

কর্ত্তব্য বোধের একান্ত অনুরোধে এ ক্ষুদ্র মাসিক পত্রিকা প্রচারে প্রবৃত্ত হইলাম। অনেকের সংস্কার যে ইংরেজদিগের এদেশে আসিবার পর ইংরেজী শিক্ষা বহুল পরিমাণে প্রায় সর্ব্বত্রে বিস্তারিত হইয়া তাহার সঙ্গে সঙ্গে ইংরেজী রীতি, নীতি, আচার, ব্যবহার, বাণিজ্য প্রণালী এবং রাজনীতি এদেশে প্রচলিত হওয়াতে ভারতবাসী দিগের বিশেষ উন্নতি হইয়াছে। কিন্তু আমাদিগের সংস্কার অবিকল এরূপ নহে। ইংরেজী শিক্ষা, আচার, ব্যবহার, রীতি, নীতি ইত্যাদি এদেশে

১

7. *Anubikshan-yantra*. The iconic microscope.

Other Illuminations

Most Kobiraji practitioners of the nineteenth century denied the colonial-modernist conceit that the insides of the human body had been opaque to them prior to the coming of new technologies. Gananath Sen, one of the keenest champions of the new technologies, posited that anatomical knowledge, or *shorir-bidya*, was of two basic types. The first type of anatomical knowledge was that which was to be "obtained by external means" (*bajhyo upaylobdho*). Sen described this knowledge as "that knowledge which is obtained through the five senses, particularly sight (sometimes aided by instruments such as the microscope), by examining a living or dead body." By contrast, the second type of anatomical knowledge was that knowledge which was "obtained by internal means" (*obhyontor upaylobdho*). Sen defined this latter as "the subtlest of subtle knowledge [*sukshmanusukshmo gyan*] about the anatomy that is obtained by the greatest seers [*maharishi*], who are possessed of divine wisdom [*dibyo-gyan*], without the use of their gross senses [*sthul-indriyo*] and merely through the operations of their [inner] eye of wisdom [*gyan-chakshu*]."[3]

Sen was not alone in arguing for the "eye of wisdom." Kobiraji authors often holding diametrically opposed views on the role of modernity and the program of modernization, still agreed on the fact that the greatest of ancient seers had had an "eye of wisdom" that had enabled them to see things mere mortals could not, and which even advanced technological devices such as the microscope could not match. Thus, Haramohan Majumdar, a Kobiraj who wrote a detailed critique of Gananath Sen's famous anatomical treatise, *Protyoksho Shorir*, still held on to a distinction between "that which is seen [by mortals]" (*protyoksho*) and "that which is seen by seers" (*rishi-protyoksho*).[4] Another noted Kobiraj, D. N. Ray, wrote, "I believe that the ancients actually saw the *Tridhatus* [the three para-humours] by some means or the other at least in their mind's eye, they formed a very clear and definite conception about [their] nature."[5] A Tamil physician, Vaidyapathi T. G. Ramamurti Iyer, writing of seers in both the Ayurvedic and the Siddha tradition, likened their "*gyandrishti*" (wise-sight) with modern X-Rays.[6] Similarly, Brajaballabh Ray, one of Sen's early colleagues at the Ayurvedic College, recalled his father, an eminent but orthodox Kobiraj, insisting that the ancient seers were able to "see" much more than mere mortals and that they had tried to convey this "fundamentally extra-sensory" (*mule jaha otindriyo*) knowledge by couching it in "signs, symbols, metaphors and similes."[7] Ray's reminiscences also proved the longevity of such an idea of divine sight. Its basis was probably much older since Majumdar frequently cited

Chakrapanidatta's dictum that the wisdom of the ancient seers cannot be judged by mortal beings.[8] Describing the nature of this divine wisdom or "sight" possessed by the ancient seers, Mahamahopadhyaya Bhagabatkumar Goswami wrote in his foreword to the *Ayurveda Ratnakar* that: "Seers who concentrated their mind upon the human constitutional aspect of Creation and received the inspiration of the first truths through devotion to and meditation upon the Source of all inspirations. Upon those soul-to-soul communications they built the Ayurveda Shastra."[9]

There is a significant body of scholarly literature on the South Asian notion of darshan, or "seeing." Such "seeing" carries within it notions of mutual transformation and forms an important part of Indic notions of piety and devotion. Much less has been written on the notions of divine sight or *dibyodrishti* that the ancient seers were thought to have possessed. Dipesh Chakrabarty, in one of the few critical engagements with the notion of "divine sight," says that, together with the better-known *darshan*, it constitutes a "family of practices" that belong to the "history of practice and habitus." These practices, according to Chakrabarty, "do not require the assumption of a subject" and in fact operate to "dissolve the subject-object distinction."[10] Chakrabarty, however, was speaking of the redeployment of *dibyodrishti* by modern authors such as Rabindranath Tagore and S. Wajed Ali.

Incisive as these insights into *dibyodrishti* are, one should be wary of drawing too sharp a contrast between Indian and British ways of seeing in the nineteenth century. As Srdjan Smajic points out, there was a robust British discourse on "spiritual optics" or "inner sight" throughout the Victorian era. Commencing most likely with Thomas Carlyle's writings, this discourse reached an apogee of sorts in the work of John Ruskin. This "spiritual optics" drew exactly the kind of contrasts and hierarchies between mundane ocular sight and a higher, morally informed, intuitive inner vision.[11] Kobiraji protestations of distinctiveness, instead of being purely derivative of old, Indic modes of seeing, were most likely the result of a creative braiding of Indic *dibyodrishti* with British "spiritual optics."

One of the marks of this new, braided regime of supervision was its curious relationship with history. Chakrabarty points out that *dibyodrishti* was a way of escaping the confines of historical time.[12] Contrarily, in the Kobiraji context, rather than being a route through which to escape historical time, the supervision underlined the Kobirajes' own entrapment within historical time. All the authors asserted that ancient seers (rishis) had had this supervision, but that they themselves did not. What separated the visually empowered ancients from the visually impaired moderns was nothing but historical time.

While Kobirajes admitted to two forms of vision, they were also clear about the chasm opened up by historical time between these two modes of seeing. Refusing to acknowledge this chasm has led many a historian astray. Malcolm Nicolson, for instance, in his oft-cited essay "The Art of Diagnosis," begins by quoting Susruta, and repeatedly through the essay draws parallels between the Ayurvedic and "Western" visual practices. Nicholson argues that the relatively unstable social position of physicians in both the Greek and the Indic world led them to have a more "well-developed interest in diagnostic observation."[13] Though Nicholson follows this up with a rider against conflating such ancient visual practices with contemporary clinical ones, his discussion in fact opens up the possibility for precisely such conflations. His discussion of Ayurvedic physicians who look carefully at the messenger bearing news of illness is a way of demonstrating subtle signs that will allow them to gauge the seriousness of the patient; it clearly depicts "vision" in its modernist guise. The Ayurvedic physician here is imagined almost in the mold of a Sherlock Holmes, as one picking up slight signs others would miss, and deducing the patient's state from them. This Holmesian vision is not the *dibyodrishti* Susruta's heirs credited him with having. In fact, if we return to Susruta's own text, his directions on how to observe the messenger make it plain that his vision is very different from the rationalized, modernist vision of a Holmes-like character. He advised the physician to notice whether the messenger "portends well" (*shubho-suchok*). He followed it up by directing the physician to notice whether birds such as ducks and objects such as a pitcher full of water—both considered good omens—are visible or not, before deciding whether or not to take the case.[14] Clearly, Susruta's vision is one attuned to portents and omens, and not the kind of rationalized Holmesian vision that Nicholson implies it to be. The contrast becomes clearer still when we see how this single element of noticing the messenger evolved in later texts. A seventeenth-century Bengali text by Sheikh Chand, *Talibnama ba Shahdaula Pir*, for instance, advised the physician to

Pay heed now to what the saints have to say,
If someone falls ill somewhere
Suddenly a messenger brings you news
From which direction [he] comes, you must enquire
[If] he comes from the right, when [your] breath is to the left
He [that is, the patient] will not die, but will suffer physical pain
Comes from the left, breath flows on the right
There is no death this time, will be fine

Gives the news from behind you
Do not turn around
News comes from the front
Nothing to be gained.[15]

While Sheikh Chand's directions are admittedly a much later development, his directions are given in a chapter titled *Rogtotwo* ("Theory of Disease") and very likely had some (however mediated) genetic affiliation to Susruta's directions. It clearly evinces what later Kobiraji authors also insinuate—that Kobiraji visual practices functioned within very different culturally coded rationalities. These older visual practices and the rationalities that underwrote them were embedded in distinctive cosmologies populated by portents, omens, and all-seeing seers. Illumination here was not impeded by the opacity of flesh, but neither did illumination simply mean peering into the gross, anatomical realities. These older illuminations escaped the confines of history itself and opened up an enchanted space of magic, portents, and transcendent forms, accessed through the miraculous powers of seers. Such visual regimes may have resembled Carlyle's "spiritual optics" or Ruskin's "moral vision," but, alas, it was partially alienated from the nineteenth-century Kobirajes by the chasm of historical time.

Microscope and Chromoscapes

Illumination, Chris Otter has recently pointed out, has too often been adopted as the iconic feature of a uniform "Western" modernity. Light and vision, he further argues, have been misleadingly and rather restrictively hitched to just two modalities: the panopticon and the flaneur, or, put another way, discipline and spectacle.[16] These two modalities, whatever their original analytic power, have now, Otter argues, been overused to the point of having become clichés. As clichés, they have been divested of historical detail and specificities and been rendered a wholly abstract phenomenon that can be found everywhere but located nowhere. Instead, Otter urges for a new history of light and vision that is attentive to historical detail: one that takes technologies, politics, cultures, and bodies seriously and engenders more archive-embedded histories of vision. I take my cue from Otter's observations.

The microscope was possibly first introduced into India in a publicly visible way in the course of the early nineteenth-century cartographic operations. Captain Everest's reports on the measuring apparatus used by the Great Trigonometrical Survey of India mentioned the use of several microscopes.[17] By the middle of the nineteenth century, however, Indian medical students at the Calcutta Medical College, including those studying in Bengali or Hindustani streams, were regularly exposed to microscopes. The famous Anglo-Indian anatomist, Allan Webb, in delivering the General Introductory Lecture to the students of the Calcutta Medical College, called chemistry and the microscope the two "powerful levers of modern medical science . . . [that] have but lately burst open the hidden secrets of physical existence and finally dispelled the elemental theory."[18] John Harley Warner points out that toward the middle of the nineteenth century, the microscope's "stature as an instrument of observational science" increased sharply on both sides of the Atlantic. The founding of the Microscopical Society of London in 1839 and the publication of their journal provided a locus for much of the British enthusiasm. Anglo-Indian scientists such as Webb were obviously drawing upon those distant energies when they championed microscopic visuality in India.[19] Holding the microscope in such high esteem, Webb naturally congratulated the students on their good fortune since, "Microscopical Anatomy is now [being] most fully elucidated, both in diagrams of great beauty and fidelity, as well as by the daily use of the microscope."[20] Since numerous stalwarts of refigured Ayurveda, including such pioneers as Surendranath Goswami, Nagendranath Sengupta, Jaminibhusan Ray, Gananath Sen, and many others, held some allopathic qualification or the other, usually a Licentiate of Medicine and Surgery, they were all exposed to microscopy at medical school.

Despite the undoubted influence of allopathic medicine in introducing Kobirajes to microscopy, the instrument was put to distinctly different uses in the two traditions. In allopathic medicine, the main emphasis, as can be seen from Webb's own work, was on postmortem analysis and the examination of excised tumors.[21] The earliest Kobiraji report of the use of a microscope, however, was related to neither postmortem nor post-excision examinations. Instead, Gopalchandra Sengupta's *Ayurveda Sarsamgraha* mentioned the *onubikshon-yantra* in the context of urine examination.[22]

Joel Howell, in his excellent study of how technology changed medicine in the modern American hospital, points out that urinalysis was not new to the nineteenth century. It had a long and hallowed presence in the Western medical tradition and had even become ensconced in art and literature as a feature of medical life well before the nineteenth century. Yet, "throughout the latter part of the nineteenth century and the first part of the twentieth century, people writing in medical and scientific literature described many new physical, chemical and microscopical tests for the examination of the urine."[23]

In the Sanskrit Ayurvedic tradition, on the other hand, urine analysis was a very late introduction. In the classical *Charaka Samhita,* there are some scattered references to changes in urine mentioned in the course of discussing specific diseases, but these are not gathered together into any systematic diagnostic method. Later, important works such as those of Susruta and Vagbhatta have even less to offer on the subject. Even important medieval authors such as Madhavakara, Sarangadhara, and Bhavamisra had little to say on the matter, though the latter did for the first time include urine analysis in a list of diagnostic practices. The details of the practice emerged first in a handful of relatively minor late-medieval texts such as the *Yogaratnakar,* the *Basavarajyam,* and Bangasena's *Chikitsasarasamgraha.* These texts introduced a new method known as the "*toilo-bindu poriksha*" or the "oil drop examination."[24] The method involved dropping a single drop of oil into the collected urine and observing its behavior. Though its source remains obscure, it is worth noting that the Unani Tibbi tradition of Islamic medicine had a long tradition of urine examination.[25] In Bengal, more heterodox Islamic medical traditions from the turn of the twentieth century also gave a prominent role to urine examination.[26]

Sengupta was thus confronted with two very distinctive traditions in the "West" and the "East." In the "West," urine examination was a hallowed diagnostic tradition, whereas in the "East," it was a lately developed and somewhat heterodox tradition. In fact, Dominik Wujastyk cites seventeenth-century British traveler, John Fryer, as observing that many Indian physicians preferred pulse examination over urine examination and therefore neglected the latter altogether.[27] To complicate matters further, neither tradition was homogenous. Just as there were numerous different types of urine examination in the "West," in Bengal too, a variety of opinions coexisted. Hesitantly braiding selective strands of these knowledges and practices together on the surface of the page, Sengupta sought to develop a new Ayurvedic urine analysis premised largely upon the new capacities of the microscope. Sengupta split the page in half in just the way he did with regard to *nadi-pariksha* and pulse examination in chapter 3: the upper half presented an "indigenous" view, while the bottom half presented an avowedly "Western" view.

In the former case, urinalysis involved the addition of a drop of oil to the urine and observing its scatter patterns. Thus, if a patient suffered from a *vayu*-related affliction, the drop of oil would swim or float on the surface freely. If, on the other hand, the patient suffered from a *pitta*-related affliction, the drop would break up into numerous smaller drops and sink. For a *kapha*-related affliction, it would stand still and become dense or concentrated.[28]

The direction in which the drop of oil seemed to travel within the container holding the urine was also significant. Thus, if the drop traveled eastward, the disease was to abate quickly; if it traveled southward, the disease would only abate gradually; a westward direction would indicate a minor affliction curable by minimal medical intervention, whereas a northward trajectory would indicate a relatively more obstinate disease that would take much effort to cure. If the drop of oil traveled toward the northeast corner (*ishan kon*) or the southeast corner (*agni kon*), however, the patient was sure to die of the illness.[29] The color of the urine was also noted in these tests, but was of only marginal importance. The main emphasis was clearly on how the drop of oil behaved within the urine. Its scatter pattern, its trajectory, its floatation, or sinking—all contributed toward making a diagnosis.

Side-by-side with these older tests, Sengupta also introduced a wide variety of new urine tests in the bottom half of the page. These included "Millon's Test," "Pettenkofer's Test," "Haller's Test," "Meline's Test," "Moore's Test," "Fermentation Test," and the "Maumene's Test," besides some unnamed tests.[30] These tests operationalized the same conjunction between modern chemistry and the microscope that Webb had identified. In conducting the Fermentation Test, for instance, one had to fill an apparatus with urine, add some yeast to it, and invert the apparatus into another container holding some more urine. The entire setup was then to be left in a warm place for twenty-four hours. This would result in the separation of sugar, alcohol, and carbonic acid. The last of these would rise to the top in a reddish color. Gradually, above this reddish hue, a thin skin, similar to milk skin, takes shape. This skin has to be observed under a microscope for the detection of what Sengupta called the "uritley."[31]

These tests subjected the urine to a number of new and carefully calibrated manipulations. The urine was mixed with a range of different chemicals, heated, cooled, and looked at in new ways and through new devices. In the process, the urine manifested new physical and chemical properties that strained not only the conceptual apparatus of Kobiraji, but also the language available to communicate these new realities. One of the first problems arose with the color-coded tests. Once again, Howell's studies of urinalysis in America are instructive. He points out that the color of the urine had long been regarded as a diagnostic indicator and the new chemical tests fitted into this older style of analysis. Thus, any "change in the normal pale-straw tint of the urine could indicate a morbid condition. Darker colors suggested disease in general; bile gave an olive-green tint; blood gave a smoky or frankly reddish appearance; pus could lead to a milky coloration."[32] By contrast, since those strands of Ayurveda that Sengupta drew upon emphasized the scattering of oil rather than colors, the color orientation itself was a novel emphasis. More significantly, the very colors evoked by the exotic chemical tests were new to the Bengali eye.

Color perception, it has long been believed, is connected to the linguistic categories available to express it. It has been experimentally determined that "if two colours are called by the same name in a language, speakers of that language will judge the two colours to be more similar and will be more likely to confuse them in memory compared with people whose language assigns different names to the two colours."[33] Such differences, it has been found, develop early in children alongside their acquisition of color terms. Moreover, such native color judgement can be destabilized by what is termed "verbal interference": that is, when observers are "indirectly prevent[ed] . . . from using their normal naming strategies" for the colors they see.[34] Thus Russians, whose language automatically distinguishes between different shades of blue, are much better than English speakers at distinguishing different types of blues. Yet, if they are indirectly prevented from using their Russian color categories, their powers of discrimination are in turn adversely affected.

Likewise, Otter points out that as new techniques and technologies of illumination proliferated in the late nineteenth century, color discrimination came to vary significantly. The yellowish illumination of gaslights and the whitish illumination of electric lights, for instance, produced very different types of color perception. In fact, color discrimination came to be seen as one of the three key visual capacities against which new technologies of illumination were evaluated.[35] Clearly, then, so far as the lighting technologies under which an urine sample was analyzed varied, the colors observed would also vary—just as different microscopes would also likely produce slightly distinctive colorations.

These findings underline the fundamental problems confronting physicians like Sengupta, who sought to translate color-based tests for urinalysis into the Bengali-speaking world of refigured Ayurveda in the late nineteenth century. The success of the tests depended crucially upon the observer's ability to distinguish different hues of the chemically treated urine, but the color terms themselves were linguistically and culturally specific, and in turn shaped color perception. What was necessary therefore was the constitution of a substantially new chromoscape—that is, a recalibration of the perceptible color spectrum for Bengali-speaking physicians.

In one test, intended to test the presence of bile in the urine, the urine was to be mixed with a small proportion of sugar and treated with dry sulphuric acid. The presence of bile would be indicated by the sample turning violet. The difficulty was that the Bengali language did not have a word that was an exact homologate of "violet." "Violet" as a particular shade of color did not culturally exist. Sengupta had to thus write that the sample would turn *bhayolet red-borno* (violet red color).[36] Sengupta's hesitation showed in the fact that he even left the word *red*, for which there was a well-known Bengali homologate, *lal*, untranslated. A similar problem arose in another test where the necessary color was a deep "amber."[37] The untranslatability of colors was perhaps the most striking aspect of the effort to render modern urinalysis in an Ayurvedic garb.

The problems of constituting a new chromoscape were further exaggerated by the fact that the colors were often connected to microscopic entities whose very existence was new to the Kobiraji lexicon. These objects and substances remained equally untranslatable. *Naitret ob Silbhor* (silver nitrate), *Mutrer Ekstryaktib o Kolaring Myatar* (coloring and extractive matter of urine), *Aksaid ob Kopar* (copper oxide), and *Merunophlyanel* (Merino/Maroono flannel) were some of the numerous substances and objects that were intimately connected to the changing colors of the urine sample subjected to chemical and microscopic analysis. None of them had any easy translations into Bengali or the classical Sanskrit. This untranslatability again demonstrated how the simple tests and the instruments introduced into refigured Ayurveda created a farrago of ontologically novel entities that needed to be conceptually accommodated within the lexical and conceptual folds of refigured Ayurvedic medicine. Their relationship to the body that seemed to be producing them or reacting to them, however, remained wholly unexplained.

The lack of explanation meant that the body's visual qualities were increased without being mutually reconciled. Thus, one could see a drop of oil scatter and sink in a bottle of urine as well as that urine in turn take on a completely novel hue, namely violet-red. Both eventualities allegedly happened because of the same underlying feature: the excess of *pitta* that was now homologated with bile. Yet no attempt was made to reconcile the extrasensory and transmaterial dimensions of *pitta*, not to mention the existence of five distinctive subtypes of *pitta*—*pachok pitta, alochok pitta, bhrajok pitta, sadhok pitta,* and *ronjok pitta*—with the resolutely singular, homogenous ontology of bile. Notwithstanding the forced homologation and the inexact translation, the newly refigured Ayurvedic body that was made accessible through the new urinalyses clearly had many more material properties than it did before.

Annemarie Mol points out that medicine knows the body precisely through acts like slicing, coloring, seeing, counting, etc. The body, she points out, does not preexist these acts of knowing, but is rather materialized through these enactments of knowing.[38] Hence it is clear that as new ways of enacting the body were developed by physician-authors like Sengupta, and as these new enactments illuminated new properties and material entities in the body, the body itself changed. A new body, both more colorful and more materialized (in the sense of producing new material ontologies in the course of the enactments), emerged through these enactments.

Sengupta's early experiments with microscopes and urinalysis were not the only such efforts in the early history of Ayurveda's modern refiguration. Unlike American hospitals, where, as Howell points out, urinalysis finally began to wane by the third decade of the twentieth century, the fate of microscopic urinalysis in refigured Ayurveda was quite distinctive. After a couple of minor and marginal attempts, the next major attempt to redefine microscopic urinalysis within refigured Ayurveda was undertaken by Siddheswar Ray in 1924. Ray was a professor at one of the newly established Ayurvedic Colleges, a member and librarian of one of the main professional associations, the Ayurveda Sobha, and a frequent contributor to the Ayurvedic journal *Ayurveda*. In short, he was very much part of the mainstream of Ayurvedic modernization at the time. Ray's reimagination of urinalysis was summed up in a nearly four-hundred-page book titled *Mutro Tottwo* (Urine science). In *Mutro Tottwo*, the chromoscape, even after half a century of Sengupta's efforts, was still unstable. Ray deployed strategies similar to Sengupta in transliterating English color terms in the Bengali script. But in Ray, there was also an attempt to develop some locally recognizable allegories to convey color terms. A comparison with washings from *bichuli* (the chopped hay cattle are fed on), for instance, was used to convey the "pale straw tint" of normal urine.[39] Likewise, the color of the urine in the "late stages of diabetes" was conveyed through comparisons with the color of sugarcane syrup.[40] For another condition, the characteristic color of the urine was compared to the washings of the tiny fruits of the Trewia plant (*pituli*).[41]

More remarkably, Ray introduced a much larger range of new microscopic entities ranging from blood corpuscles, bacilli, urea, etc. These were entities "so small that they had eluded even the imagination," said Ray, but the microscope had made them all easily visible.[42] Most of these entities had absolutely no conceptual or ontological precedent in Sanskrit Ayurveda or traditional Kobiraji practice. The new refigured Ayurveda, by embracing these new entities, was once again enacting through these tests a very distinctively new body. These entities themselves had physical properties such as shape, size, weight, and color that needed to be elaborated in great detail.[43] Unlike Sengupta's earlier work, Ray's more voluminous work therefore carried illustrations of a number of microscopic entities in order to clarify their shapes and relative sizes. Once again, these attempts to communicate physical properties revealed the limits of translation. Not only colors such as "Red Sand," but also certain shapes such as that of the "Mulberry calculi" presented translation problems. In general, we notice the same trend that was witnessed in Sengupta's work. A selective braiding of distinctive strands of knowledge and practice allowed them to redefine the new entities produced by microscopic urinalysis. Just as Sengupta had braided together "oil drop examination" and tests like "Millon's" or "Pettenkofer's," Ray braided together elements from a work called *Nadipariksha* (possibly a text attributed to Ravana) with the writings of "Western" physicians such as James Meschter Anders (1854–1936), a famous Philadelphia physician. These intricately braided knowledges produced the new physiograms that underwrote the refigured Ayurvedic therapeutics.

Worming the Germ Theory

Despite the significant and creative deployments of microscopic urinalysis by Sengupta and Ray, those seeking to modernize Ayurveda did not consistently discuss urinalysis. Other sites of microscopic investigation soon overtook urinalysis. Preeminent among these was the investigation of "worms," or *krimis*.

Jashodanandan Sarkar's brief introduction to the *onubikshon-yantra* in the prefatory note appended to his translation of the *Susruta Samhita*, for instance, made no reference to its role in urinalysis. Instead it said, "It is true that this machine [*yantra*] is not mentioned by Susruta, but without the help of this machine, all of Susruta's claims cannot be verified." Having thus elevated the instrument to the status of a touchstone upon which the words of the hallowed seer, Susruta, could be verified, Sarkar added that, "for instance, he [Susruta] says in one place that 'there are seven types of worms [*krimi*] in the blood. None of these are visible to the eye.' Susruta mentions numerous other small insects [*kshudra kit*]. Their bites, etc. are also mentioned. It is astounding to think how their impact on hundreds of human bodies had been examined."[44] Sarkar's brief comments laid out the future development of microscopy within refigured Ayurvedic circles. Instead of the urine, it tended to focus on the blood. It progressively imagined larger and more diverse sets of internal parasites that it generically dubbed *krimis*. The generic terminology, however, was somewhat misleading since it included within it both microscopic entities such as the malaria parasite, *Plasmodium*, and entities as large as the tapeworm. *Krimis*, in the first three decades of the twentieth century, became one of the most productive sites upon which to elaborate the Ayurvedic version of the germ theory.

It is both impossible and misleading to speak of the "germ theory" in the singular. Even in allopathic circles, its coherence, history, and ambit remain matters of historical debate. As Nancy Tomes and John Harley Warner point out in their introduction to a collection of early articles on the reception of "germ theory," "there was no 'germ theory of disease' transcendent over time, but rather many different germ theories of specific diseases being debated in specific communities, times and places."[45] David Barnes, in his study of the transition toward germ-based etiologies in late nineteenth-century France thus writes with some frustration and much truth that "the history of aetiology is deceptively, maddeningly unstraightforward." On a more hopeful note, Barnes adds that "the perplexing aetiologies of the past reveal in remarkable depth and detail the cultural landscapes out of which they emerged."[46] The Kobirajes did not therefore plagiarize from a single, stable script. There was no single coherent doctrine that could be "copied." Their peculiar version of the "germs" was in fact refracted through their own hyphenated Bengali-Victorian cultural landscape.

Culture, however, is grounded, articulated, and reproduced through institutions and by specific agents and actors. In order to delineate the cultural specificities through which modern Ayurvedists refracted their versions of "germs," it is useful to begin by delineating the institutions and agendas they acted through. This is particularly important in view of Michael Worboys's recent definition of the alleged "Bacteriological Revolution" that led to the invention and popularization of the germ theory. Worboys outlines four key components that constituted the alleged Revolution (whose reality he doubts, at least in the case of Britain). The following are the four components outlined by Worboys:

1 A series of discoveries of the specific causal agents of infectious diseases and the introduction of Koch's Postulates.
2 A reductionist and contagionist turn in medical knowledge and practice.
3 Greater authority for experimental laboratory methods in medicine.
4 The introduction and success of immunological products.[47]

Of these, numbers three and four were almost completely absent in refigured Ayurvedic medicine. Though Ayurvedic hospitals had commenced in Calcutta by 1916, rising up to three by the 1930s, laboratory work for refigured Ayurvedic medicine was extremely rare. The newly college-trained Ayurvedists who began practicing by the 1920s as well as the many Ayurvedists who held parallel allopathic qualifications did have some exposure to laboratory work, but the medicine they practiced on an everyday basis did not include any laboratory component. Neither did the refiguring Ayurvedists—even the most committed modernists among them—devise or champion immunological products of the sort envisaged in Worboys's definition. The only immunological products they marketed were medical formulations, such as the hallowed *Mokordhwoj*, that were said to boost general immunity, they were certainly not specific vaccines or sera immunizing the patient against particular targeted diseases. In fact, modern Ayurvedic authors repeatedly emphasized that they wrote for the benefit of the thousands of rural practitioners. These latter treated using a relatively simple pharmacopoeia and would have had no access to laboratories or vaccines.

Even the first two constituents of Worboys's definition, though clearly present, were present in a form much altered and refracted through the local cultural and institutional realities. Koch's famous postulates, for instance, had three components to it. First, a specific micro-organism must be constantly shown to be present in diseased tissue. Second, the specific micro-organism must be isolated and grown in pure culture. Third, healthy animals inoculated by the pure culture must produce the disease. Each of these three components once again required elaborate and sustained laboratory work. Authors seeking to refigure Ayurveda, with few exceptions, had very limited opportunity for pursuing such laboratory work. Thus, while they often cited Koch and embraced some version of his ideas at an epistemic level, they almost never actually followed his postulates in practice.

Moreover, in British India, as elsewhere in the colonial world, germ theory had come to be developed largely through Tropical Medicine. This meant, in addition to the search for specific micro-organisms, an interest in the external vectors of communication of disease.[48] Whereas in metropolitan Europe, germ-based etiologies led to a decline in medical geography, in the colonial world, Tropical Medicine promoted a new form of ecologically territorialized thinking in medicine that on occasion came to be remarkably close to medical geographies.[49] Such ecologically informed, broad-based studies of vectors and transmission required data to be gathered over a large swathe of land. This required both manpower and resources, not to mention centralizing institutions for collecting, tabulating, and analyzing the gathered data. None of this was available to the refiguring authors of modern Ayurveda.

The Ayurvedic germ theory that developed was thus a curious one. It was a product of the laboratory, but often without any immediate access to or use for the laboratory. It relied crucially on the new visions made possible by the microscope, but it did not advocate a rigorous or regular use of the instrument. Frequently, the new microscope-enabled visions were mere memories the author brought alive in words or at best through the reproduction of illustrations copied from other published works. The level of reproduction too remained poor, comprising generally of simple block prints rather than the reproduction of photographs that were to be seen in English bacteriological works of the time.

Ayurvedic discussions of the germ theory began appearing in the latter half of the second decade of the twentieth century. One of the first authors to undertake this discussion was Ramsahay Kabyatirtha. Writing in the journal *Ayurveda* in early 1918, he sought to answer the question as to whether "Aryan seers had known about Germs." Predictably, he answered in the affirmative. But in so doing, he also made certain key intellectual maneuvers. First, he advanced two words, *jibanu* and *bijanu*, as interchangeable designations for "germs." Second, despite principally using these two words, he also homologized certain other words used in older canonical works, such as *krimi*, *kit*, etc., with "germs." Finally, he sought to straighten out certain philosophical impediments in the way of reconciling a germ-based pathology with Ayurveda's traditional emphasis on regimen. In this latter regard, he made the rather innovative suggestion that "germs" or *jibanu* could arise sui generis within the body owing to lapses in proper regimen. He recognized that his contention seemed to suggest that inanimate objects in the body could become animate and thereby violate the cardinal boundary between the "living" and the "dead." Yet, he answered such objections by stating that since God inhered in every smallest particle of matter, the animation of a previously inanimate piece of matter was quite permissible.[50]

Kabyatirtha's position on sui generis germs was far from unique or unusual. In the 1860s, Louis Pasteur had himself had to fight a dogged battle against the doctrine of "spontaneous generation" in France. In the subsequent decade, the battle carried on across the channel in Britain, led by eminent British scientists such as Charlton Bastian, John Hughes Bennett, and others. Opposing them was an equally eminent group of scientific men such as John Tyndall, John Burdon Sanderson, and others. The positions on both sides were complex and did not "neatly divide between vitalist versus materialist, religious versus non-religious scientists and biologists versus physical scientists."[51] Kabyatirtha was drawing selectively upon this older debate and resurrecting "spontaneous generation" in an Ayurvedic context.

Where Kabyatirtha was unique, however, is that his position was not connected to any experimental results or microscopic analysis. The "spontaneous generation" debate whether in Pasteurian France or Listerian Britain had always given a prominent role to the interpretation of experimental results. But the Ayurvedic discourse, while constantly notionally invoking microscopes and laboratories, was wholly disconnected from any experimental setup. "Germs" to Kabyatirtha were simply an epistemic category in a philosophical debate. There was no materiality to them.

The distinctively discursive nature of Kabyatirtha's germs becomes clearer once he is situated within other larger context of germ debates in India during the time. Just as he was writing his piece in the mid-1910s, biomedical researchers in India like David Semple were discussing the same question about the boundaries of life at the level of microbes, but doing so in relation to laboratory experiments aimed at developing new immunological products. Pratik Chakrabarty, writing of Semple's research on rabies vaccines at the time, categorically states that "'living' and 'dead' were essentially ideological categories." But, most importantly, these were ideological categories marshaled within the context of laboratory experiments, and that is why eventually "out of these ideological categories and intangible constructs" a valid, effective and hugely popular immunological product in the form of Semple's rabies vaccine could still emerge.[52] By contrast, Kabyatirtha's *jibanu* never entered the laboratory.

The following year, the issue was once again taken up on the pages of the same journal, *Ayurveda*. Jabadeshwar Tarkaratna set out to explain the "Theory of Germs in Ancient India." Mockingly, Tarkaratna spoke of how the "germ theory" was spreading faster than "germs." Having commenced in Germany, he felt it had overwhelmed not only European medicine, but literature as well. While mocking its excesses mainly in its attempts to attribute all diseases to the action of "germs," Tarkaratna, like Kabyatirtha before him, accepted the capacity of "germs" to trigger disease. What he disputed strongly were European claims to have discovered "germs." Instead, he said that "the foremost and the oldest book in the world—the Vedas—make reference to the theory of germs." Interestingly, while Kabyatirtha had mainly used the words *jibanu* (life atom) and *bijanu* (seed atom) for "germs," Tarkaratna referred to them as *kitanu* (insect atom). He did not give any specific reasons for his divergent vocabulary, nor did he refer to Kabyatirtha's essay. Tarkaratna then went on to cite a large number of Vedic hymns addressed to invisible and frequently malignant creatures, arguing that this proved that the authors of the hymns had known of the existence of microbes that were invisible to the naked eye. Tarkaratna also claimed that many rituals, such as the practice of certain purificatory rites by people returning after having cremated a dead body, were actually designed to ward off invisible microbes.[53]

Another article appearing in the same year in *Ayurveda* took a slightly different route to microbes. Instead of seeing microbes as a special class of creatures, Saradacharan Sen reclassified them as a subordinate group within a larger group of entities known as *krimis*, or worms. Saradacharan opined, "It is unseemly that India's traditional physicians will remain silent, while the independent nations of the West through their practice of science charm the world by their many inventions." Instead, he felt traditional physicians in India should try to contribute to the discussion about new diseases "without the help of new scientific machines and by reliance on what is available in the *shastras* (ancient disciplines of textual and oral knowledge)." Proceeding along this line, he felt that the modern scourge of the hookworm closely resembled classical Ayurvedic descriptions of *krimis* or worms. He classified these *krimis* into two groups, external and internal. Commenting upon the internal *krimis*, Saradacharan mentioned that many of them reside in the blood and travel through the veins that carry blood. He pointed out that some of these blood-dwelling *krimis* were so small that they "escaped the power of sight."

> Today the *onubikshon-yantra* enables us to see these subtle *krimis*. The ancients had managed to become aware of these *krimis* that elude human vision by their yogic strength. We cannot say with certainty whether the *onubikshon-yantra* had been known to ancient Indians, but from . . . [their statements] we know that subtle *krimis* of the blood were not unknown to them.[54]

These discussions of *krimis* would immediately have resonated with generations of traditional Kobirajes, the backbone of whose education was Madhavakara's *Nidan*, which had devoted an entire chapter to *krimis*. The robust commentarial tradition attached to the *Nidan*, as well as its role as a structural exemplar whose chapterization was repeatedly copied by later collections, ensured that the discourse on worms had become part of the intellectual milieu of erudite physicians. The *Nidan's* division of *krimis* into "internal" and "external" can also easily be seen to resonate with the frameworks of modern physicians like Saradacharan.[55] The continuing importance of *krimis* within the main stream of erudite Ayurvedic learning is also attested by the elaboration of these themes in Gobindadas's eighteenth-century text, the *Bhaisajyaratnabali*. The latter, which was said to be the single most popular text studied by eighteenth- and early nineteenth-century Bengali physicians, while written according to the plan of the *Nidan*, further expanded the section on *krimis*.[56]

It was Saradacharan's formulation that seemed to catch on. Ayurvedists in the 1920s came increasingly to speak of microbes as *krimis*. Germs and worms thus came to overlap. One of the reasons that made this, rather than the other uses, catch on may have been the continued absence of any visual or practical support for the discourse on microbes. In the absence of actual microscopic experience or even illustrations, the early Ayurvedic discourse on microbes was purely a rhetorical affair. By linking it to the discussion of worms and parasites, it was possible to at least provide some experiential support to the discussion. Everyone knew of and had seen one type of worm or another. In most cases they also would have had some experience of internal and/or external parasites. These lived experiences could therefore support the abstract discussions around microbes that otherwise lacked any empirical or experiential support.[57]

It is interesting to contrast the Bengali situation with contemporary China. Sean Hsian-lin Lei has recently argued that practitioners of traditional Chinese medicine, faced with the diagnostic and political power of germ theory chose to avoid ontological debates altogether and focus instead on therapeutic merits. As biomedicine struggled in the inter-war years to repeat the dramatic success of Salvarsan in curing syphilis, Chinese physicians dwelled on the therapeutic successes of their medicines. As a result, they remained agnostic about the ontological reality of "germs" and developed an alternate approach that eventually came to be called "pattern differentiation and treatment determination" (*bianzheng lunzhi*).[58]

This stark contrast between the Chinese and the Bengali responses to germ theories seem to arise partly from the political profile of the microscope itself. Lei points out that the microscope became intimately tied up with Chinese sovereignty during the Manchurian plague of 1910–11. The politically charged climate of Manchuria and microscope's ability to deliver effective public health surveillance bolstered a tottering Qing monarchy and therefore acquired a significant political backing.[59] This political status also clearly identified the microscope with biomedicine and likely made it difficult for contemporary practitioners of Chinese medicine to vernacularize it.

By contrast, in Bengal as I have shown, whether Kobirajes had access to a real microscope or use for it, they were quite willing and able to notionally adopt it and the visions it produced. Such appropriations were admittedly often merely discursive, rather than grounded in laboratory work, but that did not detract from their conceptual value or intellectual creativity.

More interestingly, even in China, not everyone was agnostic to germs. Among those who did embrace them, the strategies of adaptation were not dissimilar to those seen in Bengal. Bridie Andrews mentions a few such adaptations. As in Bengal, here "germs" were often identified with "worms." Debate raged between whether these were the classical "wasting worms" (*laochong*) or a new type of "worms" that could only be described by newly imported Japanese terminology. Once again, the debate distinguished those using the microscope from those not using it.[60] By the 1920s like Lei, Andrews too describes a certain wariness among practitioners of Chinese medicine toward conflating "worms" and "germs."

Both the similarities and the contrasts are instructive. They demonstrate how the specific history of the microscope and political attitudes toward the instrument, at least partly, shaped the way "germ theories" were either adopted or eschewed by non-biomedical practitioners. This in turn underlines the need to revisit the histories of "germ theories" in light of the histories of the microscope. Guy Attewell's account of the Unani Tibbi physicians adopting germs also points toward a comparable role for the microscope. Attewell points out how turn-of-the-century *hakims* like Hakim Kabiruddin posited the compatibility of Unani humoralism and biomedical germs on the grounds that, from "Kabiruddin's perspective the invention of the microscope had allowed Western doctors to see deeper, "in what was previously called matter," but not to discover a fundamentally different reality."[61]

By the 1920s, as both the Chinese traditionalists and modernizing Unani *hakims* were eschewing the conflation of "germs" and "worms," the discourse on *krimis* in Bengal was in full bloom in the hands of Girindranath Mukherjee. Mukherjee was himself a qualified *daktar*, but he had been intimately connected to the world of Ayurveda. His two-volume *History of Ayurveda* and his lengthy essay on the surgical instruments of the Hindus are classics that remain popular, and more cogently in print, till this day.[62] He was close to Gananath Sen and even served as editor of Sen's English periodical, *Journal of Ayurveda*. It was in the *Journal of Ayurveda* that Mukherjee wrote a series of essays in the mid-1920s extending and establishing Ayurvedic discourse on *krimis*.

Mukherjee went beyond the classical Ayurvedic texts and commenced his interpretation from the *Atharva Veda*. He also had a keen interest in helminthology, the study of worms and particularly parasitic worms. In order to draw out equivalences between the insights of helminthology and his interpretation of ancient Sanskrit works, Mukherjee had recourse to etymology. He vigorously pursued etymological derivations from classical Sanskrit and aimed for the precise identification of a wide range of parasitic worms which, he argued, were mentioned in the *Atharva Veda* and other classical sources. Mukherjee went to great lengths, for instance, to prove that the

Proglottides of *Taenia Saginata* in various conditions of contraction (After Leuckart).

Head of *T. Solium* from the intestine of a rabbit, in different stages of motion (×25). After Leuckart.

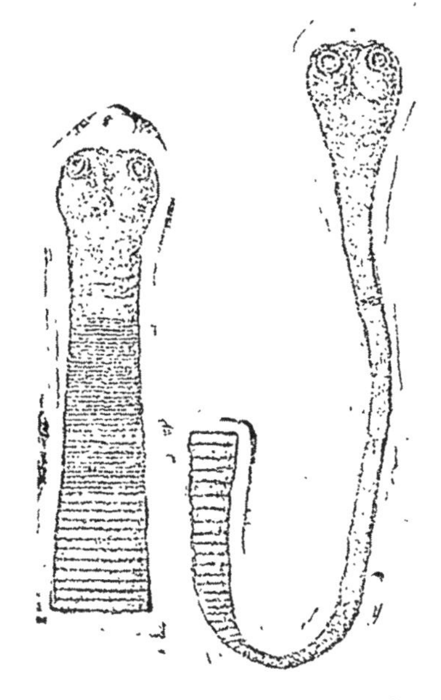

Head of *Taenia Saginata* in contracted (A) and extended condition (B). After Leuckart.

8. Girindranath Mukherjee's reproduction of Leuckart's images of the bladder worm.

tapeworm was mentioned in *Atharva Vedic* sources and that the word *solium* in the name *Taenia solium* was originally derived from the classical Sanskrit *solun*. He described a wide-ranging and evolving classificatory system of *krimis* discernible in various classical Sanskrit texts on Ayurveda that encompassed both the visible and the invisible parasites. In line with his efforts at achieving greater precision, Mukherjee included a large number of illustrations in his articles. In the majority of cases, these illustrations were of entities seen through a microscope. Thus, his illustrations of hookworms in situ in the human intestine or that of the bladder worm from the brain were all shown as they would appear through an *onubikshon-yantra*. In several cases, the illustration also stated the scale of magnification. The bladder worm, for instance, was depicted in "x 12" magnification, the head of the tapeworm from a rabbit's intestine was presented in "x 25" magnification, and so on (see fig. 8).[63] His commitment to precision and the use of illustrations put him in a different league from the early discussions of Kabyatirtha, Tarkaratna, or even Saradacharan. Yet, in one crucial way, nothing had changed between those early writers and Mukherjee. Neither of them had any direct, everyday use of the laboratory. The laboratory, with its *onubikshon-yantras* and its enhanced visuality, was an iconic and imaginary space that did not really form part of the everyday world Mukherjee inhabited, just as it had not formed any part of the everyday worlds of the earlier authors. Mukherjee's seemingly greater reliance on microscopic visibility turned out, on closer inspection, to be a red herring. The slides with due magnification that he reproduced were not things he had himself observed or collected. They were all derived from other published works. His bladder worm in "x 12" scale thus turns out to be derived from Rudolf Leuckart's *The Parasites of Man*, published in 1886, just as his "hookworms in situ" turns out to be derived from Charles A. Bentley's *Hookworm Disease*, published in 1919.

Not only are the illustrations not original, but their reproduction actually required many more mediations than are openly acknowledged. In 1886, Leuckart had relied mainly upon engravings for his illustrations. By 1919, Bentley depended upon photogravures. Both techniques relied upon the transfer of the image from the original to a copper plate that was then used for printing. While in the case of the former the transfer to the copper plate was done by hand via a careful process of controlled etching, in the latter case the image was transferred directly from the photograph to the copper plate by using certain acids. In contrast to both these methods, Mukherjee's illustrations appear to be lithographic prints that involve drawing the picture onto a special stone and using it for printing. The level of detail reproduced is significantly lower than even Leuckart's engravings, not to mention Bentley's later photogravures. Since Mukherjee usually indicated the source of the image by writing "After Bentley" or "After Leuckart" in parentheses, it is relatively easy to compare the original to the copy. Such comparisons leave little doubt about the significant loss of detail in the process.

Not only does microscopic visibility then become attenuated by the limitations of cheap reproduction technologies, but it also raises other important questions. First, the introduction of a number of intermediary stages and actors between the object observed by the *onubikshon-yantra* and the illustration seen by the readers forces us to revisit any assumption of objectification through verisimilitude. Second, we are forced to wonder what after all the use of such illustrations was, since there is very little chance that any of the readers would ever be required to detect these parasites under an *onubikshon-yantra*.

The issue of verisimilitude is clearly not a high priority since none of the readers are actually expected to make real-life decisions based on these visualizations. As a result, any argument about the objectifying power of realist illustrations should also be tempered. Yet the illustrations clearly have a role. Following Joseph Dumit's observations in the context of recent brain-imaging technology, I argue that the images of the *krimi*s must be interrogated from the perspective of how "cultural identification and intuition coincide with these representations of reality so that we are persuaded to take them as true."[64] Mukherjee's images, not to mention the very term *krimi*, a common enough word for any kind of "worm," reinforced common-sense perceptions about both the reality of such creatures and their often negative—that is, allergenic or parasitic—relationship to the human body. By an easy slippage from that, it generalized this common-sense understanding of the unhealthfulness of "worms" to posit a swarming world of diverse worm-like creatures within the cavernous insides of our bodies. The images therefore played an important persuasive part in communicating and establishing the refigured Ayurvedic version of germ theory but did not do so by establishing verifiable visual cues. Instead, they operated by functioning as suggestive hints that extended already available knowledge about "worms" into imaginary spaces within the body.

A Vaishnava Cell Theory

As the *onubikshon-yantra* focused increasingly on blood rather than urine, it was not just *krimi*s that began to emerge into a new Ayurvedic view. Blood cells too emerged. As Warner points out, German biologist Theodor Schwann's cell theory was one of the most significant consequences to follow from the enthusiasm for microscopical studies.[65] But cell theory was no more a single coherent theory in the middle of the nineteenth century than was germ theory. As Andrew Mendelsohn points out, nineteenth-century debates raged not only between those who accepted and those who rejected the theory, but even among those who accepted distinctive versions of it. Some suggested that the different versions were not simply evolutions out of Schwann's original theory but a repudiation of it.[66] Gradually, and relatively speaking rather belatedly, an Ayurvedic cell theory began to emerge approximately around the same time as the Ayurvedic germ theory made its appearance. So intimately were these two emergences connected that they began to overlap in curious ways.

I have already noted the debates over microbial life and its eventual conceptualization in the 1920s through a comparison to visible parasitic worms. Interestingly, at its origin, the Ayurvedic cell theory displayed very similar concerns. In fact, Hemchandra Sen, an eminent Ayurvedist who pioneered the Ayurvedic uptake of endocrinology, used the same word for the "cell" as Kabyatirtha would deploy only a few years later for microbes, *jibanu*. Sen argued that "every cell [*jibanu*] in our body is floating in the body's juices [*rasa*] like small islands. With the help of their powers of attraction [*akorshon-shokti*] they draw necessary substances from the body's juices towards their core and expel unnecessary material away from the core and towards the periphery."[67] As proof of this mechanism, Hemchandra urged readers to take a slice of a clean water plant and observe it under an *onubikshon-yantra*. He claimed that one would be able to see the cells full of bluish chlorophyll and by turns attracting and repelling necessary and unnecessary substances by their operation of "centripetal" and "centrifugal" forces. Going further, Hemchandra sought to map these cellular forces onto Ayurvedic descriptions of the attractions between the different types of *vayu* within the body. Dwelling particularly upon the mutual attractions of *pran-vayu* and *opan-vayu*, he argued that these two *vayus* pulled each other using a "latent force" that was akin to the electromagnetic forces operating in a telegraph wire. Playing on the words *sushumna* (the name of the principal *nadi* in Tantric lexicon) and *sushupta* (hidden/latent), he argued that this "latent force" was indeed the mysterious *sushumna-nadi*.[68]

Hemchandra was not alone in subscribing to this atomization of life at the level of cells. In 1915, the newly formed Ayurveda Sabha that sought to consolidate and centralize attempts to revitalize Kobiraji medicine held its first general meeting. The meeting was presided over by the redoubtable Shyamadas Bachaspati, a renowned Kobiraj and heir to one of the most distinguished familial legacies of medical practice. At the meeting, Kobiraj Haramohan Majumdar, a staunch traditionalist usually opposed to rampant borrowing from the "West," delivered a lecture on the nature of bodily *vayu*. Majumdar argued that "just as this [*vayu*] brings blood and semen together, it also divides them [semen and blood] into diverse forms and structurally builds them up by residing in them in the nature of a force." Following on, Majumdar continued that, "upon discussing their actions, we cannot but think of them as some form of intelligent life. Is it possible that what Western science has called a 'cell' is actually an atom of tissue mixed with *vayu* [*vayu mishrito dhatob onu*]"?[69]

Here the lines between a microbe and a cell are clearly blurred. Is Majumdar speaking of a microbe or a cell? The confusion was not his alone. As tiny microscopic entities endowed with "life," both cells and microbes seemed to share much. Moreover, both of them "inhabited" the "insides" of our bodies. The distinction was further blurred because many authors on germs, such as Tarkaratna, also drew attention to the existence of helpful germs permanently resident in the body. Putting all these elements together, it is clear to see why and how cells and germs could be confounded.

As a staunchly orthodox Kobiraj, Majumdar's acquaintance with "Western" theory was still shaky and he repeatedly misspelled "cell" as "cen." He had most likely read of "cells" in translation since in Bengali the letters "1" and "n" are easily confused. Yet it is significant that, despite his attitude and his limited familiarity, he chose to develop this line of argument at such an important speech, attended as it was by many leading Kobirajes of the day. It suggests that by 1915 such theories were becoming fairly attractive even among the more reticent and conservative circles of Kobiraji medicine.

Fascinatingly, almost three decades later when Tarashankar Bandyopadhyay wrote his novel, *Arogya Niketan,* in close consultation with his friend, Bimalananda Tarkatirtha, an eminent Kobiraj and the son of Shyamadas Bachaspati,[70] one still hears an echo of this early version of the Ayurvedic cell theory. In a poignantly eloquent passage, Tarashankar wrote of the aging Kobiraj, Jibon Moshai, meditating upon the *nadi* of an old woman afflicted by many ills. His reverie proceeded thus:

> Diseases have riddled the body. Anxiety plagues the mind. Every pulsation of the *nadi* says so. In every cell of the body, like the millions of stars in the sky, the tiny flames of life that continually worship the god of life (*pran-debota*), who anoint *pran* [life, but also possibly *pran-vayu*] with sweet warmth in order to keep him awake in the body, those flames are burning low. Some have even gone out.[71]

While the particular formulations, especially through its mapping onto ideas about *pran-vayu,* are quintessentially Ayurvedic, the roots of the cellular level vitalism were possibly not Indian. Hemchandra's discussion had arrived at the formulation through his discussion of the nervous system. Katja Guenther points out that with "the work by microscopists Theodor Schwann, Matthias Schleiden and Rudolf Virchow in the 1830s and 40s, the cell had become widely accepted as the structural unit of the organism. But the question of whether nervous tissue was made of independent cells remained contentious until after 1900." She further points out that while "reticularists" emphasized the interconnected unity of the nervous system as a whole, "neuronists" thought that neurons were independent functional units.[72] The nature of Hemchandra's discussion and the metaphors he used made it clear that he was borrowing from elements of this European debate, but unfortunately, in the absence of clear citations, it is difficult to trace exactly which authors he was drawing on.

Another, and possibly even more likely, European route to cellular vitalism lay through chemistry rather than neurology. Robert Brain has recently pointed out that while energy and evolution are generally seen to be the two major themes of nineteenth-century science, by the last couple of decades of the century ideas about protoplasm were equally expansive and popular. He calls this popularity "protoplasmania." A central aspect of this protoplasmania, as Brain describes it, was the so-called ensouling of cells by scientists such as the German naturalist, Ernest Haeckel. By seeing "cells" to have individual souls, the path was opened up for the development of an entirely new branch of inquiry known as "cell psychology."[73] Both the above-mentioned entanglements of chemistry and microscopy in the Indian medical milieu as well as Hemchandra Sen's discussion of the "bodily juices" (*rasa*) seem to draw upon ideas of "protoplasm."

Whatever its roots, this cellular vitalism raised a crucial problem for Ayurvedists. It undermined the overall unity of the self. As Dominik Wujastyk points out, Indic traditions have long conceptualized selfhood in ways that do not forestage embodiment.[74] Anthony Cerulli similarly states that "Ayurvedic literature portrays the human body as entirely detachable life-trajectory of the person to whom it belongs."[75] The self stands independent of the body. Within such an orientation, the challenge of a cellular vitalism that distributed "life" to individual cells was doubly problematic.

Surendranath Goswami sought to solve this problem. Goswami was one of the most creative and original thinkers among the Ayurvedists of his era. Many of his peers acknowledged him as the very first biomedically trained physician to return to the Ayurvedic fold. He hailed from one of the most respected Baidya Vaishnava families that had been part of the inner circle of the medieval saint, Chaitanya, and had consequently come to enjoy almost quasi-Brahmin status. Goswami's creative solution to the disaggregation of selfhood, following from the dual assault of the cell theory and the germ theory (which too had shown the body to be the possible abode of more than one living entity), drew heavily upon his own Vaishnava intellectual inheritance. Goswami argued that "either the ancestral spirits [*pitrigon*] or the attendants of the planets [*groher onuchar*] by being attached to various grains descend onto our blood cells [*roktokona*]. And there, residing as protectors [*rokshok*] they frustrate the attacks of the assassins [*atotayir akromon*]."[76] Selfhood was now no longer a matter of an individual person, but rather a community of cells that somehow carried the substance of the ancestors in their material being. Unity, at the cellular level, was reimagined in the image of a community working for a common purpose.

Emily Martin has drawn attention to the "ideological work" done by metaphors of attack and defense through which modern notions of immunity are articulated.[77] The ideological work in this case is even more complex and polyvalent. Goswami's communitarian outlook was clearly reminiscent of Vaishnava ideas about the unity of the devotional community. But it also went further. In a fascinating recent article, Tony Stewart has argued that Vaishnava communities continually sought to reproduce the mythic time of their god's worldly sojourn by replicating the architechtonics of the original *mandala* within every historically existing devotional community.[78] In a sense, a more materialist attempt to achieve exactly the same end at the level of cells and microbes can be discerned. Goswami's arguments posited that, at the microscopic level, our very existence as a unified person was always already constituted through the reconstitution of the material being of our ancestors. Since both the self and the anti-self were present within a single body, what constituted the self was the moral and material unity that tied not only the cells together, but tied them together with the reality of our ancestors. Hence, just as Vaishnava devotees see themselves as connected to, or indeed as avatars of, specific individual compatriots and companions of Krishna, so too did Goswami imagine individuals as being constituted through the cellular level re-assembly of the moral and material substance of those ancestors.

So successful was this Vaishnava interpretation that when late in the 1920s Ashutosh Roy sought to define "Cellular Philosophy of the Ancient Hindus," he baldly stated that though in Europe contemporary cell theory was often discussed through the metaphor of the solar system, among the Hindus it was "typified by the allegorical picture of 'Rasalila.'"[79] *Rasalila* is a quintessentially and exclusively Vaishnava term recalling Krishna's playful dalliances on earth in human form. Roy gave no further clarification about this image or why suddenly a festival such as *rasalila*, which is clearly stamped by its sectarian affiliations to Vaishnavism, should be taken as the preeminent metaphor. Roy did, however, add a strangely moralizing gloss on the metaphor that stands out amidst his otherwise scrupulously reductionist and scientific prose in his influential book. He wrote that

> the Human particles are constantly oscillating between sin and righteousness, between vice and virtue. While the Almighty is attracting them towards Him, the beast in man is constantly trying to prevent this holy union by arousing material desires in our breasts. Once you can completely control these beastly desires, the primitive human instinct, by making it "sublime," you are on the road to divinity.[80]

This strange description of cell theory only makes sense in light of the "cell psychology" developed along the blurred boundaries between microbes and cells that had colored Ayurveda through Hemchandra's intervention well before Roy wrote his book. In fact, Roy's book is ill-equipped to address the residual Vaishnavism injected into modern Ayurveda through Goswami, and hence somewhat incongruously inserts this moralizing image and the metaphor of the *rasalila* without much context or explanation. But his having done so proves not only the Vaishnava legacy, but also its continuing use, however ham-handed, to paper over the complex problem of a disaggregated selfhood that arose through cellular vitalism.

Roy's book, however, did not stop at this. Having inserted these incongruous elements into his discussion, he moved quickly—almost too quickly—onto what he prefaced as the "physical aspects of the cellular theory." Under this ruse and keeping more in line with his general contributions, Roy sought to reaggregate the dispersed selfhood by asserting the unity of the physical bodily mechanisms. He alleged that all cells were in effect supplied with three common structures: a "tiny nerve artery," a "tiny vein," and a "lymphatic." Through a seemingly forced set of correspondences, he argued that the nerve was "the tissue [that] possesses regulative force in the cell (Vayu)," and the artery likewise "supplie[d] food material and blood" and was controlled by an "inherent force" that was *pitta.* The vein and the lymphatic "carry away waste material" and was controlled by the force, *kapha.* [81] Hence, by redefining *vayu, pitta,* and *kapha* as "forces" aligned to the nervous, arterial, and lymphatic systems, the autonomy of the cells that had proved so problematic was undermined. The body's unitary character was reasserted by the further development of the *Snayubik* Man physiogram wherein the body's unity was premised upon a network of flows through which the three *doshe*s flowed.

The Chiaroscuric Man

Goswami's legacy was not confined to the resolution of the problem of unitary selfhood. Arguably, a more important legacy of Goswami was a historico-rational cosmology and its congruent physiogram.

In his short book titled *Ayurveda o Malaria-jwor,* Goswami attempted to overturn established Kobiraji understandings of etiology, fever, and physiology. The *bhutobidya* (knowledge of ghosts) in Ayurveda, he declared, was identical to bacteriology. The "treatment of afflictions resulting from *bhuts,*" he held, "was the resolution of Antiseptic treatment and *suchi* [Indic notions of ritual purity]."[82] In the face of generations of Kobiraji belief, he argued that the preeminent medical textbook used by erudite Kobirajes for almost the past thousand years, Madhavakara's *Nidan*, had misled Kobirajes about the true teachings of the ancient seers. Instead, drawing upon scant Vedic references and working them through imaginative interpretations of Puranic myths, he proposed that there was only one type of fever recognized in Ayurveda that was then divided into eight subtypes.[83] Denying the long-established primacy of a classificatory system for fevers based on the three para-humors, he argued that actually a radically different etiology of fevers had been proposed long ago in the classical age by a small group of radical physicians who opposed the priestly domination of medicine. Arguing counterintuitively, Goswami opined that these anti-priestcraft radicals had been the ones to develop the ideas about "ghosts" causing fevers.

Where generations of Kobirajes had erred, Goswami claimed, was in the belief that the category of "ghosts" or *bhuts* stood for those creatures of popular imagination that were "tall as palm-trees, with huge arms, deep-set eyes and a nasal voice."[84] In reality, *bhuts* were allegedly "vegetable and animal parasites (*udbhidanu* and *jibanu*)." Different types of *Bishom-jwor* (the one principal febrile category according to Goswami), such as *Sototo-jwor, Sontoto-jwor,* etc., were causally distinctive and triggered by different types of "parasites."[85] *Bhuts* and the older term *Noirjhoteyo*—designating a variety of spectral or immaterial beings recognized in different periods of history—were all collapsed into a lexicon that could articulate a sufficiently pliable theory of "germs." Jworasur, the fever demon himself, with his three heads, six arms, etc., was said to be nothing more than the description of a particular type of stained microscopic plate—namely, the "Colourdel plate."[86]

There are two important formulations here. First is the historical narrative that is staged as an epic battle between a superstitious priestly class and a radical, empirical group of physicians. This narrative is premised upon a stable and anachronistic dichotomy between the categories of "empiricism"/"rationalism" and "superstition." Second is the physiogram produced through this historico-rational argument.

At the heart of Goswami's argument is the issue of empiricism and rationalism. Goswami connected the former to the latter and opposed it to blind faith and superstition. He attributed the former positive traits to the group of "radical physicians" he claimed to have identified, while the latter negative traits were assigned to the "priestly class." Describing the conflict between empiricism and superstition, Goswami wrote that "Vedic seers feared only two things. First, empirical facts [*protyoksho*]. Second, the words of Lord Krishna. Having torn the Vedas asunder, Lord Krishna took on his earthly avatar to establish the truth. . . . Lord Krishna was an empiricist [*protyokshobadi*]—he adopted 'vision' [*darshan*] as his *chakra* [a disc-shaped throwing weapon] and thus triumphed over the unseen [*oprotyoksho*], imaginative [*kalponik*] Vedism [*Boidikbad*]." Continuing further along the same vein, Goswami stated, "He is the Lord of the Empirical/Visible [*protyoksher debota*]—He worships the visible [*protyoksher pujok*]. Upon the visible [*protyoksho*], He has established His religion [*dharma*], actions [karma], etc."[87]

This is an unexpected definition of empiricism. It plays on the multiple meanings of the word *protyoksho* in Bengali. One of the meanings of the word is "to see." Another is empiricism. A version of the first—that is, "to see"—is related specifically to Vaishnava bhakti. In this specific lexicon, "to see" becomes related to the visual manifestation of the divine. As dualists (opposed to the non-dualist divinity of Advaita), Vaishnavas, when they speak of worshipping the *protyoksho*, mean worshipping the guru or the idol of the divine. It is in this specific sense that Goswami speaks of Krishna as the Lord of the Visible. Vaishnava bhakti, in its simplest form, eschews complicated theological disputation and privileges personal devotion to the image of Krishna, either in the form of the guru or an icon. Goswami's empiricism then is a very specific type of empiricism. It is a type of seeing that is entangled with the moral universe of Vaishnavism and where seeing becomes a way of accessing the transcendental divinity within the corporeal bounds of history.

But Goswami, while drawing on these patently Vaishnava intellectual resources, also sought to curb them. By gliding subtly from the Vaishnava *protyoksho* to the secular rationality of the "Colourdel plate," he nipped the transcendental possibilities of his formulation. Instead, he recreated Vaishnava *protyoksho* in the image of contemporary ideologies of illumination. The mythic time of Krishna's avatar-hood, far from remaining a luminous space beyond history, became an episode in the ancient history of India. Shorn of the magicality and enchantment of the divine, it was reduced to being a mere clash between professional lobbies. Krishna became the leader of a radical faction rather than a god.

It was within this disenchanted cosmology that *bhuts* went from being malignant supernatural agencies to lowly microbes. Gods and demons alike were cut to size. Gods were recast as inspired forbears and happily exiled to the remote past, while demons were imprisoned on the slides of microscopes. Drained of all magic, enchantment, and divinity, the world finally stood ready to be girded by regular scientific laws.

The physiogram of the chiaroscuric man was located within this historico-rational cosmology. By speaking of the chiaroscuric man, I wish to designate a physiogram marked by its interplay of light and dark, visibility and opacity. Such a physiogram was a long time in the making. Elements of this bodily chiaroscuro were assembled well before Goswami produced the cosmology that would allow its fulfilment, just as elements were finetuned in the chiaroscuric man long after Goswami had taken his leave from this world.

One of the earliest chiaroscuric elements one comes across in refigured Ayurveda is in the writings of Binodbihari Ray. While developing his electromagnetic interpretations of the *dosh*es, Ray stated that, above and beyond the different types of *pittas* mentioned in the Ayurvedic literature, *pitta* could also be divided into two further categories: namely, *oindrik* (sensory) and *otindriyo* (extrasensory). The former, *oindrik*, as the name suggested, stood for empirically perceptible types of *pitta*. While the latter, *otindriyo*, were those forms of *pitta* that could not be perceived by the senses.[88] Hence of the five types of *pitta* mentioned in most classical Ayurvedic texts, *bhrajok, alochok,* and *sadhok pitta*, by nature "invisible," were classified as *otindriyo*; whereas the remaining two, *pachok* and *ronjok pitta*, were classified as *oindrik*. This was a minor move on Ray's part, but a telling one. The ability to perceive by one's senses had become a basis for categorizing and redistributing the elements that the classical works had assigned to the superhuman powers of seers.

Ray was not denying the existence of what he could not see. Committed to upholding the authority of classical Ayurveda, neither Ray nor any of the other Ayurvedists who followed him ever refuted the insights of the ancient seers. Yet at the same time they were also committed to the new authority of empirical demonstration. The chiaroscuric dynamic between the visible and the invisible developed at the crucible of these two contesting paradigms of authoritative knowledge. In certain ways, Ray's move was similar to the ways in which modern practitioners of Chinese medicine have sought to distinguish between the intangible "phlegm" of the classical canon and the tangible forms in which such "phlegm" becomes visible to modern scientific technologies.[89]

The compromise that resulted from this clash of authorities was to recode the invisible as the not-yet-seen. The chiaroscuric man was therefore a latticed body intricately carved with areas that were seen and those that were not-yet-seen. For those parts of the body that were not-yet-seen, the possibility of seeing was confirmed but then repeatedly deferred. Readers and audience were told that what was not-yet-seen could indeed one day be seen; only that the state of knowledge or technology was not yet at that stage.

This deferred visibility had many consequences. It meant that no agency that could willingly remain beyond our senses could be admitted. Hence, gods, demons, ghosts, and the like were recast into forms where they could potentially be seen in the future. What was vigorously denied was their power to permanently elude our grasp. Similarly, another consequence of this deferred visibility was the constant rearticulation and upgrading of the promise of visibility. Thus, as knowledge and technology progressed, new grounds were found for justifying the faith that there were elements in the body that, though invisible, existed and bore out the truth of the ancient seers. In a sense, there was always a sense of "catching up" in these upgraded promises of vision. One was always trying to catch up with the fuller, richer vision of the ancient seers.

In the decades following Ray, there were repeated attempts to assert and upgrade this promise of future vision. Initially, the electromagnetic theories of Ray began to be gradually eclipsed by the protoplasmic theories of physicians like Hemchandra Sen. But the dialectics of light and shade, visibility and invisibility, that had animated Ray's theories continued to surface in the protoplasmic ones too. Right next to the blue-colored, chlorophyll-like substance in the cells, Sen also spoke of "vibrations" traveling along universal "rivers of ether" that were much "subtler than the ends of hairs" and therefore naturally invisible.[90] Such wave- and vibration-based theories of protoplasm, as Brain shows, were common in late nineteenth-century Europe as well.[91] What is noteworthy in the context of modern Ayurveda is how these familiar European theories were selectively braided with Indic notions to produce a patchwork of the visible and invisible. Thus, in one instance, an anonymous author—who may have been Sen—recounted at length the theory that all life was essentially constituted of "lively Albuminous matter," which was variously called "sarcode" in lower animate forms, "protoplasm" in plants, and "blastema" in the higher animate forms. But this, the author hastened to add, could not account for the reasons why the cellular forms aggregated and worked with a common purpose. An invisible single cause, he posited, must be admitted.[92] The visible protoplasm had to have an invisible complement.

It was through these earlier and not altogether satisfactory attempts to resolve the dialectics of visibility that Goswami arrived at his formulations of cellular identity. Moreover, Goswami, who was a prolific author, also kept tweaking and updating his theory of cellular identity over the years, but what remained unchanged was the chiaroscuro that always played off the visible and invisible. When he first developed his theories regarding cellular identity, he drew upon the "phagocyte theory" of Edward Berdoe. He depicted the body as a battlefield where cells worked in tandem against the demonic foes or germs. The whole tenor of the serialized essay sought to show that existing science had made visible, and thus confirmed, exactly what the ancients had described. Implicit was the promise that further scientific progress would provide further insights into the battle between the *bhuts* and the *pitri* (ancestor)-derived blood cells.[93] Goswami rarely referred to actual microscopic work and published no slides or illustrations. Yet the cellular body he described was an intricate interplay of vision and opacity, all spun around a series of interconnected hypotheses. Six years later, in the same periodical, *Janmabhumi*, Goswami adopted a new formulation. Instead of Berdoe, he now cited Herbert Spencer and a whole slew of electromagnetic theorists. Instead of the cellular protoplasm variously organized as "germs" or "cells," he now spoke of mysterious hidden forces that were akin to electricity but yet unknown to science. In the same essay, he also spoke of the reclusive "etheron," a hypothetical particle that was the fundamental unit of "ether" and remained "beyond motionless vision" (*gotihin protyoksher ogochar*).[94]

Although they are often difficult to tell apart, Goswami's later formulation might in retrospect be seen as a move away from the more biochemical theories of protoplasm toward older physical theories about electromagnetic forces. Yet the chiaroscuro survived the shift. In fact, the chiaroscuro even survived Goswami's death and the eclipse of his ideas by yet another set of new biochemical formulations organized around endocrinology. It was the endocrinal enthusiasts who fulfilled the potential of the chiaroscuric physiogram. In 1916, on the occasion of the founding of the Benares Hindu University, Gananath Sen delivered an address in which he revived the old terminology of *oindrik* and *otindriyo*. He strongly objected to the tendency of European authors to translate *dosh*es as "humors." "Humors," Sen argued, were only the sensible or *oindrik* products of the *dosh*es which were themselves *otindriyo* processes.[95] Later, this was taken up and more fully developed by Ashutosh Roy.

In Roy's book, he laid out a detailed map of the chiaroscuric man. He fleshed out Gananath's formulation by stating that "at the outset we must distinguish between two entirely different kinds of Vayu, Pitta and Kapha, as described in Ayurveda."[96] The first kind was the *oindrik* type. Quoting Gananath, Roy said that this kind was "a crude visible form, the products (secretion and excretion) of those processes." To equate these "crude forms" with *dosh*es, as had been done (allegedly and wrongly, only) by European commentators, "reflect[ed] to the discredit of the interpretor (*sic*) as shallow and superficial." By contrast, the *otindriyo* type was "a fine, invisible and or essential form which mainly guides the physiological processes."[97]

Gananath's formulation might well have solved the need to perpetually upgrade and defer visibility. He was arguing that the crucial components of the chiaroscuric man's body were simply ideal types. They could never be fully visualized. But ideal types were not demonstrable. And perhaps it was this that impelled Roy to reopen the promise of deferred visibility through neurology and endocrinology. Roy concluded refiguring Gananath's ideal types into meta-material terms. *Otindriyo vayu* thus became the "nerve-force of the cerebro-spinal and sympathetic system of nerves."[98] This force was not something that was concretely known to contemporary neurology, but it was sufficiently like the ontological categories that were then in use in contemporary neurology to be within the realms of future possibility. Roy similarly redefined several other components of the chiaroscuric man that were not-yet-seen, but potentially visible in the future.

What was remarkable was not simply Roy's having reopened the issue when Gananath seemed to have settled it with his idealization of the extra-sensory dimensions of the chiaroscuric man, but rather that Roy seemed to do it with the support and encouragement of Gananath himself. In an editorial in the fourth volume of the *Journal of Ayurveda*, both Gananath's contributions to it and Roy's efforts were openly lauded.[99] Clearly then, the idealization of the categories had not been satisfactory. The promise that they would soon be demonstrable was an urgent need, even for Gananath Sen himself.

Roy's reopening of the matter therefore emphasized that the commitment to future demonstrability triumphed over any permanent idealization. It was also proved that the not-yet-visible could never again escape from history into the realm of the never-visible. Modern Ayurveda's visual regime thus engendered a historico-rational teleology that promised to one day make the chiaroscuric man fully transparent in the way he had been to the ancient seers equipped with *dibyodrishti*.

FIVE

Endocrino-Chakric Machine: Hormonized Humors and Organotherapy

Sometime around the early 1920s, Dr. Ardhendu Sekhar Bose was confronted with a curious problem. Having completed his medical education, he had recently returned from England and was struggling to make a mark in the medical profession. In an unexpected response to his many advertisements, he was introduced to a rich patient. The gentleman was both kind and generous, aside from being exceedingly wealthy. He was just the sort of patient whose social connections and financial backing could set Dr. Bose up for good in his profession. The gentleman's problem was that he had remarried late in life in the hopes of producing an heir to his fortune, but realized the physical impossibility of doing so. He consulted Dr. Bose in the hope that the physician might be able to rejuvenate his generative powers. Dr. Bose assured the gentleman that he could do this. What would be necessary was the "gland" of an "anthropoid ape" and though these were rare, they could be procured through dealers in Germany. The only problem in all this was the gentleman's lecherous, good-for-nothing brother-in-law. The latter, a lean man with a mean visage, was not only a constant source of embarrassment to Dr. Bose's patient, but repeatedly insulted Dr. Bose himself and even went so far as to make unwelcome passes at the young doctor's beloved wife.

The inventive Dr. Bose solved both his problems in a single masterstroke. Under the guise of treating the brother-in-law for a bacterial infection, the good doctor removed the necessary "gland" from the brother-in-law and implanted it into the aged gentleman, while keeping up the fable that the gland actually came from an anthropoid ape via German dealers. At the end of this hilarious comic story by Amulyakumar Dasgupta (better-known as "Sambuddha"), the characters appear one last time two years after the operations. The aged gentleman is now the proud father of a bright young boy, while the rambunctious letch has been transformed into an excessively corpulent, calm, soft-spoken man whose devotion to the doctor went so far as to make his wife jealous.[1]

In Sambuddha's story, we are told that the all-important "gland" that led to these marvelous transformations was the "thyroid." Yet the effects of its removal described in the lecherous brother-in-law are strongly reminiscent of castration. Castrated animals are usually thought to put on weight, lose their sexual appetites, and become calm and tame. Neither the conflation of the thyroid with the testes—both of which were beginning to be recognized as "ductless glands" forming part of the endocrine system—nor the therapeutic practice of inserting glands and tissues from "anthropoid apes" into humans was entirely fictitious in the 1920s. In fact, Sambuddha was not alone in finding literary inspiration from such therapies. A darker narrative using the similar therapeutic techniques can be seen in Arthur Conan Doyle's slightly earlier narrative of *The Creeping Man.* Highlighting the ubiquity of interest in the ductless glands as well as the exaggerated hopes they generated, Chandak Sengoopta writes:

> The ductless glands and their still mysterious secretions came to acquire an air of omnipotence in the 1920s. "We know definitely now," announced a popular medical work of the 1920s, "that the abnormal functioning of these ductless glands may change a saint into a satyr; a beauty into a hag; a giant into a pitiful travesty of a human being; a hero into a coward, and an optimist into a misanthrope."[2]

Usually termed "organotherapy," this therapeutic technique had its origins in the late 1880s. Pioneered by a brilliant but eccentric physician and physiologist Charles-Edouard Brown-Sequard, its original motivation—as depicted in the literary pieces—was the rejuvenation of the aged. Nelly Oudshoorn points out, however, that Brow-Sequard's work built on earlier "prescientific" ideas widely available among farmers.[3] Brown-Sequard's technique comprised principally of the injection of "testicular emulsions" made from the testicles of various animals such as dogs, guinea pigs, bulls, etc., directly into the blood stream. Somewhat inadvertently, organotherapy—namely, internal glandular secretions secreted directly into the blood—led to the discovery of hormones and, in time, the endocrinal system.[4] Eventually, the thyroid gland was identified as an important endocrinal gland. By conflating the thyroid and the testes, Sambuddha gave us a telescoped history of organotherapy.

Humors and Hormones

The *Journal of Ayurveda*, a journal closely associated with Gananath Sen, carried an editorial titled "Humours vs. Hormones in Ayurveda" in February 1928. The editorial asked whether *dosh*es were most accurately translated as "humors" or "hormones." The pithy question received an unambiguous answer. "The pioneer annotators of the last century, mostly Europeans," the editorial clarified, "literally translated 'Doshas' as Humours and were naturally puzzled by Charak's physiology." By contrast, it continued, authors "of the present century, mostly Indians, by a better and more intimate study of the various attributes of 'Doshas' came to the conclusion that they are closely analogous to Hormones."[5]

The editors were somewhat disingenuous in suggesting that it was only European authors who had translated *dosh*es as "humors." Eminent Indian authors such as Bhagvat Singhjee, Udoy Chand Dutt, and Binodlal Sen, among others, had accepted and used the translation of *dosh*es as "humors." More importantly, as I have shown in the foregoing chapters, there had been repeated attempts to interpret *dosh*es in light of the electromagnetic and protoplasmic biologies of the nineteenth century. The hormonal interpretation emerged on the back of a long history of translations that scrambled such neat divides between "European misunderstandings" and "Indian accuracy."

More usefully, the editorial acknowledged a clear origin for the hormonal interpretation. It pointed to Hemchandra Sen, MD, writing at the turn of the twentieth century as the first person to have denied the equivalence between "humors" and *dosh*es by equating them instead with "hormones." Hemchandra is known to have authored at least a couple of pamphlets and numerous articles in the medical periodical press of the day. In fact, his books, which unfortunately I have not been able to access, seem to be mostly a collection of his published articles.[6] Not only did these articles clarify the way in which he developed his interpretation of *dosh*es as hormones, but they also plainly described the central role of organotherapy as a therapeutic technology in instigating this reinterpretation of the *dosh*ic physiology.

Hemchandra Sen seemed to originally have been attracted to organotherapy in his quest for therapeutic techniques that could bolster immunity. Elsewhere I have argued that, following the plague scares of the 1890s, Bengali *daktar*s began to emphasize the body's internal "disease-preventive powers." I have also shown that the conceptualization of such internal powers of resistance was strongly resonant with the emergent nationalist ideas about *swadeshi atmashokti* (national soul-force). Though a version of the germ theory of disease was accepted at some rudimentary level by these Bengali *daktars*, they insisted that the crucial factor between health and illness was not the presence of germs but the body's power to fight such germs. Metaphorically, the need to fight foreign invasions by building up inner strength naturally resonated with an emerging anticolonial public and its anxieties of weakness.[7]

Hemchandra drew directly upon the work of Brown-Sequard (whom he inexplicably referred to as "Brown-Shepherd") and the latter's loyal assistant, Jacques-Arsene d'Arsonval. Moreover, while he gave pride of place to the testicles in his scheme, he acknowledged the existences of a complex endocrinal system of internal secretions. He wrote that

> having read of the astounding success that Dr. Brown-Shepherd [*sic.*] has been met with in his attempts at removing weakness by injecting testicular emulsions has convinced this author that the male sperm is the principal support for the disease-preventive power. Like the thyroid gland, pancreas, etc. the male testicle too has a normal internal secretion."[8]

These injections of "testicular emulsion" were mainly of interest to Hemchandra as a simple therapeutic technology rather than as a cue for rethinking the nature of *doshes*. Pointing to the numerous diseases such injections promised to cure, he wrote that "the famous French physician, Arsonval, has produced a testicular emulsion that gives excellent results in cases of *bohumutro* ("diabetes"), neurasthenia, neuralgia, *kompon* (shivering) and ataxy."[9]

Despite Hemchandra's pioneering efforts, the hormonal idiom was slow to catch on in Ayurvedic circles. It was, in fact, almost a decade after Hemchandra that hormones and *doshes* finally came to be related to each other consistently.[10] In the meanwhile, electromagnetic interpretations continued to dominate Hemchandra's largely protoplasmic theories. The person who was most influential in reviving Hemchandra's theories was Ashutosh Roy. Like Hemchandra, Ashutosh also had a *daktari* degree—though the relatively lower, Licentiate of Medicine and Surgery (LMS) rather than Hemchandra's MD. Initially, Roy made his case through a number of journal articles. A good many of these articles appeared in the *Journal of Ayurveda*. Eventually, the latter journal published his book, *Pulse in Ayurveda* in 1929. In this work, he laid out his theory in detail. It was through Roy that other Ayurvedists picked up the hormonal theory of *doshes*. Among those who were inspired by Roy to adopt this position were the two giants of modern Ayurveda, Gananath Sen and Shiv Sharma.

Even as Ashutosh Roy's writings proved most effective in establishing the hormonal theory of *doshes*, organotherapy significantly was no longer the key topic of discussion in his writings. In fact, the entire discussion of the *doshes* in Roy had taken on a reified, philosophical tone. Explaining the alignments of the individual *doshes* with "Sympathetic Endocrinology," he described *vayu* as the "nerve force manifested in action through the Nervous tissue"; *pitta* became "the force of Katabolism manifested through the Katabolic Endocrines"; and *kapha* became "the force of Annabolism manifested through the Annabolic Endocrines."[11] Reviving Hemchandra's protoplasmic ideas, Roy further laid out a "Cellular Philosophy of the Ancient Hindus." According to this, the ancient Hindus had conceived of the "cell of the organic body as a miniature solar system." The "great macrocosmic phenomena" of the solar system and the "great microscopic phenomenon" were allegedly perfectly homologized. The cell was in reality a "mass of protoplasm surrounded by an envelope."[12] The "physical aspects of this cellular theory," he held, consisted of visualizing the cell as being served by "a tiny nerve artery, a tiny vein and lymphatic." In these, "the Nerve is the tissue which possesses the regulative force in the cell (*vayu*)"; while the artery which "supplies food material and blood" is controlled by an "inherent force" called *pitta*. The *pitta* is said to "destroy and reconstruct." Finally, "the vein and the lymphatic carry away waste material" under the control of another "force corresponding to *kapha*."[13]

Roy's schema proved immensely influential and was adopted by mutually opposed intellectual camps of Ayurvedists. Shiv Sharma, whom Charles Leslie rightly depicted as a trenchant and politically savvy proponent of *Suddha Ayurveda* or "Ayurveda unsullied by western imports," explicitly cited Roy's work and adopted his homologies. Sharma wrote that "this nerve-force is not the sole connotation of *vayu* as some people understand. It is a differentiated and more crystalized form, as it were, of *vayu*. But the original meaning of *vayu* is not nerve-force, but cell-force."[14] He concluded by saying that "it is this cell-force, or rather, correlated cell-force or Vital Force [which is] *vayu*. . . . The true conception of *vayu* is not incompatible with any of the teachings of modern physiology."[15] Sharma then went on to devote an entire section to sympathetic endocrinology and rounded it off by stating that it was Roy's "erudite article on the subject" which "tempted" him to include the discussion of endocrinology.[16] "How can such interpretations be considered 'fanciful,' the 'desperate attempts' of 'Pro-Ayurvedists' to show 'the teachings of Ayurveda square with teachings of Western medicine,'" Sharma wondered, since the analogy is "unmistakably evident by comparative study."[17] Jean Langford, with some surprise, points out that Gananath Sen, who championed *Misra Ayurveda* and is therefore in the opposite camp from Sharma, held almost the same opinion on this particular matter.[18]

Hormonizing Rasayana

The *Journal of Ayurveda*, as I have noted above, was the most prominent and powerful platform for the popularization of hormonal ideas among those refiguring Ayurveda. Notably, a substantial part of the *Journal*'s revenue seemed to have come from "hormones." By the 1920s, hormonal products had become a major commodity and a number of national and international brands advertised their hormonal products regularly in the *Journal*.[19] For instance, Martin & Harris of Waterloo St., Calcutta, and Parsi Bazar, Bombay, advertised "Colwell's Pure Liquid Hormones." The product was described as "a palatable liquid preparation of *pure* hormones in natural, pluri-glandular combination—free from all proteins and toxins"; it was moreover "always pure, sweet and effective." The Colwell Pharmaceutical Corporation of New York, USA, manufactured the product.[20] Another advertisement advertised one Santosh Kumar Mukerji's *Elements of Endocrinology*. Priced at Rs. 3, the book was described as "essentially a clinical one . . . contain[ing] many useful prescriptions." Physicians were told that in order to "be up-to-date in treatment," they had to read the work.[21] The following month, G. W. Carnrick & Co. of Canal St., New York, advertised their product: Hormotone. Touted as a cure for dysmenorrhea, the advertisement said that "dysmenorrhea is due to the disturbance of the endocrine mechanism regulating menstruation—thyroid, pituitary, adrenals and ovary. Hormotone is composed of the active substance of these glands—the first successful combination of endocrinal glands used in this field." The advertisement ended by stating that Carnrick's sold "dependable gland products."[22] In the following month, Carnrick & Co. advertised another product: Viriligen. Given "hypodermically and orally it provides a treatment for the sex neuroses and lowered functional capacity," said the advertisement. It also asserted that "endocrines therapy is generally accepted today as the rational treatment in those states of failing functional activity due to the deficiency of internal secretions."[23] Next year, in 1928, a new company Reed & Carnrick of New Jersey lavishly advertised a product named Testacoids (see fig. 9). The full-page illustrated advertisements claimed that "clinical results indicate that many disabilities of old age can be modified, or long postponed, through the employment of Testacoids . . . [which] give to the glands those endocrinal elements which naturally decrease as age goes on apace." The

9. Advertisement of endocrinal product in the *Journal of Ayurveda.*

product's massive popularity was hinted at by the fact that importers from whom the product might be bought, along with their addresses, were listed throughout Asia in Honolulu, Hong Kong, Shanghai, Manila, Bombay, Karachi, Rangoon, and Calcutta.[24] These advertisements appeared even as older companies and their products, such as Hormotone and Mukerji's *Elements of Endocrinology*, continued to advertise as before. A few months later, Reed & Carnrick added another new product, Protonuclein, said to be a combination of hormones and proteins.[25] Most of these advertisers were still advertising their hormonal products in the *Journal* in the early 1930s.

Such advertising would have fitted into an older and, by the 1920s, a well-established tradition of advertising restorative medicinal products. "Commercial notices for bazaar products claiming to restore or enhance male sexual capacity—tonics, aphrodisiacs, potions, lotions and ointments—had long made up a major portion of newspaper advertisements since the late nineteenth century," writes Haynes.[26] After the Great War, global corporations entered the fray and marketed their products side-by-side with the older bazaar products.[27] By the mid-1920s, therefore, there was a thriving and well-recognized culture of advertising medicines that claimed to rejuvenate the aged, and the new hormonal products easily fitted into it.

The lavish advertising, the exotic provenance, the global markets—all contributed to the branding of hormonal products as the future of modern medicine. Refigured Ayurveda naturally sought to appropriate this promised future to its own ends. Even if we eschew a more cynical interpretation of the money generated by the advertisements as actually driving the intellectual embrace of hormones, the former surely helped create an awareness and an aura around endocrinology that would make their adoption attractive. It is also significant that, in the vast majority of instances, neither the foreign products advertised in the *Journal* nor the foreign authorities cited in the discussion on hormones in the learned articles were British. It is also worth accenting that the majority of the hormonal products advertised in the *Journal* were American. Douglas Haynes mentions the availability of German hormonal products such as the drug Okasa whose advertisements detailed the endocrinal system through texts and illustrations.[28] Markus Daechsel also takes note of the advertisements of the German-made hormonal product, Okasa.[29] These accounts fit well with the historiography of hormonal pharmaceuticals which highlight the role of European drug companies. That the *Journal* yet came to be dominated by American advertisements suggests that there was some process of selection underway. Unfortunately, our present understanding of the world of medical advertisements will not permit a fuller exploration of the motives behind such selections, or indeed, of whether the advertiser or the *Journal* played a greater role in such selections. All that can be said for certain is that it seems to be American, rather than European, hormonal products that formed the immediate context within which the *Journal* embraced endocrinology. Yet this embrace of hormones did not happen in a vacuum.

Numerous commentators have noticed that one of the eight branches of classical Ayurveda is devoted to rejuvenation therapy (*rasayana*). Lawrence Cohen has gone a step further and pointed out that "Ayurveda literally suggests the authoritative knowledge of longevity; as such, not only *rasayana* but all *astanga*, all eight branches of medicine, are seen as critical to a clinical practice preserving and extending one's years."[30] Yet this crucial dimension of Ayurvedic medicine has largely escaped careful historical scrutiny. Classical scholars and ethnographers have both tacitly agreed to interrogate only the canonical works without attending to the historical evolution of these therapies. Thus, Cohen, for instance, despite his many brilliant insights into the cultural logic of *rasayana*, chose to limit his historical inquiry to the six texts currently recognized as canonical.[31] This is unfortunate because this core canon itself is a fairly recent invention. Nineteenth-century reports are explicit in stating that classical Sanskrit works such as *Charaka* and *Susruta* were not read by even the erudite stratum of Ayurvedic physicians.[32] Hence, in order to understand the range of native and exotic rejuvenation therapies, it will not suffice to fall back on the hallowed classics.

Looking at *rasayana* historically, a number of new insights emerge. Foremost among them is the fact that Madhavakara's *Nidan*, the key authoritative text for about a thousand years before the advent of colonial modernity, did not include a chapter on *rasayana*. It was reintroduced into the mainstream of scholarly medicine in Bengal through the writings of Chakrapanidatta in the eleventh century. Chakrapanidatta's usage, while remaining strong, continued to evolve and change in the centuries that followed. The eighteenth-century physician Gobindadas provided the most authoritative view on the Ayurvedic *rasayanas* in the immediately precolonial period in his book, the *Bhaisajyaratnabali*. While *rasayana* remained focused on rejuvenation, the actual appetites that it sought to reinstate changed over time.

As Judith Farquhar's fascinating exploration of appetites in modern China has shown, bodily pleasures and the appetites they seem to satisfy are both historical.[33] Cohen makes a comparable point in stating that "the personal and social distress of old age is embodied, and embodied in different ways at different times and for different communities."[34] A comparison of the actual appetites that *rasayanas* (i.e., the recipes that effect *rasayan*ic rejuvenation) sought to revive and strengthen is productive.

For Chakrapanidatta, *rasayanas* were not simply for old people. He even spoke of *rasayanas* that made bodies of teenage boys (*balok*) grow bigger.[35] Anything that stopped the "birth of the disease called old-age," was a *rasayana* for Chakrapanidatta.[36] Upon looking at the actual goals of the recipes he listed, however, a more specific image of old age can be discerned. Some recipes aimed to remove specific ailments such as *kash* (cough), *khoyrog* (wasting/consumption), *shwas* (breathing difficulties), *orsho* (piles), *gulmo* (swelling) etc.[37] A large number of the recipes aimed to revive powers of intellection (*medha*), mnemonic powers (*smriti-shokti*), and *buddhi* (intelligence).[38] Usually the same recipes also promised to remove wrinkles and gray hairs (*bolipolit*).[39] The largest number of recipes combined these goals with the promise of restoring virility, sometimes even promising superhuman sexual capacities whereby one could satisfy hundreds or even thousands of sexual partners.[40] But interestingly, there were other superpowers too that were promised by Chakrapanidatta's *rasayanas*. One recipe, for instance, promised the ability to "sing like heavenly angels" (*kinnorer nyay gan*).[41]

By the time of Gobindadas, these appetites had shrunk. Practically all his recipes, though a good many of them derived from Chakrapanidatta, aimed to enhance *bol* (strength) and *birjo* (potency). Wrinkles and grey-hairs now became almost synonymous with old age and Gobindadas's nineteenth-century translator, Binodlal Sen, often used the terms interchangeably.[42] The number of prescriptions directly addressing either *medha* or *smriti-shokti*—that is, prescriptions promoting mnemonic power—also seemed to have grown.[43] But both the body of the teenager and the angelic voices that were seen in Chakrapanidatta had disappeared. *Rasayana* had become exclusively about removing wrinkles, blackening hair, reviving one's mind and, preeminently about sexual potency.

Other minor texts, such as a couple of small compendia of prescriptions devoted to *rasayana* by an obscure author called Bhubanchandra Basak, also exhibited similar embodiments of old age and suitable aspirations for *rasayana*. Essentially, old age here is clearly seen in strongly sexualized terms. The promotion of sexual appetites, capacities, and potency as well as the enhancement of personal attractiveness, particularly by the blackening of gray hairs and removal of wrinkles (which too one might argue is eventually part of the quest for greater sexual engagements), all point toward an increasing embodiment of old age as de-sexing. Somewhat unexpectedly, this understanding is remarkably similar to the endocrinal understandings of the 1920s. As Sengoopta points out, leading figures of the 1920s such as

Dr. Steinach came to see old age as de-sexing. When intellectuals like W. B. Yeats chose to undergo the Steinach Operation, they did so in search of a "second puberty."[44]

On the other hand, remarkably, classical Ayurvedic *rasayana* therapies such as *Chyavanprash*, an iconic preparation that Cohen found still widely marketed in various forms, were packaged in the 1920s as a treatment for phthisis rather than as a rejuvenator. The drug was marketed by Bisharad's Ayurvedic Laboratory and carried multiple testimonials. The expensive half-page advertisement, appearing regularly between the mid-1920s and the early 1930s next to the advertisements for various hormonal products, boldly declared *Chyavanprash* to be "the unrivalled preventative and curative agent in phthisis."[45] This is distinctly different from the guise in which Cohen encountered *Chyavanprash*—as a rejuvenating product that emphasized its role in the enhancement of memory.[46] It would be futile to try to figure out whether *Chyavanprash* really is a rejuvenating medicine or a cure for phthisis. In fact, *Charaka*'s original description of it, read selectively, permits both identities. What is crucial, however, is that in the 1920s, side-by-side with the hormonal products promoting rejuvenation, *Chyavanaprash* was clearly not seen in the light that it was in the 1990s. Thus eschewing the barren debate of which identity was closer to the classical identity of the recipe, it becomes patently obvious that *rasayana* therapies, though seemingly derived from age-old classics, are not unchanging entities. As their identities evolve, they are braided with different strands of "Western" science and acquire distinct resonances. In the 1920s, *Chyavanprash* was braided with phthisis. In the 1990s, it was resonating with Alzheimer's.

Organotherapy, on the other hand, in the 1920s, came to be braided with the then extant tradition of postclassical Ayurvedic *rasayana*. The imagination of old age as de-sexing was already available in texts such as the *Bhaisajyaratnabali*, and such imaginaries became easily braided with the insights of Brown-Sequard and his followers.

Indigenizing Organotherapy

As with each of the foregoing technological objects I have investigated, in the case of organotherapy, too, efforts were made to indigenize it. These efforts took two forms. First, Hemchandra Sen significantly modified the method for the production of testicular emulsions; second, he sought to relate the therapy to preexisting Ayurvedic therapeutic techniques. Both these efforts gave Ayurvedic organotherapy a distinctive profile. Brown-Sequard had ardently refused to profit from his discoveries and therefore plainly communicated his methods for others to follow. His method of producing testicular emulsions were described in his own words:

> We procure testicles of bulls at the slaughterhouse. Just after the killing of the animal a ligature is placed as high as possible on the whole mass of the spermatic cord, so as to get at least a certain amount of blood contained in the veins. When the organs reach the laboratory their coverings are at once cut away with scissors sterilised by heat. The organs are then washed in Van Swieten's liquor, and afterwards in recently boiled water. That being done, each of the testicles is cut in four or five slices, which, with the piece of cord, are placed in a glass vase, in which is thrown, for each kilogramme of the organs used, one litre of glycerine marking 30°. The vase is covered, but it is essential, during the next 24 hours, to turn over a good many times the slices and other pieces of organs. At the end of that period an addition of 500 cubic centimeters of freshly boiled water, containing 25 grammes of pure chloride of sodium, is made. The liquid so obtained is then made to pass through a paper filter. . . . [T]he paper filter and the glass funnel in which it is placed must be thoroughly washed with boiling water. The filtered liquid is slightly rose-coloured. To hasten the filtration it is well to raise the temperature of the glyceric solution to 40 ° C. (104 ° F.) . . . Of the various means we have made use of to obtain a liquid absolutely free from microbes or other dangerous pieces of solid matter, the most important is the sterilising d'Arsonval filter.[47]

By contrast, Hemchandra described his method as follows:

> To produce Liker Testicularis take the testes of any animal (the testes of Guinea Pigs are preferable) mixed with glycerine and macerate and filter it, all the while continuing to follow Dr Arsonval's [*sic.*] method of applying pressure with carbonic acid.[48]

The most striking difference between the two is of course the replacement of bovine testes with those of a guinea pig. The avoidance of bovine testes, it is fairly safe to assume, is an allowance made to accommodate Hindu religious sentiment that considers the cow a sacred animal. Though the guinea pig had indeed been used by Brown-Sequard himself in his initial experiments, its choice by Sen as the source of the therapeutic hormones is somewhat perplexing.[49] After all, if he wanted to return to Brown-Sequard's original experiments, the latter had also used canine testes in his original researches. Besides, dogs were much more easily found in colonial Calcutta than guinea pigs. Moreover, a range of other slaughterhouse animals such as goats, sheep, and even buffaloes, not to mention the many other easily available non-slaughterhouse animals such as cats and monkeys, were overlooked in choosing the guinea pig. No reasons for the choice were given. One possible reason for choosing the guinea pig might be the animal's clear associations with cutting-edge experimental laboratories.

It is also noteworthy that because the actual nature of hormones remained fairly mysterious until the very late 1930s, bioassays came to play an important role in identifying and standardizing them. By the 1920s and 30s, it was the bioassays rather than the chemicals themselves that were standardized in a bid to stabilize the identities of hormones. These bioassays usually involved guinea pigs and mice much more frequently than other animals.[50] As a result, guinea pigs and mice certainly appeared far more intimately connected to the world of cutting-edge hormonal research than other animals more easily available in India.

That the guinea pig was not an animal easily procurable in the late nineteenth-century Calcutta is attested by the fact that multiple guides to the Calcutta zoo from the 1880s and 1890s mention the animals being one of the attractions at the zoo.[51] While one of these guides did mention their use in laboratories, the fact that they were exotic enough to be zoo attractions suggests that the laboratories that used them were not in the colonies.[52] Hemchandra's preference for guinea pig testes, despite their rarity in the local context, clearly suggests that it was their symbolic associations that informed the preference.

Besides the source animal, another major difference in the two descriptions is the level of detail. Whereas Brown-Sequard gave copious details, specifying quantities and measures precisely, Hemchandra provided only the briefest summary. Clearly, while the former wanted others to use his outlines to reproduce the material and the tests for themselves, the latter had a different end in view. Since Hemchandra's descriptions would not suffice for anyone to actually create the emulsion and use it, it is plain that his intention in providing the brief description was something else. The essay in which Hemchandra provided the description was about the body's "natural disease-preventive powers." While it was clear from the details he provided that he himself had probably had some practical experience of organotherapy, he was not really writing to popularize organotherapy as such. He was writing to establish the validity of the body's internal powers of resistance and in so doing was using the latest scientific data to support his case. Interestingly, he used Brown-Sequard's experiments along with the Russian scientist Prof. Pohl's insight that the active substance in the testes was a chemical called "spermine." Hemchandra used the discussion of spermine to argue that the human sperm was the source of the body's vigour and energy.

This line of argumentation resonated with older Ayurvedic ideas. Semen, for instance, was supposed to be the most refined product of the process of "digestion" and produced only at the seventh or last stage. It was therefore the most sublime of all the bodily tissues as well as the rarest of them.[53] In the Sanskrit alchemical literature, which was also called *rasayana*, mercury was imagined to be Shiva's semen. Between the human and the divine semen thus, *rasayana* had come to attribute a substantial and prominent role to a polyvalent notion of "semen."[54] Among many religious heterodoxies of the region, this powerful somatic characterization of semen had led to its elevation to the status of a sacral substance that was the seat of life itself.[55] Hemchandra clearly reinforced these popular notions of seminal value, while also elaborating upon the contemporary *daktari* emphasis on the body's powers of immunity.

This tendency to disengage hormones from the actual therapies through which Ayurvedists first confronted them was a trend that would mark the eventual emergence of a modern Ayurvedic discourse on hormones. After Hemchandra, endocrinology progressively grew further and further disengaged from its actual therapeutic and laboratory reality. By the time authors like Ashutosh Roy or Shiv Sharma were writing about it, hormones had ceased to be connected to organotherapy or indeed to any other active therapeutic or experimental milieu (with the minor exception of mass-marketed hormone supplements). Thus, in Roy's book, *Pulse in Ayurveda*, where he greatly elaborated his endocrinal ideas, while he repeatedly cited passages from the *Journal of Organotherapy*, he never did so with a view to discuss actual therapies. He merely used the extracts to prove theoretical points about the nature of *dosh*es.[56] Essentially, by the 1920s, hormones in the Ayurvedic world had become a new idiom through which to imagine the *dosh*es and the body itself. Interestingly, Oudshorn argues that the moment when the idea of a "hormonal body" emerged in European biomedical thinking was in the 1920s.[57]

It is doubtful whether Hemchandra, hailed in the 1920s as the founding father of the new idiom, had fully intended this pure idiomatization of hormones. Hemchandra's own writings were more equivocal about his interest in hormones. In the essay on "disease-preventive powers," it was clear that his main interest was not so much in the therapy as in the theory. But elsewhere his interests were often reversed. He sought ways to adapt and indigenize the therapeutic rather than simply assert the theoretical potential of organotherapy and endocrinology.

One essay where this reversed interest is clearly manifest was published in the *Chikitsa Sammilani* the year before his essay on the powers of disease prevention. In this essay, Hemchandra drew attention to the copious evidence of the use of a variety of animal bile in Kobiraji practice. He cited the *Rasaratna Samuccaya*, a well-known Ayurvedic work attributed to Vriddha Vagbhatta. Hemchandra interpreted this use of bile in light of the emergent technology of organotherapy. Instead of ingesting "tabloids" made in Europe from organic bilious extracts, he argued it was better to use fresh bile in the form the Ayurvedists had always used it: by mixing the bile with certain herbs and then sun-drying it. He went on to describe the nearly miraculous success he had himself witnessed when he treated an asthmatic patient on the verge of fatal collapse with freshly procured goat bile. Extrapolating from this, Hemchandra suggested the elaboration of the practice. He outlined several specific diseases where the use of fresh bile according to Ayurvedic modes of administration would be preferable than existing treatment. He then went on to further state that the thyroid gland, too, exuded useful secretions and hinted therefore at its deployment along the lines of bile.[58]

In essays such as the one cited above, Hemchandra attempted to extend Ayurveda's therapeutic repertoire by conceptualizing existing practices in light of organotherapy. At the same time, he was also seeking to subtly indigenize organotherapy. Obviously, these more therapy-oriented engagements with organotherapy also altered Ayurvedic ideas about the body in subtle ways, but in the long run, these experiments failed to produce the kind of legacy that Hemchandra's more theory-oriented engagements produced.

Neurologizing Chakras

The key features of the physiogram of the endocrinal man, literally awash with endocrinal fluids, were the chakras. As Dominik Wujastyk points out, "The concept of the chakras has, of course, now entered public consciousness and media worldwide, usually in an inflated version which enumerates seven, not six, chakras. The doctrine of the chakras is widely viewed as an ancient and immutable element of the Indian worldview." Yet, Wujastyk hastens to helpfully add that "the chakras make no appearance whatsoever in Ayurveda, the classical medicine of India. Notwithstanding the contemporary growth of various forms of massage and therapy focused on the chakras, there is no such theme in the classical Sanskrit literature on medicine."[59] It is useful therefore to trace the origins of this modern insertion of yogic chakras into Ayurvedic writing.

One of the earliest hints of proximity of chakras is to be found in Binodbihari Ray's electromagnetic theories about *doshes.* Though he avoided directly mentioning chakras, he invoked both the Tantric vocabulary that describes *vayu* as "electrically charged" (*torinmoy*) as well as, more cogently, the idea of a hidden inner force known as the *kundolini.* "It is this *kundolini shokti* that is defined as *vayu* or the electrically charged substance. It resides in the spine (*merudondo*) and is divided into knowledge (*gyan*), will (*iccha*) and action (*kriya*)."[60] Ray's hesitant mention, however, was not immediately taken up. Contemporaries and later authors continued to eschew any explicit discussion of chakras.

Having said this, chakras and a largely Tantric anatomy remained inescapably part of Kobiraji—or postclassical Ayurvedic—practice in one specific context. This was *nadipariksha. Nadipariksha,* as a later accretion to Ayurvedic literature, remained in some minor senses an ill-digested appendage. It had its own specific figuration of the body that was fairly widely known and learned by Kobirajes, but that did not always square-up with the rest of their textual education. These *nadipariksha* texts, while seldom invoking the full characterization of the six chakras, frequently delineated an intricate anatomy of *nadis* where many of the specific *nadis*—including the three main ones, *ida, pingala,* and, the foremost, *sushumna*—had names identical to the names in the more canonical Tantric literature. In fact, much of the anatomical description of the *nadipariksha* texts was identical to heterodox religious texts such as the *Shiva Samhita* and the *Gheranda Samhita.* Besides describing a broadly similar anatomy and sharing the names of the *nadis, nadipariksha* texts also occasionally spoke of chakra-shaped cave-like structures where a tiny, anthropomorphic life creature known as *jib* found rest and respite from his continuous travels through the body. Thus, Gopalchandra Sengupta's textbook mentioned that "*Jib* rests at [his] abode. In some places [the abode is] *chakra*-shaped [*chakrakar*] [,] in some places [the abode is] cell-shaped [*koshakar*]. That is, at this abode-like *chakra* (*grihokar ei chakre*) virtue and sin (*punyapunyo*) are assigned."[61] The incorporation of such enigmatic passages in medical works blurred the boundaries between the medical and the heterodox religious ideas about the body and its resident soul-like life creature.

Joseph Alter points out that the somaticization of the chakras was enabled by the very logic of yoga which eschews stark contrasts such as matter and non-matter, gross and subtle, etc., operating instead on a "plane of paradox" where the gross and the subtle bodies constantly blur into one another.[62] In the context of Ayurveda, I argue, it was *nadipariksha* that provided the conduit through which the yogic logic of paradoxes entered Ayurvedic thinking and laid the groundwork for the somaticization to occur. It is also worth pointing out that Gopalchandra's enigmatic statement demonstrates that somaticization did not necessarily mean its realization in a "Western" anatomic body.

It was Hemchandra Sen who revived an explicit discussion of the chakras, though, like Ray before him, he largely avoided directly discussing the chakras themselves. With the partial exception of the *muladhar chakra*, he spoke only generically of the *Nadichakra* as the entire system of *nadis*. He did not identify chakras as specialized physiological entities. His main interest once again was in the *kundolini shokti*, and he developed his ideas about this inner force at much greater length than Ray had done. Hemchandra, however, as I have noted in the foregoing chapter, also invoked the protoplasmic, rather than the electromagnetic, theories of Ray as the general framework within which he conducted his discussions, including those regarding the *kundolini shokti*.

In Hemchandra's hands, *kundolini shokti* became a way to assert a robust vitalism. He conceptualized this force as one that was present since the moment of conception and progressively formed the various organs of the body in order to better express itself. This was totally in line with the protoplasmic theory of the underlying unity of all animate structures. Hemchandra's addition to the general tenor of protoplasmic theories was to advocate the sovereignty of the force (i.e., *kundolini shokti*) that was expressed through the common material form. Whereas most other (Western) exponents of protoplasmania conceptualized the vibrations that led the protoplasm to self-differentiate into different organs and species to be sui generis,[63] Hemchandra posited that the force itself was paramount and sovereign, but acted through the vibrating of the shared basic protoplasm. Since this force created both the brain (*mostishko*) and the heart (*hridoy*), he argued that it did not belong to either, but rather stood sovereign. Such was its power that it could destroy and recreate any or all parts of the body. It was as a consequence of the presence of this vital power that nerves arising out of the "Sympathetic Nerve Centre" at the center of the body (which Hemchandra insisted was the real *muladhar chakra*) bound the entire body in a web of nerves.[64] The ability of nerves to extend when a tumor or growth occurred on the body was taken to testify to their infinite capacity to extend the body's boundaries under the command of the *kundolini shokti*.[65]

For all its vitalism, the *kundolini shokti* was interestingly likened to an electric battery that would release its energy the moment the circuit was closed. The nervous system in general was repeatedly compared to the telegraph system, though with the significant difference of having a center or battery at the *muladhar chakra*.[66] Hemchandra's description of the *muladhar chakra* is worthy of closer attention. His principal interest in the *muladhar chakra* lay in his belief that it was the fons et origo of the entire system of *nadis*. His primary intention therefore was to determine its exact location within the body. For this, he turned to the *Shiva Samhita*, an early-modern text considered canonical among practitioners of hatha yoga. Following the *Shiva Samhita* fairly closely, Hemchandra stated that the *muladhar padma* was exactly two fingers above the anus and two fingers below the penis. It was four-fingers wide. Within this *muladhar padma* was a triangular zone (*trikon-mondol*). This triangular zone was called, according to Hemchandra, the *yoni-mondol*. It is in this space that the supremely powerful goddess, Kundolini, resided in the shape of a lightening. The three principal *nadis*, *ida*, *pingala* and *sushumna*, along with all other *nadis* arose from this triangular zone. Hemchandra interpreted this description to mean that all the *nadis* in the body arose from the Kundolini.[67]

One cannot accuse Hemchandra of outright inaccuracy. He stuck close to the description of the *Shiva Samhita* and read it fairly literally. Yet even a cursory comparison with the original passage in the *Shiva Samhita* reveals numerous telling elisions and erasures in Hemchandra's description. The *muladhar chakra* is not just a localized spot within the body in the hatha-yogic tradition. It is an enigmatic space that is at once both physical and cosmic. It is simultaneously both anatomical and mystical. In order to get a sense of Hemchandra's elisions, we must have some sense of the way the original hatha-yogic traditions conceptualized the *muladhar chakra*. Upendranath Bhattacharya, a historian of hatha-yogic cults in Bengal, provides us a detailed description of the *muladhar chakra*.

> This *chakra* is situated exactly in-between the anus and the genitals. In it resides the *muladhar padma*. This *padma* (lotus) is attached to the mouth of the *susumna nadi*. This lotus is blood red, four-petalled and only partly blossomed. On each of the four petals respectively resides four golden letters, viz. "*bo*," palatal "*so*," median "*so*," dental "*so*," each followed by the letter *onuswor*, and rotating southward.[68]

Continuing further, Bhattacharya describes a square zone in the middle of the *muladhar padma* known as the *dhara chakra*. Around the *dhara chakra* is a zone surrounded by eight pikes. In the midst of this *dhara chakra* resides a yellow-hued, delicate-bodied earth seed (*dhara bij*) in the shape of the letters "*lo*" and *onuswor.* This earth seed (*dhara-bij*) is four-armed, bejeweled, and astride the divine elephant, *Oirabot.* At the core of this seed resides the supreme god, Brahma, in infant form "like a new sun." From his four heads, Brahma recites the four Vedas and his four arms respectively hold a scepter (*dondo*), an ascetic's waterpot (*komondolu*), prayer beads (*okshmala*), and gesture of blessing (*obhoymudra*). To complicate matters further, in this *dhara chakra* also resides a bloody-eyed and terribly powerful *dakini* or a female supernatural being. In her four hands, she holds respectively a pike (*shul*), the stout wooden staff (*khotango*), a scimitar (*khorgo*), and a drinking vessel (*chashok*).

Descending to another level, below the *dhara chakra,* at the point in the *muladhar padma* where the *sushumna nadi* connects with the *padma,* there is yet another "triangular machine" (*trikon yantra*) named *Traipur.* It is said to be as bright as an electric flash, delicate, and yet a place of rest. Within it resides two entities. Both are blood red and comparable to *bondhuk* flowers. They are, respectively, a *vayu* (in this case, probably breath or air) named *kondorpo vayu* and the *kam bij,* or seed of desire. At the heart of this triangular machine also resides the Lord Shiva (*Swoyombhu*). But the Lord here takes on unique characteristics. He is the color of new leaves; his body is as soft as molten gold; his brightness is comparable to lightning or to a full moon, and he takes on the shape of a winding river. Directly above Him resides the goddess Kundolini, stretched in the form of a superfine spider's web. She winds around and protects the Lord's head. "By residing at the *muladhar padma* and by evolving by Her own will, She protects all creatures of this world."[69]

The *muladhar chakra,* along with the *muladhar padma* (which is only a small component within the former), is clearly an enigmatic structure. It is situated within the body, but simultaneously also part of a cosmic space that is beyond all particularities of anatomical place. It is rich in colors, powers, sounds, and supernatural entities. Scholars of Tantrism have long explored the imaginative and artistic dimensions of these structures. But, for my purposes, what is important is that they are most resolutely not simple anatomical structures. While they are manifestly partially located in the body, one of the central characteristics of these chakras is their seemingly unfathomable depth. As one moves from the surface of the chakra into its interiors, one keeps moving into a bottomless cosmic space. This cosmic space eventually leads one imperceptibly out of the body and to the abode of powerful supernatural entities. It is almost a kind of transcendent immanence where one transcends the particularities of anatomical individuality and localized particularity by digging deeper and deeper into the same spot.

No less important than the spatial transcendence are the robust sensory qualities of chakras. As already seen, they are described in rich colors, images, and sounds. Sounds and colors are not merely symbolic registers here but are frequently seen to possess a certain concrete materiality, such as in the case of particular syllables taking on physical form and location. None of this is comparable, or indeed recoverable, through the kind of basic anatomical mapping in which Hemchandra was engaged.

By eliding these enigmatic dimensions of the chakras, Hemchandra recast them merely as anatomical structures. However mysterious they maybe, in Hemchandra's account, they are resolutely structures trapped and localized within the individual human body. While he hesitated to make them completely compatible with "Western" anatomical imaginings—even stating at one point that the *nadis* of the Hindus is not the same as the *nadi*s or nerves of the English—he eventually concluded by making them only weakly incompatible. In fact, halfway through the essay, he seems to have reneged on his earlier contention of incompatibility and gave lists of corresponding names relating *nadi*s to particular nerves. More importantly, he sheared away the spatial and sensory excess that made chakras anatomical, mystical and cosmic, all at once. As already noted, Alter points out that yogic logic allowed both the somatic localization and subtle, extra-somatic iterations of chakras to coexist. Beginning with Hemchandra one notices an attempt to entirely reduce the extra-somatic dimensions of the chakras to their discrete somatic locations.

It was these purely anatomized chakras that Ashutosh Roy inherited and developed a little more than a decade later. In clear contradistinction to Ray and Hemchandra, Roy explicitly and extensively discussed the chakras. Yet he relied almost exclusively on secondary works on Tantrism to do so. He drew most heavily upon the writings of the British judge-turned-orientalist scholar and mystic, Sir John Woodroffe, better known as Arthur Avalon.[70] Following Woodroffe, Roy divided his discussion of chakras into three sets. In the first set, he included the "six important Chakras" usually known in the Tantric literature as the *shot chakra.* In the second group, he included "several small *chakra*s known in the West as Spinal ganglia on the Spinal column corresponding to each segment opposite each vertebrae of the Spine." Finally, in the last set, he included all the chakras that were "situated inside the skull which collectively constitute the "Brain" of modern medicine."[71]

Roy's discussion of the *muladhar chakra* is a good point of comparison that clearly demonstrates how he inherited but also advanced Hemchandra's project of anatomization of the chakras. Roy introduced the *muladhar chakra* directly and unhesitatingly through the signposts of modern anatomy. "It is opposite the Perineum and as already noted it is the meeting point of the Irah and Pingala fibres and the Bajrakshya fibres of the Sushamna. It corresponds to the Superior Hypogastric plexus of modern Medicine."[72] Nowhere in this description is there the hesitation to map Tantric anatomy perfectly onto modern anatomy that marked Hemchandra's discussion. Nor is there any talk of the "goddess Kundolini" that, despite severe refiguration, had still been presented in Hemchandra's description as a mysterious and sovereign force.

Roy did admit some of the enigmatic structure of the *muladhar padma,* but then immediately proceeded to anatomize these features. "It is conceived by the ancient Hindus," wrote Roy of the *muladhar chakra,* "as a lotus with four petals." This was slightly inaccurate in that it confused the *muladhar chakra* as a whole with the *muladhar padma,* but this was a confusion already, at least partially, present in Hemchandra's writings. Moving beyond this perplexity, however, Roy robustly anatomized this four-petaled structure. Paraphrasing Woodroffe, Roy wrote that he has "described the petals as indicating the number and position of the 'Nadis' (Nerves) issuing from or going into the plexus." Modern Anatomy taught us that "each Nerve ganglia has two sets of fibres, one communicating, the other distributing."[73] Both the glossing of *nadi*s as "nerves" and the refiguring of the petals of the *muladhar padma* as sets of fibers leading out of and into nervous ganglia, went well beyond Hemchandra's anatomizing.

Finally, Roy's most telling act of anatomizing involved the *kundolini.* Unlike Hemchandra, Roy made no mention of its status as a "goddess." Nowhere did Roy acknowledge any kind of agential characterization of the *kundolini.* Instead, he redefined her in purely anatomical terms as "a triangular ganglia situated at the tail end of the plexus and corresponds to the Ganglia Impar."[74] It was through these bald anatomizations that Roy was able to finally conclude that chakras were simply "gangliated plexuses."[75]

From their hesitant insertion into refigured Ayurvedic writings by Binodbihari Ray to their eventual centrality to modern Ayurvedic anatomy through the writings of Ashutosh Roy, the chakras had been radically transformed. From enigmatic structures that braided together anatomical and cosmic space with medical, ritual, and mystic figurations, they had been reduced to purely anatomical structures. Their sensory excess had been sheared away in favor of a new precision that was empirical, but also mundane and unspectacular. In short, the chakras had become central to modern Ayurvedic anatomy by being deprived of their magic and sensory brilliance.

Radiating Metaphysical Federalism

Despite the careful and complete anatomization of the chakras in his earlier writings, Ashutosh Roy insisted in a new series of essays in 1930 that the nature of the connection between chakras and nerves was "metaphysical and not physiological." [76] In the initial installments of this new series of essays, Roy was careful to distinguish the "Ayurvedic books" from the "books of the Hindu Yoga System."[77] Roy implicitly suggested that his earlier writings, where he had so carefully anatomized chakras, had mainly addressed the former, whereas his new writings would take up the latter. Such a distinction between Ayurvedic and Yogic corpuses is not entirely convincing, for Roy's earlier discussions were also heavily influenced by Tantric or Yogic elements. The chakras themselves, for instance, were clearly of Tantric or Yogic origin and had interested Roy from the very beginning. Moreover, in the later installments of the new series of essays, Roy dropped the distinction and started referring to a common "Hindu anatomy" or "Hindu medicine."[78] What led Roy to appreciate the importance of the metaphysical aspects of chakras seems to have been "the successful practice of *samadhi* or voluntary suspension of animation, of which we have authentic records, [that shows] that the involuntary, organic, autonymic portion can be controlled at will, a fact undreamt of by us."[79] Later he explained more clearly that it was apprehending the distinction between the metaphysical and physiological connections that allowed "the *yogi* who has controlled these chakras attain superhuman powers."[80]

What is curious is why Roy, after having tried so hard to disenchant modern Ayurvedic anatomy and anatomize the enigmatic chakras, suddenly became so interested in superhuman powers, *samadhis*, etc. The reason seems to lie in the new prominence that he gave in his writings to the writings of a young Marathi doctor called Vasant Gangaram Rele (see fig. 10). Roy repeatedly drew upon a small corpus of authorities, comprising mainly of the philosopher, Sir Brojendranath Seal, the Tantric author and judge, Sir John Woodroffe, and the theosophist, Rev. C. W. Leadbeater. In the new series of essays on the *Nervous System of the Ancient Hindus*, however, Roy included Rele as a fourth authority who was cited as copiously as the much better known previous three. It is therefore safe to assume that it was Rele's writings that had largely instigated or shaped Roy's new interest in the metaphysical. This is indeed borne out upon comparing Rele's and Roy's writings. Rele's first book, *The Mysterious Kundalini*, appeared in 1927. In introducing the book, Rele gave a detailed explanation of how he became interested in the subject. As a young biomedical doctor, Rele had been a

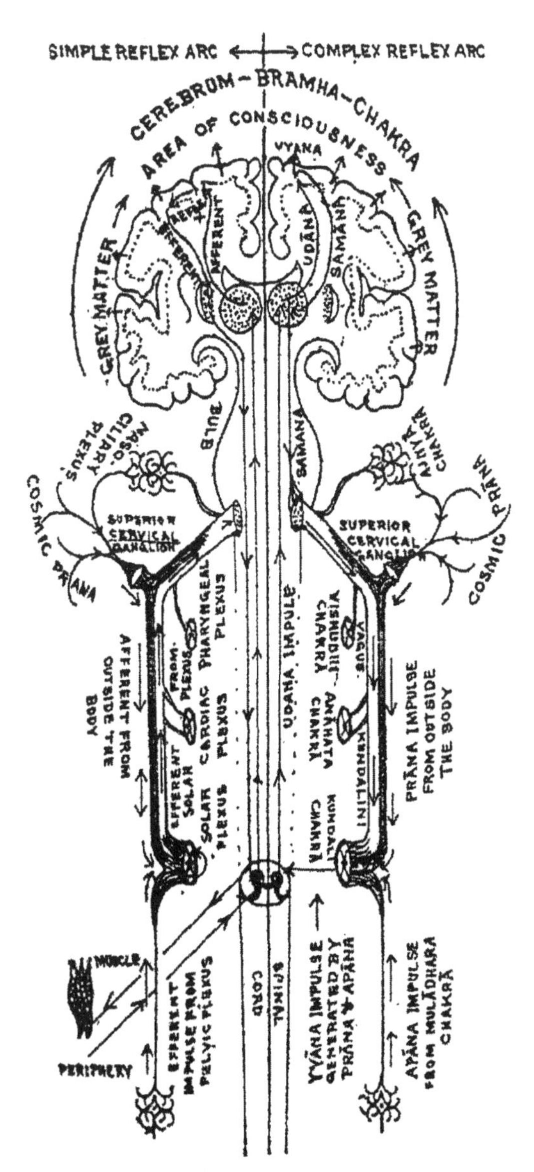

Diagram showing simple (unconscious) and complex (conscious) Afferent and Efferent Channels of the Autonomic nervous system.

10. Rele's neurologized *chakras*.

member of the Bombay Medical Club. It was at this club that a young yogi only identified by his first name, Deshbandhu, had been invited to exhibit the powers of yoga. One of the most striking feats Deshbandhu performed was to voluntarily stop his pulse. As Rele and a colleague sat next to Deshbandhu feeling his pulse and looking at their watches, the young yogi was able to completely stop his pulse for almost three minutes at one wrist and then, with equal ease, do the same at the other wrist. Rele and his colleagues were so taken by the performance that they had Deshbandhu perform under every conceivable testing apparatus, ranging from a simple stethoscope to the then-novel X-ray machine. Finally, this group of doctors were utterly convinced that Deshbandhu was in fact able to control bodily functions that were thought to be involuntary nervous acts.

What also helped Rele come to terms with this striking experience was the other major Bombay novelty of 1920s—namely, the radio. It was the radio that provided Rele with a model through which to make sense of the way Deshbandhu might have achieved his feats. Rele, following the *Bhagvad Gita*, defined *yoga* as "the science which raises the capacity of the human mind to respond to higher vibrations, and to perceive, catch and assimilate the infinite conscious movements going on around us." Drawing on his new familiarity with the radio, Rele wrote, "In fact it [yoga] makes one a broadcasting as well as receiving station of radio activity with the mind as the aerial. . . . It is claimed that all the miracles performed by the long list of saints, saviours and sages, of all times and in all climes, was due to the knowledge of this, the grandest of all sciences."[81] The radio thus provided Rele with a concept metaphor through which to imagine both action at a distance without direct contact as well as a medium through which an "embodied [individual] spirit" (*jibatma*) might remain connected with the soul of the world or the "Universal Spirit" (*paramatma*).[82]

Despite the fact that there is some evidence to suggest that the Bengali scientist, Sir Jagadish Chandra Bose, had in fact developed the radio independently of Marconi, the invention was slow to catch up in India.[83] Partha Sarathi Gupta suggests that it was as a consequence of an earlier personal scandal involving the Viceroy Lord Reading and the Marconi Radio Telegraph Company in Britain that during the Reading administration failed to support the radio in India during its crucial early decades.[84] Thus deprived of official support, the development of radio in India was left largely upto local, private radio clubs. Of these, the Bombay Presidency Radio Club was the very first. Its pioneering broadcast was soon followed by another Bombay-based experiment carried out by the *Times of India* newspaper.[85]

Rele, a Bombay resident, was thus exposed to the very first and most robust stream of enthusiasm for the radio.

As I have shown in the preceding chapters, modern Ayurveda repeatedly drew upon the image of the telegraph to create a viable new body image. In so doing, it also remained deeply entangled in a network of "wires" that it reimagined as *snayus* or "nerves." Even after hormones had displaced electromagnetism as the major conceptual resource with which to braid ideas about *doshes*, the dependence on "nerves" did not disappear. This is an interesting point of contrast between the European move from a neural to a hormonal anatomy and the Ayurvedic one. In the former case, as Chandak Sengoopta shows, there was a clear break between the two (even though they did later comingle once again). Sengoopta points out that much of the research that helped establish the endocrinal anatomy was actually based precisely on the observation that the transplanted glands continued to function when neural networks were obviously sundered.[86] By contrast, in Ayurvedic circles, nerves and hormones remained deeply dependent on each other. All the major authors on hormones in modern Ayurveda, from Hemchandra Sen to Ashutosh Roy, continued to map the endocrinal anatomy in relation to a complex nervous anatomy. While the power and potency of the hormones were widely recognized, and the chemical, rather than electromagnetic, basis to their action somewhat understood, the idea that they were "ductless glands" and did not need conduits to effect distant regions of the body was only hesitantly and belatedly recognized.

In the absence of such recognition, the reticulate anatomy of *snayus* remained central. It was only with the emergent conception of radio waves that could effect or communicate with things at a distance that this reliance on *Snayubik* networks began to wane. Yet modern Ayurveda until the 1930s was never able to wholly eschew its reticulate nervous anatomy. In fact, both Rele and Roy, despite recognizing the potential for action at a distance, continued to try to map out specific neurological routes of endocrinal action, whether it be physiological or metaphysical.

Roy's hesitancy in letting go of the nervous anatomy in the face of the possibility of radio wave-like hormonal action led him to propose a triplication of the key chakras. Slightly modifying his earlier definition, Roy argued that "*chakras* were plexuses of nerves, associated with or without ganglias."[87] But continuing further, he added that the name of each chakra actually referred to three different structures.[88] First, there were the "subtle chakras" in the spinal cord. Second, there were the "gross chakras" of type A which "were associated with spinal nerves without ganglia." Finally, there were the "gross chakras" of type B which were "associated with the sympathetic ganglia."[89] Clearly, this anatomical image was still intimately entangled with nerves and neurological structures.

In seeking to explain how the three-fold chakras worked within this intricate and complex system, Roy regularly fell back upon allegories of the state and administration. The "subtle chakras" of the spinal cord, he said, were, "the local governors of the different regions of the body, controlled by each segment of the cord." The "gross chakras" of Type B or the "gross sympathetic ganglias" were, "like so many district officers, who are autonymic in ordinary local matters, but in higher and more important matters . . . order is communicated to them." Finally, the "gross spinal chakras" of Type A, which regulated the body's interactions with its environment by assimilating positive elements and expelling malignant ones, were "like the foreign department directly under the Viceroy or the cerebro-spinal subtle centres."[90]

Roy's body image is remarkable in how closely it mirrored the dyarchic government set up in British India in the 1920s. According to this federated system, local provincial governors had substantial autonomy. They were advised, for the first time in British India's political history, by elected Indian ministries. The viceroy at the center retained a few key portfolios. Moreover, the years following the Great War had witnessed unprecedented levels of Indianization of the civil services, and hence there were many more Indian district officers who enjoyed a certain degree of local autonomy. Like Rele's radio waves, Roy was clearly drawing on the world around him to create the new body image. But what is particularly noteworthy is that, despite the striking similarities between the political realities of the day and the body image, there were also some significant differences that are made perhaps even more significant by the fact that they violated the logic of the allegory. The viceroy's office, for instance, was highly centralized and meant to provide a fairly strong unitary mold to the federalism of the dyarchy. Yet Roy was clear that there was no single central authority controlling the body as a whole. Central authority, instead, was dispersed among a number of "subtle chakras" along the spinal cord.

Roy further developed this polycentered logic of control in an intricately braided system where he sought to align chakras and nerves together with bodily functions and the cosmic whole. He did this by braiding the chakras and nerves alongside fundamental atoms called the *ponchomohabhuts*. Specific types of *shoktis* or "forces" held these chains of cosmic connection together. Table 1 gives a schematic sense of how the connections worked.

Table 1. A Schematic Representation of Roy's Alignment of Chakras, Nerve Centres, Bodily Regions, Forces, etc.

Chakras	Nerves	Spinal Region	Bodily Region	Mohabhut	Shokti
Muladhar	Pelvic Plexus	3rd sacral	Lower Pelvis & Extremity	Earth (*Prithvi*)	*Dakini*
Swadhisthan	Superior Hypogastric Plexus	5th Lumbar	Pelvis Proper	Water (*Jal*)]	*Rakini*
Monipur	Solar Plexus	8th Thoracic	Abdomen	Fire (*Tej*)	*Lakini*
Onahoto	Superficial Cardiac Plexus	8th Cervical	Chest & Upper Extremity	Air (*Vayu*)	*Kakini*
Vishuddho	Superior Cervical	3rd Cervical	Neck	Ether (*Akash*)	*Sakini*
Agna	Gania of Ribes	Cerebellum	Head	Moon & Celestial Bodies	*Hakini*
Sohosrar	(none mentioned)	(none mentioned)	Brain	Soul	*Yakshini*

What is significant in this complex and multistranded system is the suggestion that a specific type of "force" regulates each chain of communication linking a specific chakra to a specific nerve, a specific region of the body, and the cosmos. Though rendering them all into forms of a single *kundolini shokti* occasionally (especially in the concluding installments of the essays) compromises their autonomy, for the most part the suggestion clearly seems to be that these *shokti*s are independent and specific "forces" that are able to communicate between specific regions and entities in the cosmos. The model, in all likelihood, is once again derived from the analogy of radio waves.

These forces, I suggest, are imagined in light of radio waves capable of being detected at particular frequencies and not others. Whatever its source, this is clearly a new model for conceptualizing communications within the body and between the body and its environment. Whether it was the older nineteenth-century European models that emphasized the circulation or communication of electromagnetic or similar forms of energy, or the models of the 1920s that emphasized chemical interactions through hormones, they all agreed on a single type of circulating or communicating medium, a form of energy or chemical. To the best of my knowledge, no one before has sought to develop such a pluralized model where different types of "forces" would be responsible to different parts of the body or different regions of the cosmos.

Taken together with the polycentric imagination where each "subtle chakra" is a viceroy and commands his own specific type of communication channel based on a specific *shokti*, the body image that Roy and others developed was both more complex and more federal in form than contemporary European body images. This federal structure was almost certainly modeled on radio technology.

Surprisingly, however, neither the imagined plurality of radio technology, nor the political milieu of the day inspired any overtly democratic metaphors from being inscribed on to the body image. In a decade known in Indian history as the first one with significant mass participation in politics, it is surprising that nowhere in Roy's imagery of viceroys, governors and district officers does one for once hear of elected officials. In fact, while stories such as Deshbandhu's, which had led Rele, and through him Roy, to engage with the metaphysical, might well have signaled the power of the will or the ability of the humble subject to transcend the laws of everyday life to another, in modern Ayurveda it produced a well-calibrated and rule-bound idea of the "metaphysical" that barely had any space for autonomy, will or transcendence. In the hands of modern Ayurvedists, the metaphysical became what Alter dubs the "meta-material": a higher, but equally rule-bound, predictable, and mundane version of the physical.[91]

Endo-Chakric Machine

Much is made of the "fluency" with which even the modernized Ayurvedic body interacts with its surroundings. Its ability to respond more readily and sensitively to emotions, social interactions, seasonal and dietary changes, etc., has made many scholars and lay writers argue, in different idioms, that the modern Ayurvedic body is more in tune with its environment than the biomedical body. In one of the most astute as well as fascinatingly imaginative formulations of this argument, Jean Langford, describes the "fluent body" as a body

> coursing with climates and appetites, messages and passions, winds and tempers. The dosic body spans the divide between the text and the world. It is inscribed with signs that are more productively understood as versatile signifiers than visualized as definite objects. To say that it is a fluent body is to say that it is overflowing not only with dosic currents, but also with polyvalent syntax.[92]

Langford contrasts this fluency with the "docile bodies" of biomedical anatomy. She suggests further that while in the "West" the transition to docile, anatomized bodies has been fairly smooth and complete, in postcolonial

India this is being contested by the Ayurvedists who continue to course smoothly between both the fluent and the docile bodies.[93]

Despite the elegance and beauty of Langford's formulation, I find her argument problematic. What troubles me is the stark binarism between biomedical and Ayurvedic bodies. To begin with, anatomized bodies are not necessarily "docile." In fact, much of the history of dissections and anatomy today speaks to the spectacular aspects of dissection. Ruth Richardson reminds us of the spectacular aspects of public dissections allowed by the royal grants that made the bodies of executed prisoners available for dissection.[94] Closer home, the grisly scene with which David Arnold commenced his classic work, *Colonizing the Body*, wherein a poor Calcutta mortuary worker carried around a dissected arm and occasionally bit into it on the streets of the imperial capital as a begging ruse, shows us exactly how mobile, non-docile, and even bawdy, the anatomized bodies could be.[95] Similarly, Michael Sappol's riveting study of the popular culture of dissection reveals how undisciplined and untamed the anatomized body could become in the hands of radicals, entrepreneurs, health reformers, heterodox healers, and many others.[96] The tamed and docile body produced through the objectifying anatomical gaze of biomedicine that Langford purveys is unfortunately a historical caricature.

To be fair, Langford is not really speaking of actual anatomized bodies, even though she says she is. In effect, she is comparing the iconography of anatomical bodies with the practiced reality of Ayurvedic bodies. Her only source for "anatomized bodies" is textual representations of that body, not real performances of anatomic dissection. And it is largely true that since the Renaissance, the development of anatomical iconography "reveals a progressive denial of symbolic and humanist meanings of human anatomy."[97] The problem arises because she compares these representations with actual practice on the Ayurvedic side. In Ayurveda's case, her source of information is her own rich and fascinating ethnographic data. To compare this rich, situated ethnographic account of practice with the flattened representations of biomedically anatomized bodies is both unfair and misleading.

But it is not Langford's confounding of Western art and medical practice that I find most difficult to accept. It is rather her characterization of Ayurvedic bodies. Speaking of the kind of vernacular texts with which I have been dealing in this book, Langford writes that "the bodies depicted (in these vernacular texts) are not yet the passive bodies of modern anatomy. They are certainly interiorized, with organs and bones exposed. They are also, however, living bodies, not frozen in anatomical poses, but gesturing and gazing off the page."[98] Thus, in Langford's view, this "living" quality of the illustrations in vernacular Ayurvedic texts marks them off as different and distinct from the docile bodies of biomedicine.

I will not contest the fact that the illustrations do depict the bodies differently from biomedical works on anatomy. It would be completely unrealistic to do so. But I disagree that the difference is as fundamental as Langford suggests. Instead, I argue that underneath the superficial differences, there are crucial similarities in the ways in which both vernacular modern Ayurvedic texts and contemporary biomedical works understand the body. I argue that both biomedical works and modern Ayurvedic works plot the body onto a broadly compatible, even if slightly different, cosmology. Within this cosmology, the human body occupies a discrete place in space and time. Its boundaries, however porous, are clearly marked off from its environment. Internally, it forms a coherent unit that can be treated as a single whole. When Langford makes much of the images of the body and their differential representations in biomedicine and modern Ayurveda, what she entirely overlooks is the simple fact that the practice of drawing whole-body images is in itself a modern innovation in Ayurveda.

Dominik Wujastyk has commented on the lack of any significant tradition of medical illustrations in the Indic tradition until fairly recent times.[99] This is truly remarkable for a medical tradition that, in one way or another and despite all changes, breaks, evolution, etc., has been around for over two millennia. Why did such an old and robust medical tradition not feel the need to illustrate bodies? I argue that this absence, at least in part, was due to the difficulties of clearly outlining the body—that is, etching it out completely from the larger cosmos made up of both visible and invisible agential beings within which the body and the self were enmeshed. As in the case of the enigmatic chakras, so too elsewhere, the space of the body often opened up imperceptibly onto a cosmic space. Gods could literally be located within the body, not as possessing spirits taking over a mortal coil, but more enigmatically as perpetual presences. This meant that particular parts of the body always had cosmic dimensions and could not really be located strictly "on the inside" of anything. Naturally, this is not a rationalized view of (anatomical) space. But that is my point. The very rationalization of anatomic space that would make visual representation possible had not occurred until Ayurveda came into contact with "Western" medicine in fairly recent times. This is where I feel modern Ayurveda and biomedicine are in agreement. They both illustrate bodies because they both feel that bodies have sufficiently well-defined physical boundaries. Langford interrogates the different conventions of illustration without noticing that the very possibility of illustrating bodies is new to Ayurveda and is now shared with biomedicine.

What I have been tracing throughout the foregoing chapters is the emergence of these concrete images of the body—namely, physiograms. Gradually and inchoately, we have seen the emergence of a reticulate image of the body. This is what I am calling anatomization. Anatomization, in this case, is not the simple incorporation of "Western" anatomical knowledge into Ayurveda. It is the creative and historically contingent process of braiding of knowledges through which modern Ayurveda evolved a set of discrete images of the body as a whole.

This emergence was not in itself smooth or teleological. As new technological objects came to be included in Ayurvedic therapeutics, new and novel intellectual resources were necessary. Sometimes, as in the case of the chakras, these new intellectual resources also brought with them embedded elements that resolutely resisted the anatomization: namely, those elements which robustly transcended the narrow confines of a clearly defined body space. These then had to once again be refigured and shorn of their transcendent propensities in order to be made amenable to anatomization. The metaphysical had to be rendered the meta-material.

One of the most striking instances of this transformation of the metaphysical to the meta-material can be seen in the way Hemchandra treated medical astrology. The source of Hemchandra's medico-astrological views, he acknowledged, was a text called the *Shiva Swarodaya*. While the *swarodaya* texts were part of a larger corpus of Tantric learning in Bengal, it was a curious source for biomedically trained Ayurvedists to turn to.[100] *Swarodaya* texts most likely had its origins in the twelfth century. [101] The genre was essentially devoted to a peculiar species of medical astrology that related breaths in the body to the movement of planets. Rajendralal Mitra described the work as "an astrological dissertation on the indications which the human breath affords under particular circumstances and astral conjunctions of the duration of life and future events."[102] By the late-seventeenth and early-eighteenth centuries, Bengali Islamic texts, such as Sheikh Chand's (ca. 1650–1725) epic work, *Talibnama ba Shah-daulah-pir-nama*, were clearly deploying *swarodaya*-like ideas in specifically medical contexts.

It is this tradition of medical astrology that Hemchandra drew upon. His lengthy essay on the *nadichakra*, while replete with references to "nerves," electricity, protoplasm, endocrines, and so forth, also drew equally prominently upon astrological ideas. His essay began with a discussion of the cold "lunar breath" that flows through the left nostril and the warm "solar breath" flowing through the right nostril. He described in detail the internal flows of these "lunar" and "solar" breaths.[103] Even more prominently, he divided the body itself into zones marked by zodiacal signs (*rashi*). Hence the brain (*mostishko*) was under the sign of *Brisho* (Taurus), the face (*mukh*) was under *Mithun* (Gemini), the arms (*bahudwoy*) were under *Korkot* (Cancer), the stomach (*udor*) was under *Simho* (Leo), the waist (*koti*) was under *Konya* (Virgo), the abdomen (*bostidesh*) was under *Tula* (Libra), the anus (*gujhyadesh*) was under *Brishchik* (Scorpio), the thighs (*urudwoy*) under *Dhonu* (Sagittarius), the knees (*hantu*) under *Mokor* (Capricorn), the shanks (*jongha*) under *Kumbho* (Aquarius) and the feet (*padadwoy*) under *Min* (Pisces). The only non-zodiacal sign that Hemchandra included in this list was the moon that was said to preside over the heart (*hridoy*) and mind (*mon*).[104]

Having gone into such detail in assigning parts of the body to their correct zodiacal signs as well as having given a lengthy description of the circulation of lunar and solar breaths, Hemchandra then had absolutely nothing to say about astrology. He did not in any way hint at either planetary influence or prognostication. The body and the heavens were mere allegories of each other. There was not real tangible connection. The insides of the body were clearly marked off from the outside.

This was in stark contrast to the way the *Shiva Samhita*, another text that Hemchandra cited repeatedly, spoke of the body. The *Shiva Samhita* clearly stated that "in this body the Mount Meru is surrounded by seven islands; there are rivers, seas, mountains, fields, and lords of the fields too." The text continued further, stating that "there are in it seers and sages; all the stars and planets as well. There are sacred pilgrimages, shrines; and presiding deities of the shrines. The sun and the moon, agents of creation and destruction, also move in it. Ether, air, fire, water and earth are also there."[105] This is not a mere correspondence being mapped out, but a relationship of inherence that would confound any realist organization of space.

Moreover, hatha-yogic texts like the *Shiva Samhita* had tremendous popular resonance in Bengal. Besides the erudite Tantric scholastic milieu and the Tantra-inflected Kobiraji world whence Hemchandra may have picked it up, the ideas of the *Shiva Samhita* were also widely disseminated in rural Bengal through songs known as *dehotottwer gan* ("songs of body knowledge"). Sudhir Chakrabarty, who has collected and published many of these *dehotottwo* songs, writes of the centrality of "self-knowledge" emphasized in these songs. This self-knowledge translates essentially into the "knowledge of one's body and knowledge about the location and function of the *nadi*s, breaths and *chakra*s within it."[106] The debate between the bounded meta-material body and the contrasting metaphysical embodiment articulated in the *Shiva Samhita* was thus not a narrowly medical matter. It was embedded in a wider sociocultural landscape.

The two conflictual forms of embodiment were still prominent in Ashutosh Roy's redaction of Hemchandra's ideas. "The Hindus," wrote Roy, "believed . . . that the human body is a microcosm which corresponds to the macrocosm and contains in itself all parts of the visible universe."[107] "Correspondence" and "containment" are clearly not the same thing. Their juxtaposition interrupted the meta-materialization of the metaphysical. Yet Roy willfully overlooked the contradictions and interruptions and proceeded to render the metaphysical body completely anatomized.

Like Hemchandra before him, Roy identified particular parts of the body with particular celestial entities. "Cephalic extremity of the Sympathetic [Nervous System]" became the "Chandra-Mandal" (Lunar Sphere), the "sympathetic centre in the Medulla Oblongata of Sajous" became the "Seat of the Moon," the "Caudal extremity of the Sympathetic [Nerve System]" became the "Surya-Mandal" (Solar Sphere), and so forth.[108] Like Hemchandra, these meta-material entities in the body were further linked and operated by a set of hormones. Thus, the Lunar Sphere was connected to the "Kapha hormone" that soothed the organism, while the Solar Sphere was organized around the "Pitta hormone" that heated the organism and so on. Each of these hormonally organized spheres also mapped neatly onto biomedical structures. Thus, the Lunar Sphere became identical to the "Thermolytic (heat losing)" centers of "modern Medicine," while the Solar Sphere corresponded to the Solar Plexus of "modern Medicine."[109]

Yet, the metaphysical body refused to be obliterated. As I have noted earlier, by the 1930s, under the influence of Rele's writings, Roy began to develop a new appreciation of the metaphysical. In late 1930, he returned to the topic of celestial influence and penned a series of essays explicitly on the role of astrology in modern Ayurveda. As seen in table 1, he had already created a space for accommodating astrology by acknowledging a channel of communication linking the *Agnachakra* or the "Gania of Ribes" with the celestial planets through a *shokti* called *Hakini*. His essays built on this to advocate the therapeutic value of specific gemstones on specific organs.[110]

Instead of dismissing astrological influence altogether, Roy's efforts in the 1930s were directed toward the meta-materialization of the celestial influence itself. By providing a detailed model of action of the planets, he sought to expunge the mystic or occult nature of planetary influence implicit in astrology. Roy argued that just as the alchemical pursuit of the "Elixir of Life" had been modernized by Claude Bernard's work with testicular extracts and Metchnikoff's work on Lactic bacilli, so too should that happen with astrology. For starters, because the "positions of the stars had considerably changed," the old Indian almanacs had to be updated with reference to the "Nautical Almanac." "Astrology like Ayurved," Roy asserted, "cannot move on old grooves entirely."[111]

By braiding older Indic strands of knowledge with strands of modern European knowledge, Roy endeavored to not only create an updated meta-material framework for celestial influence on the body, but also render its seemingly metaphysical nature more compatible with the hormonal body of contemporary biomedicine.

Men such as Roy tried hard to expunge precisely the sort of radical and absolute difference that Langford attributes to Ayurvedic bodies. Yet, they did not seek to make Ayurvedic bodies entirely indistinguishable from contemporary biomedical bodies. Indeed, they actively cultivated a certain level of distinctiveness and distance from biomedical anatomies. What they aspired for was a form of mutually intelligible difference. While they extinguished more outré forms of embodiment such as those found in the *Shiva Samhita*, they resisted the complete collapse of difference by engendering entities such as the "Kapha hormones" and astrologically informed gemstone therapies.

It was in this anxious space between identity and difference that a specifically Ayurvedic version of the hormonal body emerged. This latest physiogram built upon the earlier physiograms like the *Snayubik* Man. It was reticulate and its reticulation was engendered in an intricate network of nerves. The main difference from the *Snayubik* Man was that, while the *Snayubik* Man's inner nervous networks, deprived of a brain, had lacked a controlling center, the new endocrinal man was regulated by a number of "nervous plexuses" or chakras. It was not a single unified command post, but rather a series of command stations that worked in mutual cooperation. Hemchandra had fallen back upon an updated version of the old telegraph metaphor to describe this new multicentered physiogram of an *endocrino-chakric* machine. He compared the many control stations to the "Government Telegraph, Telephone and the Commissariat Departments."[112] There was no viceroy or monarch in his system. It was simply a matter for ministries or departments functioning in tandem to coordinate among themselves. Roy, as mentioned before, developed this further into an image of dyarchy-era federalism. A complex hierarchy of partially autonomous district officers and provincial governors under a consortium of equally powerful viceroys replaced Hemchandra's ministries.

Roy's complex administrative structure was further complicated by the variety of channels that each chakra controlled. Whereas the first five channels mentioned in table 1 were relatively unremarkable and only served to consolidate the reticulated endocrinal machine, the remaining two connected the body directly to the cosmic and the spiritual domains. In a way, this might once again be taken to be evidence of the "fluency" of the modern Ayurvedic body, but it is worth noting that this "fluency" is not only constructed by drawing upon modern, transnational, and diverse intellectual resources, but it is also not identical to the way in which premodern Ayurveda conceptualized the action of planets or the soul.

Despite the novelty of the conception, it is clear that the physiogram that emerged through Roy and others by the end of the 1920s was still deeply physiospiritual. Yet this physiospiritualism was now reticulated and enclosed within a highly regulated system. Thus Roy, for instance, argued that all the distinctive *shokti*s including that communicating on the spiritual plane—namely, *Hakini*—were all versions of *kundolini shokti*, and that this latter was merely a "regulative force" within the body that balanced the creative and destructive processes within it. The much-touted metaphysical dimension of this body was nothing but a latent version of this "regulative force."[113] The spiritual, while explicitly being embodied, was thus also simultaneously rendered mechanical and largely de-spirited—that is, in Alter's words, metamaterial. The physiogram of the endocrinal body that emerged by the end of 1920s was thus an endocrino-chakric machine intercalated by a tame and deflated physiospiritualism, and immaterialized in a thickly reticulated neural network controlled through multiple "nervous plexuses" or chakras.

SIX

Baidya-as-Technology: From Diagnosis to Pharmacy in a Bottle

Gopalchandra Sengupta's pioneering book on modern Ayurveda detailed a list of prerequisites without performing which no physician could proceed to examine the patient. The evacuation of the physician's bowels in the morning, followed by a bath and the performance of stipulated ritual prayers (*ahnik*), and even dressing in a way consonant with his social identity (*samajochito beshbhusha*) were all said to be imperatives.[1] These instructions to the physician can easily be written off as therapeutically inconsequential instructions principally aimed at the enforcement of didactic social norms. It is precisely against this separability of the sociocultural and the therapeutic that Rosenberg has proposed his more expansive notion of "therapeutics." In keeping with that line of reasoning, I urge a closer examination of these instructions for the physician.

Developing Rosenberg's arguments along a slightly different line I argue that the inseparability of the sociocultural and the therapeutic was engendered by the simple fact that the body of the Ayurvedic physician functioned as a technology in and of itself. To this end I focus on a single one of Gopalchandra's injunctions, the one about the evacuation of the physician's bowels. The reason Gopalchandra was concerned with the physician's bowel was not simply in order to enforce a social more, but because he and others like him felt that a constipated physician was a blunt instrument that was therapeutically inefficient. That the injunctions about bowel movement were motivated by concerns over therapeutic efficiency, rather than any isolated concern about social normativity, is strongly suggested by the fact that subsequent authors who stripped away at Gopalchandra's other seemingly more superfluous injunctions, chose to retain the injunctions about bowel movement. Hence, Binodlal Sen, writing more than a decade later, jettisoned Gopalchandra's injunctions about dress and ritual (*ahnik*), but retained those about bowel movements.[2] Years later, Binodlal's cousins, Debendranath and Upendranath Sengupta, reiterated the injunction on bowel movements in their four-volume classic, *Ayurveda Samgraha*, while ignoring the injunctions about dress, etc.[3] Dhurmodass Sengupta, another eminent late nineteenth-century Ayurvedic physician, also retained the injunction and even attempted to provide a rationale for the injunction. Taking the Sanskrit phrase *pratah krita samacharaha* (lit. "performance of morning duties") literally, he added a footnote after the word "morning" stating that "at noon and in the evening the Pulse, having heat and rapidity of motion—characteristics common in fever—the physician cannot with certainty attain to a knowledge of the patient's health."[4] The rationale is unconvincing. Dhurmodass disingenuously breaks up the phrase that idiomatically refers to the evacuation of bowels and focuses only on the word "morning."[5] Moreover, though the Sanskrit verse he was translating referred to the body of the physician and Dhurmodass's translation too clearly did the same, the footnote seems to be referring to the patient's body. What is striking, however, is that despite such convoluted attempts to rationalize the injunction, Dhurmodass obviously felt it was too central to be omitted altogether. A clearer and more convincing reason for the injunction was advanced by Pandit Prayagchandra Joshi's Hindi commentary on a precolonial *nadipariksha* text published in 1959. Joshi, a Sanskrit scholar and a practicing Ayurvedic physician, in glossing the injunctions about the movements of bowels explicitly stated that "the *nadi* cannot be properly discerned so long as one has retained one's faeces, urine etc. (*molmutradi*)."[6] Clearly the physical conditions of the physician's bowel's mattered because his body was itself a diagnostic instrument.

The Ayurvedic physician's relationship with the small technologies he gradually came to embrace over the period of my study was not a simple unidirectional relationship. His own body was a technological object in itself, and as he embraced other small technologies, the physician's own body-as-technology was simultaneously recalibrated. In this chapter, I interrogate how the physician's body-as-technology was remade in the course of the modernization of Ayurveda.

The Baidya's Mindful Body Instrument: A Precolonial History of Practice

There is no need to read between the lines to discern the implicit logic of injunctions about evacuations to realize that Ayurvedic physicians had long figured their own bodies as medical technologies. The *Susruta Samhita*, urging the need for both textual and practical grounding of physicians, had famously drawn a parallel between a good physician and a two-wheeled chariot. Just as the chariot had two wheels, the *Susruta* had argued that physicians needed to be acquainted with both "knowledge of the shastras" and practical skills. These were his two "wheels."[7] While this metaphor is frequently cited as a testament to the importance given to practice in classical Ayurveda, what has been systematically erased is the specific genealogy of practice that was espoused by the *Susruta*.

The modern popularity of chariot metaphor begins with the British orientalist H. H. Wilson. Writing in the *Oriental Magazine* in 1823, Wilson reproduced a pithy translation of the chariot analogy and followed it up by lamenting that "it is much to be regretted that these aphorisms have so little influenced Hindu practitioners."[8] In 1837, another British orientalist, John Forbes Royle, again repeated the comment.[9] Thence it gradually evolved into a cliché that is always rolled out to strengthen Golden Age narratives about the precocious advances of "ancient Hindu science."[10]

Lissa Roberts and Simon Schaffer point out that there is a long and hallowed tradition dating back to Aristotle that distinguishes episteme and techne by rendering manual and mental labor as two totally different categories of action. No matter how positively repurposed in the writings of Wilson, Royle, and their Indian inheritors, this breach between episteme and techne sustained a whole range of social, moral and scientific hierarchies which had its roots in early modern Europe.[11] But Roberts and Schaffer do not simply show up this polarity to be false. Instead, they espouse a historical approach to what they call the "mindful hand." They urge an exploration of the actual historical spaces, disciplines and discourses through which such contrasts and hierarchies were operationalized.

As Pamela Smith points out, in the Aristotelian tradition there were three hierarchically organized terms that structured knowledge: *episteme*, *praxis*, and *techne*. The first, *episteme*, was thought of as "theory," demonstrated by logical syllogisms and geometric proofs and was the grounds of certainty. *Praxis* was a collection of particular experiences and could never generate certainty. Finally, *techne* was what produced effects and thus made knowledge productive but had no claims to certainty. In the late seventeenth century, this system was overhauled and the Aristotelian schema was reorganized. Episteme, praxis, and techne now came to be aligned in a way that made the certainty of the episteme reliant upon its ability to produce effects.[12] It is this ideal that is espoused by Wilson and everyone else in his wake who has reproduced the chariot metaphor.

Instead of stopping at the metaphor, if one reads further along the section in the *Susruta Samhita* from which the metaphor was extracted, a very different epistemology of practice can be found. The section provided a detailed list of the kinds of "practice" it saw as complementing knowledge of the shastras. These injunctions might be heuristically organized into three sets. Foremost was a set of injunctions about the maintenance of ritual purity (*shuchi*). Second was a group of injunctions about personal comportment. These latter touch both upon mental and somatic states of being. For example, they instructed the medical pupil to wear an *uttoriyo* ("modesty scarf"), not be overeager in his mind, be brave, not be sleepy, etc. Finally, there is a set of injunctions about inculcation of *shastric* knowledge through memorization. Here again a number of specific directions were given as to exactly how to pronounce texts and how to memorize them.[13] Every gesture and action required by these three sets of injunctions was called a *karma* or "practice." "Practice" or karma clearly did not have the modernist designation of manual or clinical dexterity that was implied by Wilson and others. Nor indeed did this sense of karma-as-practice become the grounds for generating epistemic certainty by virtue of being able to produce effects. The karma-as-practice that the *Susruta* designated as the grounds of therapeutic prowess were not the sorts of practice engendered in Wilson's post–seventeenth-century epistemology of practice.

Yulia Frumer has recently pointed out that, when speaking of translation of technological practices between distinct historical cultures, the word *practice* often appears "to be all-encompassing and hence too coarse."[14] The problem, Frumer points out, arises partly through the collapse of distinctive notions such as "custom," "convention," and "habit" into a single monolithic concept of "practice" in our historical lexicon. In the case of the chariot metaphor, as it has been handled by modern authors since Wilson, the coarseness of the term *practice* has certainly been misleading. *Karma*, the word Wilson had glibly translated as "practice," was a much more polyvalent term. It included, and indeed continues to include, actions pursued with an eye to ritual purity, gestures aimed at aiding the memorization of textual knowledge, mental dispositions, and possibly certain specific types of clinical practice.

Moreover, the hierarchy between the mind and the hand implicit in the post-seventeenth century epistemologies of practice operationalized by colonial orientalists were difficult to sustain in pre-reform Ayurvedic therapeutics. As seen in the case of *nadipariksha*—one of the most iconic aspects of Ayurvedic therapeutics—an unevacuated bowel was believed capable of leading the physician's clinical judgement astray. Evidently, the mind did not rule the hand. Even the stomach could overrule the mind.

Throughout the section dealing with the chariot metaphor, the goal of the *Susruta* was to make the medical student fully dextrous in the shastras or *shastra-porag* in the words of the nineteenth-century Bengali translator, Jashodanandan Sarkar. But what exactly did it mean to be dextrous in the shastras? In order to fully comprehend the precolonial epistemology of practice, it is crucial to understand the true import of what it meant to become *shastra-porag*. To begin with, as Sheldon Pollock points out, *shastras* do not simply mean "texts." Prima facie, shastras are any form of verbally codified systematic knowledge. But in effect they stand for an eternal, infallible, and transcendent knowledge that precedes and shapes every possible activity—religious or secular. It "establishes itself as an essential apriori to every dimension of practical activity."[15] It is this inescapable and transcendent priority of the shastras that renders modern notions of "practice" as touchstone of "theory" utterly redundant in classical Ayurveda.

It is also thus that the karmas listed under the chariot metaphor were all aimed at producing shastric dexterity: ritual purification of the body, proper modes of reciting the verses, appropriate intonations, the right pitch of the voice, etc. These karmas were what maybe defined as the "arts of learning," including within them both "arts of memory" and "arts of utterance." The body in its totality was deployed here as a palimpsest upon which transcendent shastric knowledge is inscribed. Marta Hanson has recently commented that "Chinese publications from at least the eleventh century through the early twentieth century attest that . . . manual arts of memory practices were as varied as they were widespread throughout the traditional domains of Chinese knowledge well beyond medicine."[16] While there has unfortunately been no study of embodied learning practices in Indic medical traditions, the brief section in the *Susruta* certainly suggests that the bodily "arts of learning" in South Asia were no less rich or diverse than in China.

In this regard, it is worth underlining that the emphasis on ritual purity or *shuchi*, which is steadily elevated in postclassical texts, cannot simply be reduced as a matter of the exercise of caste power.[17] While ritual purity was most certainly caught up in issues of caste-based social authority, its grounding in medical theory could only be sustained because the physician's body was itself a key therapeutic instrument.

In fact, tracking the genealogy of the chariot metaphor in classical Indic texts is in itself revealing. Dominik Wujastyk points out that the very metaphor of the chariot is well known in a story found in the *Katha Upanishad* wherein the self is described as a "chariot owner" and the body his chariot.[18] Rather than the orientalist epistemology that clearly distinguished the "theory" and the "practice" and obscured the physician's body as an instrument of knowledge, the *Katha Upanishad* story and its popularity in the classical textual corpus suggests that the *Susruta* may originally have deployed the chariot metaphor precisely to underline the centrality of the physician's body in classical Ayurvedic therapeutics. "Knowing" in this case is similar to Stacey Langwick's description of Tanzanian healing epistemologies wherein it "is not about acquiring a store of information that a healer can draw on in order to address discomforts and disabilities; 'knowing' takes place when a healer has cultivated his or her sensitivities to a range of relations with human and nonhuman actors and welcomes useful revelations at critical moments."[19]

The physician's mindful body, though clearly conceived of as a technology—that is, a chariot—did not remain unchanged since the time of *Susruta*. In Gobindadas's *Bhaisajyaratnabali* we have an important and locally hugely influential version of how this technology had evolved since the classical period. The most obvious shift was the explicit refashioning of the body instrument within the emergent local notions of caste. As I have noted in chapter 2, the eighteenth century was a period of great social ascent for the Baidya caste under the stewardship of Raja Rajballabh. Gobindadas's reworking of the form and function of the physician's mindful body instrument was also shaped by this emergent status of the caste. Gobindadas emphasized that any medicine prepared by one not belonging to the Baidya caste was to be "untouchable" (*osprishyo*). Even if a patient mistakenly consumed a medicine prepared by a Brahmin, the consequences were dire. For the lowest castes, they would have to perform ritual penance, while the higher castes would permanently lose their caste status.[20]

Whereas in the *Susruta Samhita* a physician could achieve ritual purity (*shuchi*) by engaging in certain acts of purification and observance of set rules, by the eighteenth century in Bengal this had emerged into a much more firmly "genetic," though still equally embodied, quality. This is not to suggest that only hereditary practitioners were physicians in the eighteenth century. As I have demonstrated in chapter 1, there is enough evidence to suggest that the Baidya caste, despite its astounding social ascent in the period, was not sealed off and continued to admit new entrants into its folds. Yet the emphasis on caste identity as a marker of ritual purity and its connection to therapeutic success most likely made embodied social identity a much more conspicuous factor in therapeutic considerations. In any case, it showed how the physician's mindful body instrument continued to evolve through new figurations of the physician's body and its material role in therapeutics.

Instrumentalizing the Diagnostician's Body

The influx of small technologies produced and nurtured a new breach between "sensory data" and reasoning. Instead of the body instrument being materially entangled in the diagnostic process, it gradually became a mere medium through which sensory data was collected.[21] Ironically, it was precisely at this moment when the body was being robbed of its embodied powers of diagnostic reasoning, and thereby being reduced to the status of an unthinking tool, that its status as technology akin to any of the other technologies came to be vigorously denied. Whereas earlier texts had unabashedly equated the body to technological objects such as the chariot, now the body came to be contrasted dramatically with the various tools that the physician regularly used.

This double move whereby the physician's body was denied any power of diagnostic reasoning, and yet promoted as being much more than a tool, was most clearly visible in Nagendranath Sengupta's *Kobiraji Shiksha.* On the one hand, Nagendranath insisted that the body was merely a medium through which necessary sensory data was obtained, while, on the other hand, he also presented all technologies as mere extensions of bodily capacities, thereby in turn elevating the body to a unique and privileged position. In the process, Nagendranath also opened up a clear divide between textual knowledge and sensory data. Whereas earlier texts had constantly sought to inscribe textual knowledge through memorization into the realm of bodily instincts, Nagendranath clearly separated the sensory knowledge obtained through the body from the textual learning.

In Nagendranath's view, there were three components to diagnostic reasoning—*shastropodesh* (instruction in shastras), *protyoksho* (witnessing), and *onuman* (deduction)—and these were organized along a linear chronology. Rather schematically, he said that the physician had to begin by witnessing and then, comparing the information thus obtained with the shastric instructions he had, the physician had to deduce the illness. Through this algorithmic structuring of medical reasoning, Nagendranath not only displaced the earlier emphases on embodied reasoning, but also organized the entire reasoning process into a linear, temporal framework that was akin to the post–seventeenth-century European model. Since Nagendranath's algorithm placed the greatest emphasis on clinical witnessing, this is what he spent the greatest time developing.

It is useful here to remind ourselves that precolonial texts such as the *Bhaisajyaratnabali* had clearly eschewed any such notion of clinical witnessing. It had stated with ample clarity that, for a diagnosis to be effective, one needed to be textually learned and industrious. Skill in various diagnostic techniques ranging from *nadipariksha* and *mutropariksha* to *mukhopariksha* was strongly recommended for a proper diagnosis. But these were certainly not presented as mere passive acts of witnessing.[22] The seeing and the thinking could not be neatly separated in these acts of diagnosis.

By recoding these complex techniques which almost always involved some form of embodied reasoning with a formulaic abstract form of deduction, Nagendranath was able to engender precisely the sort of stable subject, disconnected from its object of examination, which could be the destination of empirical knowledge derived from sensory data. He therefore catalogued precisely the kind of information each sense could obtain. Whereas Gobindadas had spoken of diagnostic acts such as *mutroporiksha* as a specialized form of analysis where various senses and reasoning were combined, Nagendranath mentioned "colour, size, slim-ness or obesity, beauty, faeces, urine and eyes" as things to be observed. Visual witnessing or *protyoksho darshan* came to replace a specific form of analysis or *poriksha*. Auditory, olfactory, and gustatory forms of witnessing were similarly promoted.[23] Once this move had succeeded in separating the knowing subject from the data through which it knew and the body through which it acquired such data, it was only a matter of time before sensory prosthesis and enhancers were being advocated.

Nagendranath argued that the senses had limits and, for one reason or the other, it was not always possible to use them. The taste of urine or feces, for instance, could not be determined by the use of one's own taste buds. As a result, he recommended the use of other creatures such as ants. If ants flocked to a urine sample, one could be sure of its sugariness. Similarly, the incidence of lice or too many flies being attracted to a patient's body signaled the preponderance of sweetness in the body. Offering a blood sample as food to a chicken or dogs could likewise help discern the qualities of the blood. By dislocating the acquisition of sensory knowledge onto the bodies of lesser animals such as ants, dogs and chicken, an equivalence was set up between the body's own senses and the actions of the animals. Embodied human perception was therefore neither unique nor irreplaceable for the medical reasoning. The senses unthinkingly gathered information and it was in processing that information, through a purely abstract cognitive act, that medical reasoning emerged.

It was through this framework that Nagendranath could also seamlessly introduce a range of new small technologies as doing little more than supplementing or extending the human perceptual apparatus. Whereas the earliest modernizing authors such as Gopalchandra, as seen in chapter 3, had struggled to reconcile the contradictions generated by the use of devices like the pocket watch, Nagendranath noticed no contradictions whatsoever. In speaking of the stethoscope, for instance, Nagendranath stated that "*Akornon* is the name given to the examination of the various sounds of the chest through the medium of the auditory sense (*srobonendriyo*). This can be done in two ways, i.e. immediate (*protyoksho*) and mediate (*poroksho*) or through a machine (*jontrodwara*)." Proceeding further, he stated that these two modes were exactly alike. The physician could either put his own ear directly to the patient's chest or he could use the stethoscope. "For a variety of reasons, the stethoscope was used more often," he pointed out, though, in some special cases such as when treating infants the direct approach might also be taken.[24] While Nagendranath's terminology—especially given his slightly different use of the word *protyoksho* as witnessing only a few pages earlier—acknowledged the new layer of mediation introduced by the stethoscope, his entire discussion denied that such mediation is in any way significant. By collapsing the difference between mediated and immediate use of the sense of hearing, Nagendranath rendered the stethoscope a mere extension of the physician's own bodily sense of hearing. This can only be done if sense perception is refigured as a purely data-gathering mechanism devoid of any cognitive aspect.

From the use of ants and dogs to the use of stethoscopes and thermometers, Nagendranath promoted a vision of technology as an extension of the physician's body. In so doing, however, he also rejected the capacity of embodied reasoning in the physician's body. The physician's body, as Nagendranath conceptualized it, was a purely instrumentalized entity: unthinking, mute, and entirely subordinated to an abstract and disembodied linear reasoning mechanism he had outlined at the outset.

Once the physician's body had been thus instrumentalized, the earlier injunctions about ritual purity of the physician's body lost all therapeutic significance. Instead, developing further along the lines suggested in Gobindadas's eighteenth-century text, modern Kobirajes turned the older injunctions about ritual purity into a question of caste monopoly defended in the name of tradition. A lengthy essay written by an eminent Kobiraj, Satyacharan Sengupta, in the leading Ayurvedic periodical of the day in 1923–24, bears ample witness to the changes in the ways in which the physician's body was conceptualized. On the one hand, the author lamented that, whereas in the past "only Baidyas used to be Ayurvedic physicians . . . nowadays from Brahmins to Chandals (a ritually low caste) everyone has earned the right to practice Ayurvedic medicine . . . in Calcutta, the city of wonders, one even notices three or four Muslims who have opened up Ayurvedic clinics."[25]

The disdain and lament was backed up by copious Sanskrit quotations to justify that only Baidyas had the right to dispense medicine and that anyone who consumed medicines prepared by other castes, high or low, stood in violation of religious obligations. Despite Satyacharan's obvious concerns and his attempts to bolster the Baidya monopoly, he made absolutely no attempt to connect the embodied caste identity of the Baidya physician to the therapeutic value of his diagnosis. He defended the monopoly as a matter of religious obligation backed up by scriptural quotations, not a diagnostic necessity. Not once did he say that the Brahmin, Chandal, or Muslim Kobirajes were bad physicians. His point was simply that being treated by them would contravene religious obligations.

By thus separating religious obligation, social identity, and diagnostic value of the physician's body into separate spheres, Satyacharan could then, just as Nagendranath had done before him, speak unabashedly of the supplementing of the physician's embodied diagnostic capacities by mechanical devices. Working this argument for supplementing into a general rhetoric of decline, he said that since the modern Kobirajes no longer had the "power to diagnose by touching the *nadi*," they ought to embrace the use of the *Tapman-yantra* used by *daktars*. Similarly, lacking the powers of yore to examine the patient's chest, they ought not to be embarrassed to adopt the stethoscope. Finally, since modern Kobirajes lacked the proper knowledge of yogic purificatory practices like enemas (*bostikriya*), he felt they ought not to shy away from the use of injections.[26] Satyacharan's fundamental move was to locate historically recognized precedents (though these were occasionally quite far-fetched, as in the case of enemas and injections) for modern small technologies and then urge their uptake on the grounds that the embodied skills of the modern physicians had declined since the classical age. Yet, interestingly, this decline was not attributed to any change of the physician's body itself, but rather to the fact that they were no longer properly trained and committed as they had once been. The physician's body was merely a diagnostic instrument whose capacities depended upon erudition and commitment, not on any embodied, pre-reflexive quality as such.

The physician's body that emerged through the writings of Nagendranath and Satyacharan among others was a body in which the biological and the social were neatly, separately, and hierarchically organized. By contrast, the physician's body available in precolonial texts was much more resolutely biomoral; the biological and the moral were deeply and inseparably intertwined with each other.[27] As a result, in these earlier figurations the moral valence of the physician's body—whether calibrated to ritual performance (as in *Susruta*) or birth (as in Gobindadas)—had a direct relationship to its diagnostic function. However, in the emergent modern physician's body, social duties attached to religious affiliations and the diagnostic performance could be entirely decoupled.

In this regard, it is also worth noting Rachel Berger's argument about the communalization of biomorality from the 1920s.[28] In Bengal, as can be seen from Satyacharan's comments above, this communalization did not pit the Hindu against the Muslim. Rather, because of the entanglements between Baidya caste politics and Ayurvedic modernization, the Baidya is pitted against the Brahmin, Chandal, and Muslim together. The community that underwrote this communalization, therefore, is not the homogenized "Hindu" community, but rather the "Baidya" caste community.

Practice as Pharmacy

As small technologies came to replace the bodily practices of embodied diagnostic reasoning, a new figuration of "practice" began to emerge. The clichéd rhetoric about the need for both practical and textual learning now evolved in a new direction. At the most obvious level, discussions of "practical skill" were relocated from the realm of diagnosis to that of treatment. Practice now gradually came to be equated with the knowledge and competence in pharmacy. A lengthy anonymous article published in the journal *Ayurveda* in 1329 BE (1922) which aimed to outline the qualities necessary in an ideal physician asserted that "to be a good physician, it is not enough to have read the shastras or been well-instructed, nor is it enough to possess certain innate virtues. One must be an able pharmacist."[29] Harking back to a Golden Age, the author posited that the decline of Ayurveda had resulted from the decline in the practice of pharmacy. The lament around decline caused by the overly textual learning of Ayurvedists and their neglect of "practice" goes all the way back to Wilson's regrets. By the early 1920s, this allegedly neglected form of "practice" had been identified with the practice of pharmacy.

The anonymous author was not alone in equating Wilson's regrets with a decline in pharmaceutical practice. In fact, one of the leading Kobirajes of his day, Amritalal Gupta, writing in 1916–17 in the same journal, *Ayurveda*, had made an almost identical point. Having invoked Susruta's famous chariot metaphor, Amritalal went on to explain with a list of examples what kind of karma or actions had been implied by *Susruta*. "The preparation of medicines etc. according to the proper rules and the proper alchemical processing (*jaron, maron, sodhon*) of metals like gold and semi-metals, [as well as] the making of [medicinal] oils (*toilo*), butters (*ghrito*), confectionaries (*modok*), molasses (*gur*), distilled spirits (*asob*), fermented wines (*orishto*), powders (*churno*), pills (*botika*), etc. according to the proper rules," Amritalal clarified, were the substantive instances of the kinds of practice insinuated by Susruta.[30] In a similar vein, in the following year, Jogendrakishore Loh wrote another lengthy article in *Ayurveda* on the importance of practical training. While Loh nominally invoked a broader notion of "practice" associated with clinical training and surgical skill, in effect, like Amirtalal and others, he focused mainly upon pharmacy. Well over three of the four pages that Loh's essay ran into were devoted to the importance of the practice of pharmacy.[31] These authors were not atypical. Numerous others too developed this discourse of practice as pharmacy. For some of them, like Amritalal, Ayurvedic practice meant exclusively the pharmacy; for others, like Loh, while practice notionally suggested a broader notion including clinical experience and surgery, in actuality it came down to the pharmacy once again.

This redefinition of "practice" allowed the modernizing Ayurvedists to appropriate the regret voiced by orientalists like Wilson and mobilize it for their reformist ends without necessarily introducing any dramatic changes to Kobiraji customs. In any case, in the absence of any Ayurvedic colleges till 1916 which in turn ruled out clinical experience, and the unavailability or distaste for cadaveric dissections that ruled out regular surgical or anatomical experience of any sort, meant that there were in effect very few forms of "practice" that a modern Ayurvedist could undertake. Pharmacy was naturally one of the things he could realistically undertake, and so recoding "practice" as exclusively comprising of pharmacy allowed the modern Kobiraj to both focus upon what was realistically possible as well as present it as a radical reform that would reverse the alleged decline of Ayurveda.

This reorientation of practice as pharmacy also led to the ascendancy of a particular type of medicines known as *pachons*. The anonymous author who asserted the need for Kobirajes to be able pharmacists then went on in the same article to explain how this new "practical" orientation would lead to a new emphasis on *pachons*. He explained that the pharmaceutical imagination (*bhesoj kolpona*) of Ayurveda comprised of a wide range of different types of medicaments including oils, butters, fermented wines, ointments, etc. By contrast, *pachons* referred to "only one small limb of that vast body of medicines." Yet, despite its humble size, *pachons*, the author reiterated, were often more useful and valuable than the entire set of other Ayurvedic medicines.[32] In a similar vein, the enormously influential Kobiraj Satyacharan Sengupta, who served as an editor of the journal *Ayurveda* for many years, wrote that simple *pachons* frequently produced stunning results in whose comparison the effect of more expensive formulations would pale into oblivion.

Sengupta too thus sought to exhort modern Kobirajes to make greater use of *pachons*.[33] Indubhushan Sengupta, a *Bhisagratna* (a qualification indicating specialization in pharmacy), argued that *pachons* acted much faster than other types of medicines. He also cited scriptural authority to suggest that the ancient seers had always acknowledged the vast superiority of *pachons* over other forms of pharmacy. Unfortunately, Indubhushan felt, Kobirajes tended to avoid *pachons* as they were relatively more difficult to make and patients disliked the bitter/astringent taste. Yet he urged all Kobirajes who sincerely wanted to uplift Ayurveda to take up *pachons* as their main instrument of cure and also convince patients about its powers.[34]

While these authors did not propose any all-encompassing definition of *pachons*, they did insist that the term included within it two other popularly known categories of medicines known as *mushtijog* and *totkas*. Technically speaking, *pachon* was defined in an important thirteenth-century text, the *Sarangadhar Samhita*, as "that which causes the digestion of the chyme (*am*) but does not light the [digestive] fire."[35] They could contain from one to numerous ingredients, so long as the total weight of the entire mixture was two *tolas*. The proportions of individual ingredients were expressed in terms of this two *tola* weight.[36] This technical definition of *pachon*, however, was seldom adhered to. Contemporary Bengali-to-English dictionaries describe it as either "a digestive, gastrive" or a "medicinal decoction."[37] The Rev. Lal Behari Dey described *pachons* as "aperient mixtures" in his 1874 novel, *Govinda Samanta*.[38] This gradual blurring of the names for digestive medicines and medicinal preparations in general is interesting in itself and might well be related to the widespread belief in the central role of digestion in pathogenesis. In most cases, they comprised of secret recipes that were passed along lines of discipleship and were crucial to the reputation of the physician. Dey's novel mentioned, for example, that "that [the particular Kobiraj] had a collection of the best and rarest medicines was a fact admitted by everyone in the village."[39] Though Dey's tone was sarcastic, the link between possession of these lists and a physician's reputation was clear. Another unsympathetic observer, Herbert Hope Riseley, recorded the mode in which *pachons* were administered. "A *pat* or *pachan*," he wrote, "that comprises of from nine to sixty ingredients is considered a [good] alternative tonic. The patient being given twenty-one powders made of a jumble of herbs, takes one daily and boils it in a seer of water until only a quarter remains; then straining and putting aside the sediment, he drinks the decoction. After the twenty-one days have expired, all the sediments are taken, and the decoction drunk for eleven days longer. Finally, the sediment is put into boiling water and with it the patient takes a vapour bath (*bhapra*)."[40] Most importantly, *pachons* in these descriptions were usually not to be found in canonical Ayurvedic texts. Their very value and their connection to the physician's reputation arose from their unique and secret nature.

From the mid-1890s, a number of collections of *pachons* were published. Most of these claimed to be disclosing secret medicinal recipes passed on for generations through familial or apprenticeship chains. The authors of these *pachon* collections included some of the most eminent modern Ayurvedists of the day. Debendranath, Upendranath, and Nagendranath Sengupta, for instance, all published their own respective collections of *pachons.* Whereas Debendranath and Upendranath emphasized the esoteric familial provenance of the recipes, Nagendranath combined these with prescriptions culled eclectically from canonical Sanskrit sources. In Nagendranath's collection therefore, the non-canonical, secretive, and familial origins ceased to define *pachons.*

Nagendranath emphasized a new and alternative definition of *pachon.* He confessed that formerly the term *pachon* was reserved for astringent digestives (*poripaker jonyo projukto koshay*). But this older meaning was no longer in vogue according to Nagendranath. Instead, any astringent (*koshay*) had come to be called *pachon.* This broader set of *pachons* could be divided into five constituent categories: namely, *Sworos, Kolko, Shrito koshay, Shito koshay,* and *Phanto. Sworos* was the name given to the essential juice or extract wrung from any material. *Kolko* was the juice extracted by pasting or grinding the material with a stone mortar and pestle. *Shrito koshay* was the juice extracted by boiling the material. *Shito koshay* was the extract obtained by leaving the material immersed in water overnight. *Phanto* was the extract obtained by first immersing the material in warm water and then pasting it in a mortar.[41] The five types were progressively more and more easily digestible. Thus, the *sworos* took the longest time to be digested, while the *phanto* took the least time.

Nagendranath's definition seems to have caught on. In 1920, just nine years after the first publication of *Pachon o Mushtijog,* his book was in its fourth edition.[42] Moreover, in 1927, Ashwinikumar Chattopadhyay wrote another book on *pachons* that was essentially a paraphrasing of much of the material presented in Nagendranath.[43]

Even as these new definitions began to take root, the importance of *pachons* continued to grow within modern Ayurvedic pharmacy. Besides the regular columns that started to appear in the various Ayurvedic periodicals, the numerous advertisements for Ayurvedic products in the general press attested to the rising preeminence of *pachons* as the main form of medicine. In the early to mid-1890s, Kobiraj Lakshminarayan Ray's *Bharat Ayurveda*

Aushadhalya was one of the most regular advertisers of Ayurvedic pharmaceuticals in the *Amrita Bazar Patrika*, the leading Indian daily newspaper of its time. Their advertisements listed a number of their leading products, but none of these were advertised as *pachons*.[44] Bijoyratna Sen's much more eminent firm which also advertised its leading products regularly in the *Amrita Bazar Patrika* in the 1890s, also avoided *pachons* altogether. Their focus instead was on pills, oils, and musks.[45] Around a decade later, in 1904, however, new advertisers focused almost exclusively on *pachons*. Thus, Srischandra Dutta's *Peacock Chemical Works* advertised five products and all five were explicitly described as *pachons* or *koshays*.[46] Around the same time, the eminent Kobiraj Binodlal Sen's *Adi Ayurveda Oushodhaloy*, one of the most successful firms of its day, regularly published full-page advertisements in the *Bankura Darpan* highlighting its flagship products among which the majority (four out of six) were *pachons* or *koshays*.[47]

The earliest collections of *pachons* had also begun to appear from around the mid-1890s but their popularity only grew in the 1910s. Debendranath and Upendranath Sengupta's coauthored collection appeared in 1895.[48] Thereafter, it took almost a decade for the next collection, that of Kaliprasanna Bidyaratna, to appear in 1906.[49] In fact, despite the Sengupta brothers having been one of the most widely read Ayurvedic authors of the time, their *pachon* collection remains one of their least successful works. Thus, while some of their other works remain in print till this day, their *pachon* collection is virtually entirely forgotten. The first edition is difficult to find today even in the better-stocked libraries and it is usually the third edition, published in 1911—incidentally the same year as Nagendranath's collection first appeared—that remains more easily accessible. Both the enormous success of Nagendranath's collection in the 1910s and the numerous journals that carried columns on *pachons* clearly shows that it was from the second decade of the twentieth century that *pachons* came to dominate modern Ayurvedic pharmacy.

In fact, so dramatic and explicit was the shift that its reverberations were felt even in the religious culture of the day. Thus, the iconic representations of the mythic Dhanwantari, whom the medical orientalists had called the "Hindu Aesculapius" and who in Bengal was also considered a mythic ancestor of the Baidya caste, were radically transformed precisely at this time. In nineteenth-century lithographs and woodcuts, Dhanwantari had usually appeared in conspicuously mythic time. Perhaps the most popular visual trope showed him as emerging from the cosmic oceans during the churning of those oceans by gods and demons (see fig. 11). The image shows him

11. Dhanwantari in cosmic time.

12. Dhanwantari as pharmacist.

surrounded by a host of others divine and supernatural beings and emphasized his divine and exalted status. He was also shown sitting with a *punthi* (traditional manuscript-book) in hand. By contrast, a woodcut included as a frontispiece in a Hindi edition of the *Charaka Samhita* dating from the early 1920s shows him as a pharmacist surrounded by innumerable bottles displayed in glass-fronted almirahs (see fig. 12). Though he is still shown engaged in *nadipariksha*, it is clear that the bottles, the almirahs and the other accoutrements of pharmacy, such as the mortar and pestle, dominate the iconography. What these signposts confirm is precisely what can be surmised from the various articles in the journal *Ayurveda* that equated practice with pharmacy—namely, that *pachons* were not always equally important to Bengali Ayurvedic pharmacy. Their importance is a relatively modern phenomenon and seemed to peak in the second and third decades of the twentieth century.

By the 1910s, "practice" had come to be equated with the knowledge of *pachons*. And these extracts or astringents were beginning to displace all other types of pharmaceuticals. Naturally, it is important to ask why *pachons* emerged as the dominant form of Ayurvedic medicine at the time.

The Liquefaction of Modern Ayurveda

Satyacharan Sengupta provided a partial answer to this conundrum. Contrasting contemporary practice with a dimly remembered prior era, Satyacharan wrote that "the medicines of yore were neither stored in stoppered phials not neatly displayed in glass almirahs. Kobirajes did not usually keep prepared medicines with them. They wrote out a list of ingredients (*phordo*) [and instructed the patient's family to prepare it]. In the rare cases where Kobirajes did keep ready-made medicines, they were stored in a carpetbag (*puntli*) or an earthen pot (*handi*). But today a Kobiraj who fetches his medicines out of a carpetbag will hardly be respected."[50] In an almost identical mocking tone, another Kobiraj commented on how modern Kobirajes felt it was enough to "display medicines of many colours in bottles that were arrayed in glass-fronted almirahs."[51] Bottled medicines thus had come to represent modernity and respectability. Kobirajes who refused to deal in bottled medicines or carried carpetbags were not considered respectable.

Medicine bottles, despite being perhaps the most iconic material form in which people in the nineteenth century saw and handled medicines, have remained utterly neglected in medical history. While representing old medicine seems to almost instinctively call to mind old medicine bottles, they are seldom engaged with either as material objects or as modern packaging

G.N.ROY'S

Invaluable remedies of different diseases

PREPARED STRICTLY ACCORDING TO THE FORMULÆ OF AYURVEDA &c.

Kalpatarusuda.—The wonderful blood purifier &c, 1 Phial Rs. 2.

Kandarparasa.—A sure cure for impotency &c. 1 Phial Rs. 3-8.

Jarantakatsub.—The best anti-malarious mixture &c. 1 Phial Rs. 1-3

Chitaronjan Oil.—Specific of fits and Brain-diseases &c. 1 Phial Rs. 1.

Hair Vigor.—It restores grey hair to its natural color &c. 1 Phial Rs. 1-12.

All the above preparations are harmless.

Sir Raja Radha Kant Deb's Rajbari, SOVABAZAR.

13. Bottled medicine as a sign of modernity.

technology. Thus Kavita Sivaramakrishnan's excellent history of Ayurvedic modernization in the Punjab, for instance, is called *Old Potions, New Bottles* and yet does not really talk about bottles at all. This is of course not at all unusual. Very little attention has been paid to this humble medical packaging technology. Yet, as is clear from Satyacharan's comments, the bottles were in themselves important signifiers of modernity.

William Ward, the Baptist missionary, writing over a century before Satyacharan, had stated that "when a Hindoo doctor goes to see a patient, he takes with him, wrapped up in a cloth, a number of doses in cloth or paper. He has no use for bottles, every medicine being in a state of powder and paste, liquids, when used, are made in the patient's own house."[52] Though Ward's tone is only mildly critical, it is clear that the absence of medicine bottles were already becoming a signifier of lack in Kobiraji medicine. More than half a century later, Shibchandra Basu in 1883 still reported that the Bengali Kobiraj carries with him "different kinds of pills and powders, wrapped up in a paper, in small doses . . . he seldom uses phials; liquids, when required, are made in the patient's own house."[53]

It is difficult to accurately pin down when Kobirajes began to adopt bottling as a medical packaging technology. By 1887, however, at least one regular advertiser in the *Amrita Bazar Patrika* had adopted bottles.[54] G. N. Roy's *Kalpatarusuda*, regularly advertised in the newspaper, was sold in glass phials. Yet this was most likely a precocious and pioneering effort. For the bottle was still conspicuous enough for Roy to actually adopt it as his brand-mark. His advertisements thus always carried the image of a bottle prominently across the top with the name G. N. ROY's written on the image.[55] That Roy's move was precocious is also suggested by the fact that other regular advertisers, including the well-known firm of C. K. Sen & Co., which was headed by Debendranath Sengupta, was still advertising medicines exclusively in the form of pills and sold by weight in paper packets.[56] Even in the early 1890s, while some companies such as Roy & Co. had switched to predominantly liquid medicines sold in glass phials, others such as B. Basu & Co. continued to deal exclusively in pills sold in packets.[57] Over the next decade and a half, the use of bottles continued to grow in Ayurveda. Yet bottled medicines formed only a small part of the medicinal repertoire of Ayurveda. Usually only medicated hair oils, blood purifiers, and a handful of other, more generally used, drugs were sold in bottles. Everything else remained in non-liquid form.

Remarkably, the demand for medicine bottles peaked at exactly the same time that Ayurvedists started to emphasize *pachons* as the dominant form of medicine. Thus in 1911, when Nagendranath's important collection of *pachons* was published and Debendranath's forgotten collection republished, a new company called the Upper India Glass Works was opened up in Ambala to manufacture glass bottles for medicines in India. The nationalist newspaper *Amrita Bazar Patrika* welcomed the new effort by pointing out that India annually imported glass bottles in large numbers from Germany and Japan and hence constantly drained wealth to these countries. The "drain of wealth" had emerged as a major ideological plank of early nationalism and anything that required money to leave the country was seen to be hurtful to the nation.[58] The new company, however, lacked the know-how to produce glass bottles and had to bring in an Austrian glassblower as the director of the process while at the same time apprenticing several Indian youth under the Austrian to learn the process.[59] By 1920, Calcutta had become the biggest market for medicine bottles in all of British India. But bottles still had to be brought in from up-country. Most bottle manufacturers were located in Allahabad and the freightage and breakage added considerably to bottling costs. During the Great War, however, Calcutta suffered a bottle crisis. Companies such as Shaw & Bros., who used to manufacture a well-known brand of hair oil, found it difficult to procure bottles for their products. As a result, Birendrakumar Shaw started the first bottle manufacturing concern in Belgachhia near Calcutta. Soon they found that production costs were much lower in Calcutta and the business blossomed. In May of 1920, the glass company of Birendrakumar was turned into a public limited concern with a total capital of Rs. 15,00,000.[60] The Calcutta Glass and Silicate Company Limited continued to thrive and, with the commencement of the Gandhian national movement, began to position itself as a "national" concern.[61]

As both the supply and demand for medicine bottles grew in Calcutta, the number and range of bottled medicines grew manifold. Instead of a small range of general purifiers and hair oils, a whole host of vegetable extracts came to be sold in stoppered glass phials. In 1911, for instance, the famous Sanskrit scholar, E. J. Lazarus, was advertising the "essence of chiretta"—a classic *pachon*—at Rs. 1.8, Rs. 2.8, and Rs. 4 per bottle.[62] The year before, P. M. Bagchi advertised a "brain tonic" made of the essence of several exquisite flowers.[63] About a decade hence, C. C. Ghosh advertised a range of remedies including one containing the essences of neem, gulancha, etc.[64]

These new organic essences naturally required a fairly professional knowledge of chemistry for their distillation. Brahmananda Gupta has outlined the number of chemists who went on to found successful early Ayurvedic pharmaceutical firms. Mathuramohan Chakrabarty, a poor chemistry teacher at a school, for instance, founded the hugely successful Shakti Aushadhalaya of Dhaka in 1901. Later, Jogeshchandra Ghosh, yet another chemistry teacher, this time from Bhagalpur, followed suit and established the Sadhana Aushadhalaya.[65] There were also obvious overlaps with the powerful discourse on "indigenous drugs" and, through it, with organic chemistry once again. In fact, 1910s was precisely when the discourse on "indigenous drugs" was most prominently subverted by the creation of a common pharmacopeia applicable throughout the British Empire.[66] It would therefore make perfect historical sense to expect modernizing Ayurvedists to become more strident in their efforts to assert the indigeneity of their "indigenous plants" in the face of the imperial centralization.

Interestingly, in the context of biomedical pharmacy, this was a time of transition wherein the "figure of the pharmacist as an expert in *material medica*" was gradually replaced by the twin figures of the "apothecary turned retailer" and the more prestigious "industrial entrepreneur who owned factories." Through these new figures, medicinal plants or rather their extracts, though "less visible than hormones or vitamins," had become major mid-century biologics.[67] Jean-Paul Gaudilliere defines "biologics" as "not just therapeutics made out of biological things" but rather as "technical objects both socially and culturally associated with nature and elaborated through industrial mass-preparation processes."[68] Colonial biologics, however, Gaudilliere clarifies, were somewhat different in that they therein "standardization acquired a different meaning that was unrelated to laboratories and/or industrial plants, but was centred on the problems of plantations and local agriculture."[69] Ayurvedic biologics were still different. As technological objects, they were not completely innocent of laboratories. The involvement of a number of chemists attests to as much. But neither were they completely dominated by the industrial scale production that underwrote biologics in Europe. Ayurvedic biologics, while sharing some features with biomedical biologics, retained some of their distinctive character in the way their standardized production and distribution was organized.

To focus on these larger scientific developments, however, would risk overlooking the role of much humbler, more everyday forms of small technologies such as medicine bottles in engendering therapeutic change. The growing importance of *pachons* among modern Ayurvedists, I argue, was at least partially the result of the fact that medicine bottles had become iconic signs of medical modernity. As a result, the old pills, pastes, and powders, bundled into a carpetbag, came to be seen as backward and unrespectable. The shift toward *pachons* enabled the adoption of medicine bottles and thus the articulation of a certain medical modernity. Yet, this shift could not have taken place without the parallel redefinition of the meaning of "practice." Together, the recoding of practice as pharmacy and then the narrowing of pharmacy further to *pachons* dovetailed into the simple and conspicuous fact that Ayurvedic medicines were now readymade and bottled.

This shift toward liquid *pachons* sold in bottles naturally required a parallel recalibration of the Kobiraj's embodied skills. His body, as before, remained a crucial technology in the therapeutic regime, but now needed a range of new capacities. These new bodily capacities were only obliquely hinted at in the *pachon* collections that were published. Not only did the Kobirajes have to learn a number of formal aspects of *pachon* making, but they also had to develop new types of tacit knowledge. "Tacit knowledge," as Harry Collins points out, is knowledge that is not explicit, but whether this lack of explicit articulation is a matter of choice or epistemic necessity cannot be fully untangled.[70]

On the more formal side, most importantly they had to remember elaborate substitution schemas that told them which ingredient could be substituted by which other ingredient under what specific conditions. They also had to learn the qualities of specific plants growing under specific conditions and an entire complicated system of weights and measures. Finally, the formulas in published collections as a rule did not mention what part of a herb was to be used, and so the Kobiraj had to know what part of the plant was to be used in which preparation.

Besides this formal knowledge with its emphasis on memorization, there were also new embodied capacities that had to be developed. The lowest weights, for instance, could not actually be measured. The smallest unit of measurement was the speck of dust one can observe in a beam of sunlight coming through a crack in the window. The higher weights gradually built upon this basic unit of weight called *bongshi*. In actuality, it was clearly impossible to measure a speck of dust seen in a beam of sunlight. Neither therefore were the units immediately above the *bongshi*, such as the *morichi* (made of six *bongshis*) or *rajikas* (made of six *morichis*).[71] Clearly then, such measures when required would need a tacit visual sense on the part of the *pachon*-making Kobiraj. Similarly, since a number of the *pachons* required heating and since there is no indication of thermometers being integrated into *pachon* making at this stage, the temperature at which something is to be taken off the boil or something else added to it, relied on the physician's *andaj* (conjecture/surmise).[72] Once again, however, this surmise could only be based on an embodied tactile and visual apparatus. Finally, the much more elaborate use of various ferments ranging from sugarcane juice, to molasses and boiled rice, all required a delicate sense of timing that relied largely on visual and gustatory estimation.

Even more strikingly, *pachon* making invoked a range of biomoral aspects of the physician's body. Ritual purity, or *shuchi*, that had once been important to make the right diagnosis, was now deemed necessary to ensure the potency of the *pachon*. Loh pointed out, for instance, that if there is any kind of *oshuchi* (ritual impurity) anywhere in the body at the time of the making of a medicine, not only is its potency lost, but its taste and smell too become perverted. In fact, according to Loh, such consequences would follow even if the mere shadow of a person who was *oshuchi* were allowed to fall upon the medicine while under preparation.[73] In a statement strongly reminiscent of the injunctions against *nadipariksha* without having cleaned one's bowels, Loh mockingly said that "in the past physicians used to collect herbs respectfully and after the performance of appropriate ritual prayers, whereas today he collect herbs on his way back from the toilet."[74] Nagendranath's collection of *pachons* also recommended appropriate ritual rules for the collection and preparation of the *pachons*. It was clear that, not only was the physician's mindful body an important technological device for the production of the *pachon*, but that this device was in itself a potent biomoral entity rather than simply a material tool.

In short, then, the liquefaction of the modern Ayurvedic therapeutic repertoire did not erase the technological function of the physician's mindful body. Instead, it recalibrated it and developed new capacities that were both reflexively inculcated as well as inscribed into the realm of instincts, while also reconstructing the biomoral valences of the physician's active body.

Maestro to Gem Collector

Each of the preceding physiograms has been concerned with the way in which the body of the patient was being therapeutically understood or operationalized. In this chapter, however, we have turned the lens in the opposite direction and engaged instead with the body of the physician. I have argued that not only was the physician's body in itself a technology of crucial importance to Ayurvedic therapeutics, but also that it was the changes in the technological function of the physician's body that resulted in dramatic shifts in the actual treatment regimes of modern Ayurveda.

Broadly speaking, the practical deployments of the physician's mindful and biomorally constituted body moved from being essentially a diagnostic technology to a pharmaceutical technology. Precolonial Ayurveda privileged the physician's practical skills in the arena of diagnosis. *Nadipariksha* in particular was a "practice" that required complex and yet subtle deployments of the physician's body. By the 1910s, however, this had shifted. As the diagnostician's body changed from being a site of embodied knowing to a mere medium for collecting datum, "practice" increasingly came to be understood as acts of preparing *pachons*. This transformation of the physician's body and the shifts in its technological function from diagnosis to liquid pharmacy were book-ended by two contrasting physiograms: a maestro and a gem collector.

In the *nadipariksha* texts, which repeatedly invoked the technological function of the physician's body and the need to deploy it properly, the physician's body too, was repeatedly compared to that of a musical maestro. Thus, Pandit Salimuddin Bidyabinod in his versified *nadipariksha* text wrote that "just as all tunes are expressed with a vina in hand/ All diseases are revealed most certainly with the *nadi* in hand."[75] The patient's body was rendered as a vina, and in order for it to express itself properly the maestro must know how to play it.

This image of the physician as a vina player was not a nineteenth-century invention. It derives from a text known as the *Yogaratnakar* that discusses *nadipariksha* at length. While there is lack of consensus among scholars on the actual dating of the *Yogaratnakar,* it was most certainly written around the mid-eighteenth century, if not slightly earlier.[76] The *Yogaratnakar* had mentioned that, just "as the instrument made of by [*sic.*] the union of fine wires emits out the various melodic tunes when it is stroked," similarly the *nadi* is able to express the various diseases of the body.[77]

Interestingly, the *Yogaratnakar* was not alone in comparing the patient's body to a musical instrument and the physician to a maestro. Another equally well-known and possibly even more important text for *nadipariksha* was the *Nadi-vigyan* of Kanad. The *Nadi-vigyan,* which unfortunately remains undated, figured the body of the patient as a *mridanga,* a type of earthen drum whose surface is held together by a number of tightly wrought leather straps.[78] The *Nadi-vigyan* and the *Yogaratnakar,* two of the most important texts in Ayurvedic *nadipariksha,* both therefore conceptualized the physician as a maestro.

Even more interesting is the evolution of *Yogaratnakar's* imagery in later, modern texts. Whereas the image of the *mridanga* in Kanad was a concrete image, the *Yogaratnakara,* by merely speaking of a five-string instrument, left enough ambiguity for later authors to develop it as they wished. Bidyabinod, as already seen, chose to think of the stringed instrument as a vina. Haralal Gupta, one of the most successful late nineteenth-century authors of a *nadipariksha* text also thought of it as a vina.[79] Binodbihari Ray of Rajshahi, who repeatedly features in the earlier chapters, however, thought of the instrument as a violin.[80] Sarva Dev Upadhyay, a contemporary Ayurvedic physician and historian, on the other hand, interpreted the image as a sitar.[81]

While the concrete musical referent changed in keeping with the historically fashioned musical tastes of individual authors, the basic image remained remarkably stable and widespread. Notable in this image was the explicit articulations of embodied skill, technique, and the importance of *obhyas,* or habituation by repeated practice. As Ray put it, "One can know everything by adding *obhyas* . . . / No harm is caused by not [being able to] see the *nadis*/ Just as you play the *sitar* with its notched strings/ So can you play the *violin* with un-notched strings."[82] "Practice" in this case was tied to the perfection of embodied skill and technique through habituation and putatively connected to the act of diagnosis.

By contrast, the image of the physician that emerged once "practice" had been relocated to the realm of pharmacy and narrowed to *pachon* making, is one of a gem collector. In a rather elaborate deployment of the image, Shitalchandra Chattopadhyay compared the Ayurvedic *materia medica* to a treasure trove. Each of the medicinal ingredients known to Ayurveda, he said, was a gem and their total number so large that it was beyond anyone's capability to count them. These gems, he alleged, were scattered everywhere and often secreted away by individuals unwilling to share it. The physician who made *pachons* had to collect both the ingredients as well as the knowledge about them. For Sitalchandra, the medieval Ayurvedic author, *Chakradatta*, who compiled a large number of new prescriptions, was above all a talented gem collector.[83] Once again, this comparison of medicinal ingredients to gems was far from unique or idiosyncratic.

Precolonial texts such as Gobindadas's *Bhaisajaratnabali* had already established the association of "gems" (*ratna*) and materia medica (*bhesaj*) in the popular parlance. Authors such as Sitalchandra continued to sustain and nourish that older usage. But the older image had stopped by comparing medicines to gems. In its modern form, because the image was deployed within a larger context devoted to the redefinition of "practice," the emphasis shifted to the act of collecting.

Nagendranath's famous *pachon* collection introduced itself to its readers by saying that "God has scattered all the medicines for all illnesses all around us. If only we know of them and undertake the slight trouble needed to collect them, we could avoid disease altogether."[84] An editorial in the journal *Ayurveda Hitoishini* lamented that "owing to our neglect we have lost many of the invaluable gems we possessed. If we do not undertake the slight labour to collect the few that are left, soon all will be lost."[85] Nabinchandra Dey, in a letter to the journal *Ayurveda Bikash*, stated that "readers [of the journal] should all attempt to collect the divine medicines scattered all around us and establish medicinal gardens."[86]

Interestingly, this emphasis on collection seamlessly conflated the actual collection of the herb with the collection of knowledge about its identity, therapeutic value, etc. By thus collapsing the physical object and its knowledge into a single indivisible "gem," the act of collection became less of a physical act and more an act of social observation. As a result, the act of collection almost never dwelled upon the actual physical act of collecting the herb. Rather, it became an act of ethnographic voyeurism. Nabinchandra Dey, who urged his readers to establish medicinal gardens, in his own turn spoke of how he had acquired his knowledge of medicinal plants by observing cowherds.[87] An anonymous reporter writing in *Ayurveda Bikash* similarly mentioned how he had discovered a new therapeutic use for a well-known local weed by observing a poor Muslim fakir collecting the plant. In another incident mentioned by the same author, a gentleman friend of his discovered a medicinal plant when he, upon instructing the gentleman's upcountry manservant to cut down an unwanted plant growing in his garden, had found the manservant hesitating. Upon inquiring, the servant revealed that the plant was actually a hard-to-find upcountry medicinal.[88] Nagendranath said it was mainly old women who had possessed the knowledge of herbs.[89] It was clear that the social locations from which the modern Kobiraj had to salvage these gems were outside his usual domain. The gems lay scattered in the hands of old women, Muslim fakirs and upcountry manservants.[90] One author colorfully wrote that the poor herbs were embarrassed from "being in the hands of lower class people" and therefore were trying to "hide their faces in shame."[91]

This generalized Othering of women, Muslims, and non-Bengalis would seem to fit neatly into the Hindu revivalist cultural project that many historians have identified in late nineteenth- and early twentieth-century Bengal. What complicated this seamless Hindu-ness was the strident caste politics of the Baidya Kobirajes. Their identities were seldom devoid of caste markings. As a result, the same ethnographic gaze that represented women, Muslims, and non-Bengalis as Othered and subordinated also represented a smaller number of Brahmins as such. Like in the case of old women, the representation of Brahmins was much more respectful and appreciative, but the articulations of Otherness were still difficult to miss.

It is clear that the ethnographic discourse through which the practice of collection was presented was an important site for fashioning the social identity of the physician. It would be wrong, however, to conclude that this identity was simply an unmarked Hindu *bhadralok* identity. Though admittedly rare, occasional collections attributed to Brahmins or even Baidya ancestors were not unknown. A serialized list of medicinal prescriptions, including one for making an amulet, published in *Ayurveda*, for instance, were attributed to a departed Brahmin Kobiraj and yogi from north Bengal, Jairam Lahiri.[92] The following year, an eminent Calcutta Kobiraj, Indubhushan Sengupta, published some prescriptions in the *Ayurveda* that he attributed to his ancestor, Ishwarchandra Sengupta Shiromani.[93] The social identity of the physician that was represented in these narratives was neither stable nor monolithic. It had always had a degree of ambiguity and was a work-in-progress. In each narrative of "gem collection," the social identity of the collector was inchoately worked out through the actual details of the action described. The proverbial devil was indeed in the details of the socially situated action itself. There was no set template for elaborating a stable identity of the physician.

This partial and performative nature of the physician's social identity had two very significant consequences. On the one hand, while sharing a large repertoire of images, tropes and sentiments with the Hindu revivalist cultural project of the period, modern Ayurveda maintained a degree of what is at best described as an anxious intimacy with the politics of Hindu revival. The seamless Hindu identity championed by the revivalists often revealed their cracks in practice. Thus, for instance, many of the Baidyas spoke passionately of a "Hindu medicine" and a glorious "Hindu past"; those very Baidyas then equally passionately recalled and defended precolonial textual injunctions against all non-Baidya castes, including Brahmins, from preparing or dispensing medicines.[94] In this regard, it is also worth pointing out that one of the most ardent promoters of the Baidya monopoly, Satyacharan Sengupta, was in fact the editor of the leading journal, *Ayurveda,* and often contributed over a third of the articles published in it in a year. The argument that Baidyas alone have the right or ability to make proper Ayurvedic medicines was therefore far from being a marginal argument. At least one of its most vocal proponents was at the heart of the modernizing project. It was in the practical elaboration of social identity through *pachon* making, therefore, that the politics of caste that animated modern Ayurveda in Bengal becomes visible once more.

Interestingly, this is a complementary instance to Joseph Alter's study of the modern braiding of yoga, science, and right-wing Hinduism. In Alter's account, what interrupts the smooth assimilation of yoga into a seamless Hindu nationalist project is the articulation of a broader, post-Western universalism by scientists like Dr. Kumar Pal of Delhi who insist that yoga is not just Hindu.[95] In this case, too, a tension between the seamless Hindu identity and the project of Ayurvedic modernity that unfolds, as Alter suggests, can be found precisely around the body. Yet in this case, the tension arises not from the broader universalism of the Yogic body, but from the narrower parochialism of the Baidya's caste body and the pan-Indian agenda of the Hindu nationalists.[96]

On the other hand, it is once again this performative and ever-inchoate aspect of the physician's bodily identity that allows for a constant synchronization of the different strands that constitute therapeutic change. Rosenberg, let us remind ourselves, had defined "therapeutics" expansively as "a complex interactive system, centring on the doctor-patient relationship but incorporating the specific physiological activity of the drugs, social relationships at the bedside, and the expectations of the participants as well as the views concerning the nature of the human body and the physiological basis of health and disease."[97] So long as I had focused on the bedside interactions and the body of the patient in the foregoing chapters, the interaction of these with the actual regimes of therapy had remained obscure. But the moment one accepts that the most important technology in the therapeutic milieu is in fact the physician's own body and turns toward it, the disparate strands come together. The politics of caste, the liquefaction of the Ayurvedic repertoire, and the transformation of maestros into gem collectors altogether engender a dynamic, evolving, and inchoate physiogram that doubles up as the hub that holds the entire "complex interactive system" in place.

Yet the very fact that I had to turn the lens around and focus on the body of the physician, rather than that of the patient, to be able to see the hub that holds the system together, is in itself not unworthy of note. It signposts an alternative figuration of Ayurvedic modernity. Like the "Gandhian modernity" described by Dipesh Chakrabarty in *Habitations of Modernity*, this Ayurvedic modernity complicates the stable division between the private and the public that organizes the hegemonic versions of "western modernity."[98] There is no stable interiority here called the "doctor's mind" that can fully objectify the patient's body while reducing itself to a pure, disembodied rationality. Despite all attempts to reduce the body to a mere medium for processing sensory datum, the Baidya's body is constantly and explicitly implicated in the embedded and interactive milieu within which therapeutics unfolds. It was perhaps befitting that under the contradictory pulls toward abstract, disembodiment and trenchant embodied reasoning, acting, moralizing, etc., the very notion of "therapeutics"—its etymology, its history, and its conceptual coherence—all began to come undone. Unless an unmoved Baidya held it in place.

Writing in the *Ayurveda* in 1921–22, the eminent romantic historian, Dineshchandra Sen, who was also a Baidya by birth, asserted that the word *therapeutics* was a testament to the long interactions between "East" and "West" and had come into classical Greek from the Pali word "*thera*" (as in "Theravada Buddhism"). The word *thera*, Dineshchandra claimed, had derived from the Sanskrit term *sthavira* meaning "unmoved." It is worth pointing out here that Dineshchandra was no Hindu chauvinist. His argument should not be confused with the Hindu revivalist arguments about Europe's debt to India. Ideologically, he is perhaps best described as a romantic Bengali cosmopolitan.[99] Pride of place in his account of "therapeutics" was reserved for the Buddhist king Ashoka. The word *mahasthavira* or the "Great Unmoved One" referred to the Buddha himself. According to Dineshchandra, it was this word that in Pali and Prakrit had been vernacularized as *thera*. It had traveled to Europe with Ashoka's Buddhist missionaries who had also taken with them classical Ayurvedic medicine. Hence, argued Dineshchandra, in Europe, the medicine had come to be known as the medicine of the *thera-put* ("sons of Buddha"). The supremely impersonal medical rationality of the "West" was thus rendered as a willful forgetting of that rationality's historical embodiment in Buddha's sons. Dineshchandra lamented that, by rejecting Ayurveda, European therapeutics was acting like an ungrateful son who turned his back on his father, while still carrying the latter's name.[100] The disembodied, abstract medical rationality of European therapeutics, in the hands of the romantic Baidya, had become an embodied, post-Western universalism that rejected abstraction in the name of bodily kinship.

CONCLUSION

The Pataphysics of Cosmo-Therapeutics: A Requiem

Part 1

Ayurvedic technomodernity in its earliest decades crystallized around a series of small, humdrum technological objects or instruments. Besides such obviously medical technologies, like the clinical thermometer, the microscope, the injection, and so forth, there were also those objects that we do not usually think of as medical technologies, such as pocket watches and glass bottles. All of these, to repeat a metaphor I used in the introduction, acted like a speck of sand on the oyster's back that instigated a gradual but fundamental reimagination of the way Kobirajes understood the body.

While these small technologies drove or motored the therapeutic change in Ayurveda, they did not do so by propelling modern Ayurveda to evolve in any specific or predetermined direction. They were simply instigators. The therapeutic change (in Rosenberg's more capacious understanding of that term) that emerged in the wake of such agitation was an open-ended project heavily reliant on creative agency of the modernizing Kobirajes.

The creative agency that I have described in the preceding chapters entailed an intricate act of braiding. Instead of seeing cultural interaction as an "encounter," I have argued that it is more appropriate to see "cultures" as spools that were constituted of innumerable internally diverse and heterogeneous threads. The modernizing Kobirajes, I have argued, pulled on threads selectively from "Indic" and "European" knowledges and braided them together into a series of physiograms.

These physiograms were neither universally shared by every Kobiraj, nor were they simply individual fantasies. They were, what I have called, middle-level generalities existing somewhere in between the levels of "culture" and "individual." Moreover, the physiograms were not just explicitly described anatomical figures. They were rather images of the body that could be found embedded in the theories and practices that came to constitute modern Ayurveda in its early decades.

In chapter 2, we find the physiogram of the "clockwork body" wherein the famous watchmaker analogy of William Paley's natural theology was repurposed through an intricate braiding of older Indic ideas to fabricate a new machinic physiospiritual body. This image of the body combined the seemingly incompatible poles of vitalism and mechanistic thinking. In the following chapter, a second physiogram crystallized around the clinical thermometer. Drawing upon a range of electromagnetic theories and braiding them with selective elements of the Ayurvedic thinking on *pitta*, modernizing Kobirajes overlaid a reticulate network of *snayu*s on top of the preceding physiogram. It was this uncentered reticulation that made it possible to speak of the body as one marked by various kinds of flows. The next physiogram, discussed in chapter 4, was secreted around the absent presence of the microscope. While microscopes and their powers were widely recognized by Kobirajes, few actually used them. Yet it was this elusive presence around which the new physiogram of the chiaroscuric man was concretized. By drawing upon both orthodox and heterodox "Western" ontologies of vision and braiding them with older South Asian ways of seeing, the chiaroscuric man figured the body as a translucent space where visibility was forever limited or deferred and the visible and the invisible were constantly blurring into each other. Finally, in chapter 5, the technology of injecting organ extracts—that is, organotherapy, and the then-emergent ideas about endocrinology—engendered a physiogram that I call the endocrino-chakric machine. Here the acephalous, reticulate but discrete body of flows was equipped with a set of six control centers in the form of chakras by braiding together strands of Tantrism with endocrinology. The new body image that emerged through the early decades of Ayurvedic technomodernity was therefore that of the clockwork body constructed like a machine but made of vitalized matter and overlaid by an acephalous reticulation of *snayu*s regulated by six chakras. The physiospiritual machine was internally undifferentiated and through it coursed a number of endocrines recast of humors.

While the physiograms are what connect the chapters, each chapter also provides detailed descriptions of a number of practical and theoretical innovations that were occasioned by the birth and maturation of Ayurvedic technomodernity. Such innovations ranged from new techniques for measuring pulse without an actual watch to a Vaishnava cell theory, apart from others. These innovations were part of the kaleidoscopic developments that constituted the totality of Ayurvedic technomodernity. Each of them does not enter directly into the construction of a relevant physiogram, but then physiograms are only a small window through which to map therapeutic change.

A summary map of the direction of the overall therapeutic change is provided in chapter 6. Here the physiogram illuminated by the discussion is very different from the foregoing physiograms. Instead of the body of the patient, this last of the physiograms is an image of the body of the Kobiraj himself. As the meanings of "practice" changed, the Kobiraj's body evolved from a diagnostic tool to a pharmaceutical instrument. This change also required a change in the physical form of the Ayurvedic pharmacopoeia, mutating from powders and ointments to mainly bottled liquids. The iconic appeal of glass bottles and new biological extracts were part of the symbolic resonance of Ayurvedic technomodernity that informed the changing image of the Kobiraj's body.

The liquefaction of the Ayurvedic therapeutic arsenal also signaled a general shift from electromagnetism toward endocrinology that marked the history of Ayurvedic technomodernity in its early decades. By the 1930s, modern Ayurveda had begun to take on a distinctively new shape. A new body image, that of the endocrino-chakric machine, had taken shape and was in the process of becoming institutionalized through the numerous Ayurvedic medical colleges and professional journals. The modernization process had succeeded in establishing modern Ayurveda as a valid para-science by braiding together selective threads of Indic and European knowledges.

But all success comes at some cost. In the next part of this chapter, I provide a thumbnail sketch of those costs by mapping what was it that became orphaned or unacceptable to modern Ayurveda because of the modernization.

Part 2

Intangible Agencies

Hemchandra Sen, whom we have met in the foregoing chapters and who is a key figure in shaping modern Ayurveda, had been called upon to visit a patient on March 16, 1904. Sen arrived at the house in Jhamapukur, Calcutta, around 8:00 in the evening. Upon entering the patient's room, he noticed that one of his students from the Campbell Medical School, a man named Surendranath Das, was already present. As Sen began his examination of the patient, Das, who was sitting nearby, suddenly fainted. Surprised by the turn of events, Sen turned his attention to Das. The latter seemed to have fallen into an absolute coma at first. Though the vital signs such as the pulse and breathing were still present, nothing could revive him. Sen had Das's boots removed, pinched him, made a small cut with a knife and even held a lighted candle to his skin. Yet nothing affected him. Sometime later, Das began to mumble incoherently. After a period of such mumbling, Das regained his consciousness equally suddenly. He immediately sat up and, complaining of extreme heat, took his shirt off. But before long, he lapsed back again into a coma. This time he seemed even worse. His body became taut and bent like a bow. With his stomach on the bed upon which he had been laid, his limbs arched backward. Members of the family whose home it was wrestled with the comatose Das trying to straighten his limbs, but it was to no avail.

Even as the perplexed family and Hemchandra struggled to make sense of the situation and help young Das, things took an even more bizarre turn. Seemingly out of nowhere an amulet (*maduli*) dropped near Das. The head of the household picked it up and found it attached to a string meant for wearing it. He immediately put it on Das and miraculously, the latter's limbs relaxed. Within a few more minutes, Das had recovered fully. Once recovered, Das narrated how he had suffered from repeated possession by an evil spirit for a few years now. Each episode was physically and mentally tortuous. Finally, a few months ago, the spirit of his departed younger brother had appeared during one of the possession episodes and handed him the amulet. As long as he had the amulet, Das would be fine. But sometimes, inexplicably, the amulet would disappear and with it the possessions would recommence, until the amulet once again reappeared. Sen investigated the case thoroughly and, upon being convinced of its genuineness, presented both Das and the head of the household where Sen had first witnessed the possession at a meeting of the Bengal Theosophical Society.[1]

Sen was not alone in encountering or believing in the reality of spirits. Spirits and other supernatural entities had played an important part in Ayurvedic medicine for a long time. Even canonical works on Ayurveda acknowledge a place for both malignant and benevolent entities. Gobindadas's *Bhaisajyaratnabali*, arguably the most important medical text used in eighteenth-century Bengal, mentioned, for instance, that tertian (*tritiyok*) and quatrain (*choturthok*) fevers arose from the "anger of ghosts, etc." (*bhutadi abesh*).[2] It also mentioned that for fevers caused by spirit entities, one needed to act according to shamanic wisdom (*bhutbidya*).[3] Comments such as these were not rare either in the *Bhaisajyaratnabali* or indeed in the Ayurvedic corpus as a whole.

Such affordances present within the canonical textual corpus opened up everyday Ayurvedic practice to a number of possible interactions with both malevolent and benevolent spirit entities. In one instance, not at all dissimilar to Sen's encounter, one hears of another eminent Ayurvedist, Kobiraj Shyamacharan Sengupta, having obtained a cure for a particularly difficult illness afflicting a young girl from the patient's departed older brother who had died the previous year from the same illness.[4] Sengupta was no quack. He was an erudite Sanskrit scholar and physician who published an Ayurvedic medical dictionary and contributed to leading periodicals of his time. Yet as he confessed, like Sen, he too believed in both the benevolent spirit entity and the cure it offered. Somehow, neither Sen's nor Sengupta's beliefs in spirit entities and their capacity to cure found a place in modern Ayurveda. Despite being prolific and influential Ayurvedic authors, neither of them wrote of their encounters with spirit entities and the capacity for the latter to cure in any Ayurvedic fora. Both anecdotes have survived in isolated spiritualist memoirs, but not as part of the archive of modern Ayurveda. Yet, importantly, they survive and do so fairly publicly and without embarrassment. From the 1920s, however, such intangible agencies would become increasingly embarrassing to modernizing Kobirajes.

While both Sen and Sengupta's experiences concern spirits of the departed—namely, entities we would most likely describe as "ghosts"—there were a range of intangible spiritual agencies that were recognized in traditional Ayurvedic practice. Besides ghosts, there were natural divinities: gods and goddesses. The *Uttarakhanda* of the *Susruta Samhita*, for instance, recognized nine *graha*s. Though the word *graha* means both "seizing" or "laying hold of" and simply "planet," it is clear from the discussion in the text that they are imagined as a kind of intangible but conscious agent. Susruta mentioned nine of these *graha*s: namely, Skanda, Skandapasmar, Sakuni, Rebati, Putana, Andhaputana, Sitaputana, Mukhamandika, and Naigamesh. If the mother or the nurse of a young child transgressed specific ritual and medical rules, then these *graha*s were said to become jealous of the child and enter into its body. They were invisible to the human eye and hence could not be detected entering the body.[5] Chakrapanidatta extended this discussion further in his treatise, but redescribed these agencies as "mothers" rather than "planets." As in Susruta, Chakrapanidatta's "mothers" (*matrika*s) appeared as invisible agencies threatening the life and health of little children. Whereas Susruta had spoken of nine such agents, however, Chakrapanidatta spoke of twelve: namely Nanda, Sunanda, Putana, Mukhamunditika, Kataputana, Sakuni, Suskarevati, Aryaka, Sutika, Nirrta, Pilipicchika, and Kamuka.[6] Later still, Gobindadas developed this discussion still further. Referring to them as *graha-matrika*s (planet-mothers) in his *Bhaisajyaratnabali*, he related each of the twelve agencies to specific periods of influence upon the infant's life. Thus, Nanda threatened the infant's health on the first day, the first month and the first year of the child's life, just as Sunanda's province was over the second day, second month, and second year of the child's life, and so forth. This new schema also meant that as the child grew older, it was often simultaneously under potential threat from multiple different *graha-matrikas*. Gobindadas largely agreed with Chakrapanidatta on the names of the *graha-matrikas* except in two cases. Chakrapanidatta's Sakuni and Nirrta were replaced in Gobindadas by Sabdanika and Nirjhati, respectively.[7]

Some scholars have argued that there was a fairly linear, teleological progress from the "religious therapeutics" found in the *Atharva Veda* to the increasingly "rational therapeutics" of classical Ayurveda.[8] Attending to the development of the lore about the *graha-matrikas* within the medical canon problematizes such teleological narratives. It demonstrates not only the resilience, but also the growth of belief in intangible agencies having the capacity to affect human health. Moreover, the constant growth of the lore also suggests that the medical authorities were possibly drawing upon extra-medical sources. Interestingly, however, while the cult of the *matrikas* has a long antiquity, the medical lore is significantly different from the cult as such. Michael Meister has described the original *matrika* cult dating from Vedic times as being comprised of a "sacred heptad." Initially identified with the god of war, Skanda, the cult much later in its career became identified with Shiva. It was also extended to include an eighth *matrika* in certain places. Once they came to be identified with Shiva, the first three *matrikas* became consorts of the Brahmanic trinity and were therefore identified as Brahmani, Maheshvari, and Vaishnavi. While there was some variation in the *matrikas* in the middle of the set, the last of the heptad was also consistently represented as the goddess Chamunda.[9] Neither the individual names of the *matrikas* nor the total number match the medical lore. Nowhere does one find the set of nine or twelve *matrikas* or individual *matrikas* like Nanda, Sunanda, etc.

What likely explains the divergence is that the *Susruta* almost certainly conflated the cult of the *saptamatrika* (the "sacred heptad") with that of the *navagraha* or "nine planets." It is generally accepted that a later editor added the "*Uttarakhanda*" of the *Susruta Samhita* sometime during the first five centuries of the Common Era.[10] Meister points out that from the ninth century CE, temples in central India regularly paired a set of the "seven mothers" together with the "nine planets" atop their doorways.[11] Since the medical evidence of the conflation long predates the visual and iconic evidence, it can be surmised that the medical authors drew upon an emergent religious culture well before it had sufficiently developed to be etched in stone.

Meister does mention in the passing the existence of some known instances of later popular religious practice having confounded both sets ("seven mothers" and "nine planets") and connected them to children's health.[12] Besides their association with the cult of *matrikas*, the *navagrahas* were simultaneously also central to astrological concerns. The medical lore about the *graha-matrikas* demonstrated the intercalation of medical texts with both the emergent popular religious culture surrounding them and astrological scholarship.

This process of intercalation did not stop with the *graha-matrikas*. Throughout its history, scholarly Ayurvedic traditions remained connected to other neighboring spheres of religious, astrological, and other social activities, and through such connections incorporated newer intangible agencies into their ambit. One of the best-known of these is possibly Sitala, the smallpox goddess. She has been widely written about by scholars and is an excellent example of a divine agency that was well-integrated into traditional Ayurvedic therapeutics.[13] While clearly a goddess whose cult originated outside erudite medical circles and whose worship even in the nineteenth century had been strongly associated with the subaltern classes, Sitala was eventually included and integrated into the elite textual corpus of Ayurveda and featured in such important texts as Bhavamisra's sixteenth-century classic, *Bhavaprakash*.[14] Another such disease deity who appeared both within and without the textual corpus of Ayurveda is the fever demon, Jworasur. Once again, while in the nineteenth century he was mostly associated with subaltern orders, his presence within the elite textual corpus is unquestionable.[15] Such deities demonstrate the robust dialogue and repeated incorporations of intangible agencies into the scholarly tradition of Ayurvedic therapeutics.

But these appropriated agencies were not exceptional. Numerous other gods and goddesses had retained their places within Ayurvedic therapeutics from the very beginning. The foremost instance of such older intangible agencies that remained at the heart of Ayurvedic therapeutics was the deities associated with the proto-humors or *doshes*. For instance, Ashutosh Roy, one of the key exponents of the endocrinal refigurations of modern Ayurveda, stated that, traditionally, the three *doshes* were identified with the three dominant gods of the Hindu pantheon. Thus, Shiva or Mahadeva was identified with *pitta*, Vishnu with *vayu*, and Brahma with *kapha*.[16] There was, however, some heterogeneity of opinion on the matter. Premvati Tiwari states that each *dosh* is usually associated with two gods. *Vayu* is dependent upon Marut and Akash, *pitta* upon Agni and Aditya, and *kapha* upon Soma and Varun.[17] These prominent identifications were further elaborated in keeping with the mythologies about these deities in the postclassical period. In Gobindadas's eighteenth-century text, the *Bhaisajyaratnabali*, for instance, there is an incantation that sought Shiva's protection from Varun and beseeched Vrishaketu to protect from *vayu*.[18] Vrishaketu is a minor figure available only in certain versions of the epic Mahabharata.[19] These versions, particularly popular in the eastern parts of South Asia, represented him as the grandson of the sun god, Aditya or Surya. Clearly, the locally popular mythology had allowed the grandson, Vrishaketu, to gradually replace the grandfather as the deity presiding over *pitta* in therapeutic contexts.

The socially embedded nature of such intangible, charismatic agencies also meant that there was a degree of cosmopolitanism about such agencies. Elaborations, appropriations, and incorporations meant that a range of both benevolent and malevolent entities were often therapeutically recognized despite their provenances not strictly conforming to the denominational affiliations of the physician-authors. One manuscript collection of prescriptions dating from 1795 preserved in the Visva Bharati University's manuscript collections at Santiniketan bears witness to such esoteric cosmopolitanism. The collection includes remedies both in the form of incantations and recipes. In the incantations, a wide variety of intangible agencies were invoked, ranging from the local snake goddess, Manasa; to the preeminent Tantric deity, goddess Kamakhya; to the Vaishnava saint, Gauranga; and to prophet Mohammad and even Allah himself.[20] This seamless blending of Shakta, Tantric, Vaishnava, and Islamic sources of power engendered a robust esoteric cosmopolitanism that marked healing by intangible agencies throughout the otherwise increasingly communally divisive period of the late nineteenth and early twentieth centuries. Even as modern Ayurveda eschewed such cosmopolitanism, a number of small collections of healing incantations were published from the second half of the nineteenth century onward. Though available side-by-side, they formed no part of modern Ayurveda. It was in these collections that the continuing appeal of the intangible agents of healing can be discerned. Atulchandra Gupta's *Gupto Mantra* (Secret Incantations), for instance, was in its sixth edition by 1907. The incantations in the collection once again invoked a wide variety of intangible agencies ranging from the popular local deities, Manasa and Dharma Thakur, to more widely known deities such as Shiva, and an entire gamut of Islamic sources of power ranging from the prophets Ibrahim and Mohammad, to the Mohammad's first wife, Khadija and daughter Fatima.[21] Possibly the earliest of these collections was an anonymous collection entitled *Bongiyo Bishwasyo Mantraboli*, published sometime around the middle of the nineteenth century, and it included once again a wide range of invocations from the Hindu god,

Rama, to the locally popular Islamic saint, Ghazi Pir.[22] But numerous other collections followed in its wake. By the early 1920s, some collections were even invoking a motley crew of Biblical angels in the cause of healing.[23]

In a classic article originally written in Bengali, Dipesh Chakrabarty draws attention to the radically different ways in which the subaltern and bourgeois imaginations of the body and disease had operated in colonial India. Drawing heavily upon the works of fellow Subaltern Studies scholars such as David Arnold, Chakrabarty argues that the subaltern imagination of the body retains a distinctively premodern aspect in which the "mundane and heavenly times and worlds become entangled. Mortal beings enter into stories of immortal gods . . . deities take the form of human beings and humans act out the part of deities. . . . In the circle of another time . . . all participate in a world that is purely human."[24] Capitalist modernity seeks to break this circle and rationalize the human body. The vitality of the human body is rendered progressively more tangible by attributing it to chemical and physical mechanisms. Anson Rabinbach points out that while Descartes in the seventeenth century had distinguished the "human machine" from automata as being possessed of a soul, by the nineteenth century such a distinction disappeared. The industrial machine and the human machine became indistinguishable. Both were agents merely of energy conservation and transformation. The soul disappeared.[25]

If the soul within the body became progressively untenable, the disembodied soul was even less tenable. Yet it was precisely in this period—in the second half of the nineteenth century—that Spiritualism and Theosophy sought to carve out a space for the intangible, disembodied soul. It did so by seeking ways to render the intangible soul tangible through techniques such as spirit photography.[26] Both Sen and Sengupta were connected to the Theosophical Society in Calcutta. It is to this involvement that we owe the preservation of the traces of their own trysts with intangible agencies. But this archive remained distinctive from the archive of modern Ayurveda.

Within modern Ayurveda, as seen in chapter 5 in Surendranath Goswami's elaborate refiguration of *bhuts* (ghosts) as "germs," the trend was toward metaphorization. Intangible agencies were glossed as metaphors or some form of tangible or soon-to-be-tangible energy or substance. Goswami was not alone in attempting such metaphorizations. There are numerous equally and even more elaborate examples of metaphorizations. One of the most conspicuous of these engendered the metaphorization of the references to the fever demon, Jworasur. Numerous leading Ayurvedists including Gananath Sen developed an elaborate metaphoric reading of the Puranic lore around Jworasur. According to this reading, the entire narrative of the fever demon's birth from an enraged breath exhaled by Shiva was rewritten as an allegory for the chemical process of digestion and its occasional vitiation leading to disease.[27] Ashutosh Roy similarly metaphorized Shiva as the "Katabolic force" and Brahma as the "Anabolic force" in cellular metabolism.[28]

Historical scholarship, as Dipesh Chakrabarty points out, has often struggled to effectively write about the "supernatural" without reducing it to mere "belief." This mode of writing is similar to what is dubbed in Sociological circles as the "Thomas Dictum": that intangible entities deserve analysis because they are social facts—facts believed in and shaping the actions of "real" social actors. Such an approach, as Chakrabarty points out, always already assumes sides with the rationalizing worldview and denies these intangible entities any claim to reality outside of the minds of the actors. In a poignant recent intervention, Ruy Blanes and Diana Espirito Santo have tried to transcend this "psychologization" by arguing for an "anthropology of the intangible." Turning the question of tangibility on its head, they have argued that, instead of starting with the fixed template of reality, we should investigate how and why certain experiences come to be accepted as "tangible" while other experiences are denied such reality. By thus placing the visceral experience—of both the "tangible" and the "intangible"—at the center of analysis, these scholars follow science studies approaches in seeing why certain experiences become "knowledge" while others remain at the level of "belief."[29]

Applying such a perspective to the above anecdotes would mean that, instead of reducing the spirit entities witnessed by Sen and Sengupta to mere figments of imagination, we would need to see why they could not be accommodated within the epistemic fold of modern Ayurveda. Kristina Wirtz points out that such a strategy of mapping "spirit biographies . . . violate[s] empirical notions of historicity, requiring those of us who are partial to the academic discipline of history to hold our empiricism in abeyance if we are to understand the significance of spirits."[30]

Wirtz also recommends that, for the sake of clarity, we distinguish between two types of signs relating to spirit-like entities. These, according to her, are respectively, *signs of spirits* and *signs about spirits.* The former are the embodied experiences, such as the falling into a coma, the arching of the body or, most dramatically, the appearance of the amulet that we saw in Das's case, while the latter is the narrativization through which these accounts are further materialized and disseminated. The interplay of these constitutes, according to Wirtz, specific "imaginal realms of possibility." Particularly, the *signs of spirits* are connected directly to forms of embodied experience.

The exile of the spirit entities from the folds of modern Ayurveda thus entailed two intercalated movements. First, the "imaginal realm of possibility" that constituted Ayurvedic knowledge was altered. Its boundaries were reordered and certain types of signs and their referents made redundant. Second, this redundancy in turn altered the repertoire of bodily experiences of the physician. Certain types of experience became impossible. By denying intangible agencies, modern Ayurveda therefore produced a distinctly new imaginal and experiential map for the physician.

The Hyperphysical Body

Intangible agencies were not the only entities excluded from the new imaginal realm of modern Ayurveda. One of the most prominent other exclusions was that of what Gopinath Kobiraj has called the "hyperphysical body."[31] Kobiraj was one of the preeminent intellectual historians of Sanskrit and especially the broadly Tantric knowledge traditions in the early twentieth century. More importantly, he wrote as an insider to these traditions. He was an initiate into an esoteric Tantric lineage of bodily practice and wrote as a committed practitioner. The "hyperphysical body," for Kobiraj, was a modular type. He cited numerous specific exemplary instances of such a "hyperphysical body." Kobiraj points out that ancient authorities including Charaka, who is perhaps the foremost of classical Ayurvedic authors, admit to eighty-four lakh (hundred thousand) animate species that are all mutually and hierarchically related. Life ascends through these forms achieving greater complexity, perfection and extra-physicality. It is only the lowest insect-like forms that are constituted simply by the *Annamaya Kosa* that is a purely physical form. Subsequently, higher forms develop by turns the *Pranamaya*, the *Manomaya*, and the *Vijnanamaya Kosas*.[32] The third form is human, but none of them are purely physical. All but the lowest form is imbued with special, spiritual qualities and attributes.[33]

The hyperphysical body clearly redraws the boundaries of animacy by locating the human being within a hierarchy of eighty-four lakh animate beings constituted of different biomoral substances. Clearly, then, by repositioning human beings within this distinctive cosmology, Kobiraj was interrupting the smooth emergence of a biopolitically enscripted notion of biological "life." Instead, his hierarchy of animacies gestured toward radically othered cosmologies. Following Michael Silverstein, Mel Chen has recently repositioned the notion of "animacy hierarchies" to point out not only that there are no universally valid boundaries between the animate and the inanimate, but to also draw attention to the ways in which "animacy hierarchies" can open up new critical possibilities. According to Chen, animacy can "trouble and undo stubborn binary systems of difference, including dynamism/stasis, life/death, subject/object, speech/nonspeech, human/animal, natural body/cyborg. In its more sensitive figurations, animacy has the capacity to rewrite conditions of intimacy, engendering different communalisms and revising biopolitical spheres, or, at least, how we might theorize them."[34]

One dichotomy, missing from Chen's list but yet, directly subverted by the Bengali Kobirajes and their specific animacy hierarchy was the humanity/divinity dichotomy. An important aspect of the hyperphysical body Kobiraj described was the existence of deities within the human body. Unlike in cases of possession where an intangible agent is supposed to temporarily occupy the body of a particular human being, in the hyperphysical body, there were intangible, divine entities that constantly inhabited the body. One of the most enigmatic of these entities is one called *jib* ("creature"), mentioned repeatedly in *nadipariksha* texts. One text from 1883 described the *jib* as traveling all over the body astride the vital breaths (*pran somuho*) the way a spider travels along its web. The entity "incessantly practices the Lord Brahma in its mind" and thus is immortal. Its abodes within the body are circular or cave-like.[35]

This reference to circular or cave-like structures has often been read as referring to the Tantric chakras that, as I noted in chapter 6, Hemchandra Sen and others assimilated into modern Ayurveda. But once again, what is ignored about the chakras is that they are frequently said to be the abode of gods in a very literal sense. Though the exact structure, hierarchy, number, and investiture of the chakras vary from one tradition to another, all of them hold chakras to be the seats of intangible and divine entities (see fig. 14). The *muladhar* of the Tantrics is known as the *Adhar chakra* among the Nathists. The chakra is presided over by a deity called Ganeshanatha and his two "powers" or "goddesses" (*shaktis*), Siddhi and Buddhi. The chakra right above the *Adhar chakra* is known as the *Mahapadma chakra* and is presided over by a deity unknown elsewhere, Nilanatha. Besides these gods and goddesses, however, there are many strange creatures. One of them is a "Gayatri named Kamadhenu" who even provokes Kobiraj—perhaps the greatest modern authority on Tantric intellectual history—to comment that it is a "strange figure." The creature looks like a milch cow with four long (individually named) teats, but it also has a peacock's tail, a horse's neck, an elephant's tusk, a tiger's arms, a cow's horns and "wings consisting of Lila Brahma and Hamsa."[36]

Besides the *jib* and the entities resident in the chakras, many of the traditions also explicitly believed in specific circles of deities residing in the

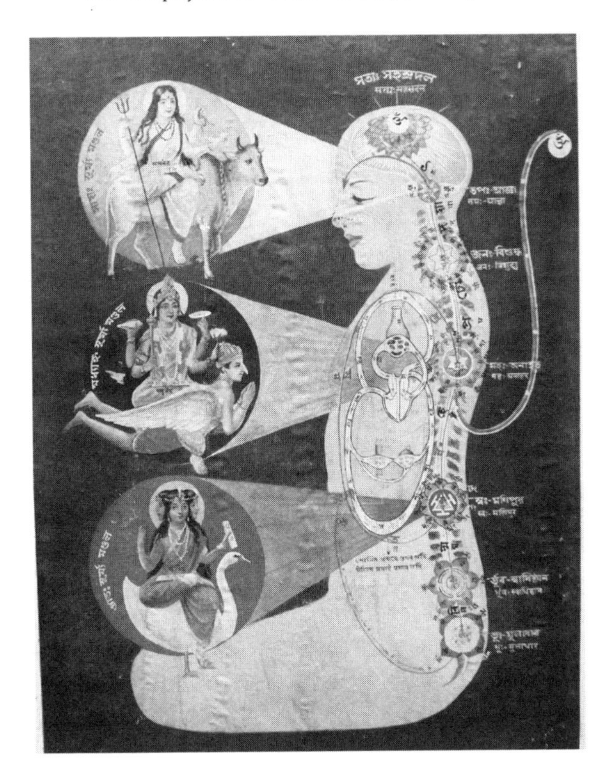

14. Macranthropic Body.

body. For instance, Shaiva-Tantrism, which clearly influenced texts such as Gobindadas's *Baisajyaratnabali*, also believed in a circle of deities within the body. Gavin Flood, describing the circle of deities within one such Shaiva-Tantric tradition, writes that "the body is inhabited by the circle of deities; this pantheon animates the body which becomes the *mandala* wherein they reside. One of the terms for the pantheon of goddesses here represented is "clan" or *kula*, a term . . . one of whose meanings . . . is, indeed, the body."[37] Complex but distinctive "circles" of deities and supernatural beings of various kinds residing within the body were also spoken of in both Tantric Buddhist traditions and the cult of Dharma Thakur. Among the Tantric Buddhists, there is a complex system of quintuples wherein five Dhyani Buddhas, five Shaktis, five Skandhas, five bodhisattvas, five human Buddhas, and the latter's five animal mounts. All reside within the body and are associated with distinctive parts of the body. Each single element in one set of five corresponds to specific individual elements in the other quintuples and together these individual elements all inhere in specific parts of the body. In other words, the Dhyani Buddha Vairochana, the Skandha Rupa, the Shakti Tara, the bodhisattva Chakrapani, and the human Buddha Krakucchanda and his mount the Dragon are all associated with the head, just as the Dhyani Buddha Amoghasiddhi, the Skandha Samskara, the Shakti Aryatara, the boddhisattva Viswapani, the human Buddha Maitreya, and the mount Garuda are all associated with the legs.[38] In a structurally similar yet distinctive alternate system, the devotees of Dharma Thakur place corresponding groups of *Patras*, *Kotala*s, *Dwara-pala*s, *Ghata-dasi*s, and mounts (*vahanas*) in charge of the different parts of the body.[39]

In Tantric circles, the sense organs too were often thought to be directly presided over or engendered by specific intangible but embodied deities. Among these are the Tantric Buddhist goddesses, Vajra, Gauri, Chauri, Vajrayogini, and Nairatmya-yogini, who are said to control the individual senses,[40] and the ten *Mahavidya*s of Hindu Tantrism who are said to respectively preside over and inhere in the eyes, the ears, the nostrils, the anus, the penis, etc.[41] To the extent that the communication of sensory information is the province of the individual types of *vayu*, we do then have a tradition of interpretations that comes close to investing subordinate *vayu*s with independent lives. An Islamic rendition of this idea relating specific supernatural beings to specific sensory organs can be seen in the following lines from Sheikh Chand's *Talibnama*:

Nasut mokam karna nashika Malkut
Jabrut chokshu hoy much je Lahut
Lahut mokame boshe name Israfil
Nuri-ferishta taye badi hi kamil

Jabrut mokame boshe mehatar Ajrail
Dono jahane kaji ferishta kamil.[42]

Incidentally, such images of a body bustling with a population of multiple animate spirit-like or soul-like entities are rife throughout Asia. In Southeast Asian traditions, for instance, such as those among the Hmong people, the body is animated by thirty-two souls (*khuans*).[43] In fact, some scholars have even explicitly suggested that the Southeast Asian belief in multiple "souls" in the body has come from India.[44] Similarly, in China, both Taoist and Chinese medical texts speak of the body sheltering within it a thriving population of entities. Chinese traditions also speak of "the multiple souls or spirits [that] represents the essences of the energies (*qi*) of the body."[45] Elsewhere, the same author, Kristofer Schipper, cites Taoist author, Ko Hung, as writing that "there are an infinite number of texts circulating among initiates that teach how to visualize all the spirits inside the body. All these methods are efficient. Some people are capable of materializing, through concentration, thousands of beings to protect themselves."[46] What Ko Hung's comments also clarify is that we cannot reduce these entities to mere symbolic fictions. The "reality" of these beings in the body is far more enigmatic whereby concentration and visualization can lead to the materialization of actual beings.

The existence of these deities within the hyperphysical body also meant that the body's spatial dimensions were neither neatly bounded nor rationalized. The *nadipariksha* texts, for instance, speak of the heart being "as large as the world and its pre-eminent location" (*hridoy bishwer ayoton o prodhan sthan*).[47] The very conception of the chakras, too, defies any rational organization of space since they open up to the cosmos. In fact, the very idea of deities permanently residing within the body confounds any notion of a limited, rational and discrete body space. Statements such as the ones affirming that the sun and the moon reside in particular *nadi*s within the body—unless metaphorized—also once again clearly defies rational, discrete physiograms.[48] The hyperphysical body thus is a body that is both constitutively and spatially transrational and supramaterial, and straddles a difficult-to-figure line between the mundane and the transcendent. The body itself was inspirited and miraculous. It formed part of an enchanted cosmos where rational causal order was not always necessary.

This hyperphysical body was progressively rendered implausible in modern Ayurveda. Upon Kobiraj's death and years after Kobiraj's coining of the term, the "hyperphysical body," the son of a close friend, who himself was also a scholar of Indic intellectual traditions, recalled Kobiraj's "implicit faith and childlike credulity" in "miraculous (yogic?) matters."[49] What even half a century earlier had appeared plausible to erudite scholars, now seemed like mere "childlike credulity" even to sympathetic scholars.

Superhuman Health

Eventually, medicine is a practical project seeking to achieve practical goals. What made the hyperphysical body look absurd and implausible was in fact the reorientation of the goals of Ayurvedic medicine. Joseph Alter has recently pointed out that Ayurveda traditionally had a very distinctive ontology of health. It did not simply imagine health to be a normal state of being free of disease. Instead, it imagined health as a perpetually augmentable state that could aspire to higher and higher levels of perfection. The goal of medicine, instead of being the remedying of a diversion from a normal state of health, becomes therefore the achievement of immortality through a reversion of time that will rejuvenate the body. At the core of this distinctive vision of health that Alter describes, is the existence of a powerful metaphoric ontology of transformations.[50]

Ayurveda, before the modern delimitations refigured it, operated to transform the hyperphysical body into an immortal, superhuman body. Dwarakanath Datta's *Chikitsa Chakrasar*, a text relatively unmediated by the reformism of modern Ayurveda, articulated some of these goals explicitly. Writing of a certain prescription, Datta states that the medicine made the patient "fair-skinned as Gourango, eloquent as Brihospoti and prolongs his longevity by five hundred years . . . this medicine makes youth stable in the body." Of another medicine, Datta wrote:

> Upon taking the medicine every morning, the following results ensue. The eyes, the shoulders and the body are strengthened and beautified. Wrinkles get no place in the body. Life is extended, the semen does not erode, the power is born to consort with a hundred women, no disease arises even after eating enormous quantities of food. Eating, sleeping and sexual intercourse all become completely subservient to desire.

Going further, about a third medicine, it is stated that "taking this medicine for three weeks gives the patient a body that cannot be burned by fire, which does not die upon drowning and one which cannot be pierced by iron or by thorns. Taking the medicine for four weeks gives one the power to leave the earth and travel to the sky."[51]

Clearly, the aim of the remedies prescribed by Datta was not the remediation of an illness that had subverted a normal state of being. Rather, in keeping with Alter's observations, the goal was to transform the body beyond its normal course and to raise it to superhuman status. But the superhuman aspirations too have some key accents. It seems organized around overlapping quests for physical beauty, sexual prowess and immortality. These three distinct aspects are not collapsible into each other and do not necessarily go together. One can easily imagine immortality without either physical beauty or sexual prowess, just as one can imagine great physical beauty without either sexual vigour or immortality. The bringing together of these three engender a particular configuration of the superhuman that is frequently invoked in much of the Bengali Ayurvedic tradition. Other superhuman qualities such as the power to levitate mentioned by Datta in the last instance, or the power to eat enormous portions of food mentioned earlier, are also occasionally present but these are definitely not as frequently sought as are the three core features of immortality, physical beauty, and sexual prowess.

These superhuman aspirations had a long genealogy in Ayurvedic medicine. As seen in chapter 6, these recipes for superhumanity were usually found in the sections on *rasayana* and there is a long history of medicines promising to transform the normal human body into a beautiful, sexually precocious and immortal entity. In Chakrapanidatta's influential and eponymous medieval work, one recipe promised to destroy

> various types of poisons—animal, vegetable and artificial, sorcery, effects of invisible organism and inauspiciousness. It promote[d] strength, sexual power, virility and lustre. If kept in mouth, it ma[de] one favourable to kings and gives victories in debates and disputes. If used regularly for a year, it ma[de] the person endowed with lustre, intellect, physical strength, immunity, corpulence, clarity of senses and energy and free from wrinkles and greying of hairs with maximum life-span. If taken for two years, it further enhances life-span.[52]

Gobindadas, in the eighteenth century, hugely expanded the repertoire of remedies promising such superhuman transformations. But most of his prescriptions continued to focus on the triangular focus on physical beauty, sexual prowess and longevity. One such prescription was to "quickly destroy all diseases and engender a new body. . . . [O]n consuming [this medicine] man will be strong, sexually attractive and powerfully libidinous."[53] Of another prescription, the patient was told to

> smell it, apply it as paste, wear it as an amulet, fumigate the home with it and keep it at home. It destroys ghosts, *Alakshmi* [a malevolent goddess], rebelliousness of others [towards you], incantations, fire, danger and enemies. Moreover [it also] destroys nightmares, addiction to women, accidental death, [as well as] the fear of water, storms and thieves. It increases wealth, grains and brings success to your work [while] improving your vision, complexion, longevity and fortune.[54]

These specific aspirations toward superhumanity are usually referred to as individual *siddhis* and seem to be connected to the alchemical traditions of South Asia. The precise doctrinal history of such traditions remains a matter of much scholarly disputation, though in Bengal much of this seems to derive from Nathist roots. In her magisterial study of the Nath cult and their traditions, Kalyani Mullick clarifies the difference of the Nathist tradition from other Indic philosophical traditions. While most Indic philosophical traditions recognize a two-stage, hierarchized process of the soul's liberation wherein *jibonmukti* (embodied liberation or living-liberation) is subordinated to the *summum bonum*, namely, *bidehimukti* (liberation from the body or postmortem liberation), the Nathists, Mullick points out, eschew the latter and take the former as the *summum bonum*: "In the Nath philosophy the ideal is the state of *jibonmukti* . . . the Naths say, it is [our] duty to make our body—by which we have attained the ultimate truth—imperishable, immortal and to travel wherever we wish in it."[55] Notwithstanding the fact that certain other heterodox traditions and saints too have on occasion recognized and elevated *jibonmukti* above *bidehimukti*, the Naths stand out by the boldness and consistency of their message. Even though written theological treatises expounding their position are rare to come by in Bengal, there is a wealth of lore relating to the Naths that consistently depict Nath siddhas as wielding superhuman powers.

While the Naths were singular in promoting *jibonmukti* above *bidehimukti*, numerous neighboring religious and philosophical traditions agreed upon the need for some sort of basic body practice and embodied liberation as a precursor to the final liberation out of the body. This meant that, despite differing about the final goal, the Nathist beliefs partially overlapped with certain neighboring traditions. Each of these latter in turn had their distinctive ways of trying to attain the elevated physical state. Hatha yogis, for instance, depended upon complex manipulations of the breath. Alchemists depended upon various mercury-based substances. Certain sects of Buddhists and Vaishnavas depended upon meditation and mantras, and finally upon specific semen-based techniques. Outside their immediate milieu, Mullick points out that the Nathist aspirations for *jibonmukti* bore some resemblance to Daoist doctrines.[56] Nathists themselves were broadly divided into two sects: while one used mercurial substances to attain *kayasiddhi*, the other pursued the same goal using meditation and incantations.[57] Outlining a somewhat different set of affiliations, David Gordon White posits that the Nathists constituted part of a larger, somewhat dispersed Tantric tradition. Their external links, he says, connected them to China, Tibet, central Asia, Persia, Europe, etc., while within South Asia their closest interactions were with Alchemists (*Rasa Siddhas*) and the cult of the goddess Kubjika known as the *Paschimamnaya*.[58]

Describing the Nathist belief in *jibonmukti*, Mullick writes that, "upon becoming *jibonmukto*, the yogi's body is transformed. Just as salt dissolves in water, so too does divinity [*brahmatma*] dissolve in the yogi's body. Once the yogi's body becomes indistinguishable with Brahma, divine consciousness permeates his/her entire body, the bodily senses too then become permeated with the same consciousness." The mortal body at this point is transformed into the *siddhodeho* and the attainment is called *kayasiddhi*.[59] The attainment of *kayasiddhi* is materialized in a transformation so radical that the new *siddhodeho* is also known as the *nirmankaya* or the built body, to distinguish it from the naturally born body. This latter is therefore also said to be *oyonijo* or not born of a womb. Interestingly, Gopinath Kobiraj's essay on the hyperphysical body had described and accepted the existence of a number of different types of *oyonijo* bodies. Such bodies were understood to be fundamentally distinctive from natural-born bodies and therefore subject to very different therapeutic regimes. *Nirmankaya* bodies, for instance, were said to be non-*ponchobhoutik*—that is, not made up of the five fundamental particles that are recognized to be the building blocks of all mundane reality,[60] thereby undermining the very cornerstone of everyday therapeutic logic that related these fundamental elements through the intermediate category of *dosh*es to various physiological processes. Kobiraj spoke of a number of instances where naturally born human bodies may have been transformed so fundamentally as to become distinctive types of bodies. Certain legendary figures such as Mandhata, Caru, Kapolamalini, Amrapali, etc., were said to be *Svedaja* or "born of sweat or moisture," while rishis or seers and "hell-beings" were asexually constituted by the "atoms combined under the influence *dharma* and *adharma*."[61]

Tarashankar Bandyopadhyay's novel, *Arogya Niketan*, which, as I have stated throughout, is much more than a work of fiction because of the author's friendship with one of the leading Ayurvedists of his day; it vividly describes a lasting faith among traditionally minded Ayurvedists that, therapeutically, urged one to treat the bodies of saints and seers differently. Describing a clinical encounter between the hero of the novel and an old sadhu or holy man, Bandyopadhyay wrote that

> the substance-temperament [*dhatu-prokriti*] of *sadhu-sonnyasi*s are distinctive. It is quite different from the bodies of common people. The effect of following different rules of etiquette and leading a highly regimented life is to enormously raise the body's tolerance. Medicines react in surprising ways. They behave like seeds sown on virgin soil that has never been ploughed before. Therefore its results cannot be predicted. Sometimes, even when death is near, it retreats due to the strength of the vital-force of these *sadhus*. Jibon

> Dutta has seen many instances of this throughout his career. His father too had spoken of it. Diagnosing by *nadipariksha* too was difficult for such patients. His father had advised him never to foretell death by the *nadi* for such patients. Instead he had said one must first ask the patient what he desires. Men often become despondent, but *sadhus* do not have despondency. Their will is strong and their desires unalloyed. They can only die if they willingly desire to forsake life.[62]

These broad similarities between medical and antinomian religious ideas about superhuman health, however, did not mean that there was complete correspondence between medical and Nathist traditions. While the Nathists were explicit in seeking an embodied immortality alongside a lustrous, beautiful body that was indestructible, they did not seek unbridled sexual appetite. While rites involving sexual contact with women possibly formed part of their religious rituals, they seldom explicitly sought to enhance their libidos.[63] In fact, one of the most common tales associated with the cult in Bengal, *Goroksho Bijoy*, tells of one of the key Nath siddhas, Gorakhnath, having to go and rescue his guru, Minanath, after the latter had been seduced and entrapped by a bevy of beautiful women and had lost his powers.[64] The positive appreciation of a superhuman libido, which I discussed here and in chapter 6, then seems to be distinctive about the medical iteration of superhuman health. Yet, being different does not disallow overlaps and conversations, and the medical position on superhuman health almost certainly evolved out of such discussions and conversations with antinomian religious traditions such as the Naths who also sought superhuman bodies, even if they did not end up being identical.

Such superhuman health became impossible to imagine in modern Ayurveda. Here, gradually, health was redefined along the more familiar biomedical terms. Alter agrees that today the "remedial bias" that undergirds biomedical understandings of health were all too common in Ayurvedic circles as well. The distinctive superhuman aspirations were already in the process of excision and refiguration at the end of the nineteenth century when modern Ayurveda itself was born. What eased the marginalization of the ontology of superhuman health was the rise and prominence of Vaishnavas within those circles that sought to modernize Ayurveda. The *siddhi*-based superhuman health, though far from being exclusive to any one theological tradition, was far more closely identified with the Tantrics than the Vaishnavas. In fact, key modernizers like Gananath Sen explicitly identified mercurial prescriptions (i.e., the main repertoire of *rasayanic* therapies aimed at *siddhi* acquisition) with Tantrism.[65] While the mercurial remedies were too well integrated into everyday Bengali medical practice for the Vaishnava modernizers like Surendranath Goswami and Gananath Sen to exclude them completely, they stripped away the cosmological frameworks that aligned these therapies to *siddhi*s and Tantrism.

Aural Therapies

One of the most significant excisions from the repertoire of modern Ayurveda has been that of aural therapies. Both religious hymns and mantras or incantations played an important role in Kobiraji therapeutics prior to the modern refigurations.

Mantra use was widespread in premodern Kobiraji practice but constituted only one part of the total range of aural therapies that were in vogue in premodern Bengal. Other aural therapies included the communal recitation of holy names: namely, *zikr* in Islam and *sonkirton* in Vaishnavism; the chanting of particular scriptures, namely *Chondipath* in Shaktism; and the singing of devotional hymns, namely *Kalikirton* in Shaktism. Together, these diverse aural therapies had constituted a robust, plural and important component of the larger landscape of traditional medicine. Many of these therapies had survived well into the period of colonial domination and continued to proliferate, even though they were increasingly marginalized within the folds of modern Ayurveda.

Mantras or incantations were regularly included in even classical medical works. Kenneth Zysk has argued that the use of mantra*s* in early Ayurvedic texts can be localized at five distinct therapeutic sites: namely, for the treatment of swellings and wounds, treatment of mental disorders, treatment of poisons, treatment of fevers, and the collection and preparation of certain medicines.[66] There are, however, a few other sites too where mantra*s* have been deployed in both classical and postclassical texts. The *Susruta Samhita*, for instance, advocated the use of specific mantra*s* to treat attacks on children by the various *grahas*.[67] The *Chakradatta* and the *Bhaisajyaratnabali* also followed suit in this regard.[68] *Kumaratantra*, as this specific area of therapeutics was called, thus became a major site of mantric therapies beyond the ones recounted by Zysk. The preeminent domain of mantric usage, however, was undoubtedly *Bishchikitsa*, or toxicology. The *Susruta Samhita*, in discussing the treatment of snakebites, for instance, mentioned that "the *mantra*s uttered by the gods and the divine seers are true and powerful. Such *mantra*s do not fail, indeed they destroy the poison forthwith. The speed with which *mantra*s that are true, divine and charged by meditation destroy the poison, cannot be rivalled by the application of simple medicines."[69] Later, postclassical works bore witness to a further expansion of the application of mantras beyond the traditional domains of infant health and snakebites. Both the *Chakradatta* and the *Bhaisajyaratnabali*, for instance, deployed new mantras against fevers that do not derive from the earlier *Atharva Vedic* tradition that Zysk identifies.[70]

Beyond these texts, there was an even larger domain of diseases and illnesses to which mantric therapies were deployed. A manuscript belonging to an unnamed Bengali physician simply called Samar and dating possibly from the turn of the eighteenth century, included mantra therapies for stomach aches, a variety of different types of regular and irregular fevers, suppurating sores, serious burns, elephantiasis, a type of skin disease, different types of diarrhoeas, indigestion and wounds. It even had one mantra for driving away feelings of despondency and depression. Besides human health, Samar's collection also contained some mantra therapies for cattle diseases.[71] The many collections of mantra therapies that appeared in the nineteenth century as cheaply printed tracts also attest to the widespread applicability and use of such mantra remedies. Sivaramakrishnan has rightly pointed out that at least partially, the opposition to mantras by men such as Gananath Sen was driven by their class anxieties, since mantra use was more prominent in the therapeutic arsenal of lower caste and lower class physicians such as Ojhas.[72]

Aural therapies had received a major fillip in their popularity through the emergence and popularity of important religious movements that emphasized personalized devotional religiosity and mysticism toward the end of the fifteenth century. In Bengal, the Gaudiya Vaishnava movement inaugurated by Chaitanya Mahaprabhu in the sixteenth century, ushered in a veritable social and cultural revolution. At the heart of the changes was the Mahaprabhu's encouragement of communal singing of religious hymns. At its simplest, it could simply entail the religious community going through the streets of the town or village loudly chanting the name of god: *nogor sonkirton*. The word *kirton*, which originally meant any kind of utterance or exposition, was transformed through the Gaudiya Vaishnava movement into a specific and preeminent devotional activity where the name, descriptions or exploits of Vishnu and his avatars were sung out loudly. When done collectively, it was called *sonkirton*, whereas individually it was referred to as *kirton*. The emphasis on singing aloud meant that all those around the devotee could hear the hymns and be healed or saved by it. With the rise of the devotional, mystic religious movements together called the Bhakti movements since the fourteenth century, *kirton* was institutionalized as an important devotional activity.[73] The Mahaprabhu told his followers that "the supreme healer in this iron age is *sonkirton* of the Name"; that is, in this sinful age, the communal intonation of the name of Krishna is the supreme healer.[74] Lochan Das's versified biography of the Mahaprabhu, *Chaitanyamangal*, mentions an incident eliciting the healing power of *kirton*. One day, when the Mahaprabhu was out singing *kirton* along the streets of his native Nabadwip, he encountered a badly disfigured leper. While the hapless leper fell at the Mahaprabhu's feet and sought to be healed, he also pointed out that the reason he had been afflicted was that he had gone around abusing and maligning some of the Mahaprabhu's closest disciples. The saint explained that he would have happily forgiven the leper had he maligned the saint himself, but he could not abide by his disciples being mistreated. When the leper repented and persisted in seeking his mercy, the saint first sought the permission of Sribas Pandit, the disciple who had originally been maligned. As soon as Sribas had indicated his willingness to forgive the errant leper, the saint began to loudly utter the holy name of Krishna (*e bol suniya Probhu kore Hori-nad*). At that instant, the leper obtained the "supreme medicine" (*porom-oushodhi*). He was instantaneously healed and obtained a new, divinely blessed body.[75] The same incident is also described in Krishnadas Kobiraj's biography of the Mahaprabhu, the *Chaitanyacharitamrita*, with some minor differences in detail.[76] It is also noteworthy that both these biographers, Krishnadas and Lochan, were physicians in their own right.

The closely related practices of *sonkirton* and *nogor-kirton* remained popular in Bengal until the very early twentieth century. Tarashankar's novel *Arogya Niketan*, set at the turn of the nineteenth and twentieth centuries, for instance, mentioned collective *kirton* singing at times of cholera epidemics.[77] Moreover, there were actual news reports of *nogor-kirton*s being organized during the plague epidemics of the late 1890s and early 1900s in Calcutta. These news reports were also backed up by a serious discussion in *daktari* circles about the possible positive effects of such *sonkirton*.[78]

It was approximately two centuries later, in the eighteenth century that *Sakta podaboli* began to emerge as a musical genre in Bengal. Yet the genre shared a strong "family resemblance" with the Bhakti music of the Vaishnavas. Rachel Fell McDermott points out that both shared "attitudes toward the divine [that] are emotive, highly personal, pervaded with an undercurrent of love and adoration . . . expressed in an earthy vernacular that abounds with regional words and metaphors; grace and self-surrender are emphasized." Both genres also frequently allude to "the teacher's guidance" while invoking "the philosophical language of absolutes, superlatives, and coincidences of opposites . . . to substantiate claims about the Goddess's pre-eminence."[79] Just as the Vaishnavas had spoken of the *kirton* as the highest medicine, the Shaktas too spoke of the emergent *Kali-kirtons* and *Shyama-songit* in the same vein. Ramprasad Sen, the son of a physician, and unquestionably the preeminent author of *Kali-kirtons*, sang thus: "Kali's name is a powerful medicine, the best prescription/drink it with devotion. Oh sing! Drink"![80] Once again, there is clear evidence that this faith in aural therapies that began to develop in Shakta circles in the eighteenth century had survived into the twentieth century. Around 1912, a well-known and respected Ayurvedic physician, Kalidas Bidyabhusan, published a small book entitled *Chikitsa-tottwo o Bhaboushodh*. The book included a lengthy essay on the reasons for the decline of "indigenous medicine" and the need to revive it, followed by an exposition of Ramprasad Sen's philosophy and a collection of songs mostly, though not exclusively, from Ramprasad Sen's oeuvre. These last were entitled "Wordly Medicines" or *bhaboushodh*. The author used this enigmatic phrase to deliberately blur the lines between "medicine for the soul" and "medicine for the body." Given the overall presentation of the book, the topics covered in it, and its title, it was clear that the author considered Shakta music to have healing powers of one sort or another.[81] In the preface to the second edition, the author further clarified that "I suppose many would have witnessed for themselves that the moment *Prasad-songit* [Ramprasad's compositions] are sung, they instantaneously inaugurate in the audience a state of enlightenment and bliss. What should I call songs which quell obstinate diseases and desires but 'earthly medicines'"?[82]

Finally, besides mantras, *sonkirtons*, and *Sakta podaboli*, there was a fourth important stream of aural therapies available in Bengal. Mystical strands of Islam had developed a strong tradition of *jikir* (from the Arabic *dikr* or Urdu *zikr*) that also entailed the communal chanting of holy names. Jagadish Narayan Sarkar noted the similarities between *kirton* and *jikir* in 1907. Sarkar observed that "the Sufi practice of *zikr* (remembering God or reciting His name) when accompanied by violent actions, became similar to *nam-sonkirton* or loud Vaishnava recitals of the Name accompanied by song and dance."[83] But *zikr* had its own Islamic genealogy that stretched much further beyond either Bengal or India. T. P. Hughes, writing more than a century before Sarkar, had noted that "*zikr* is the religious ceremony, or act of devotion, which is practiced by various different orders of Faquirs . . . the performance of *zikr* is very common in all Muhammadan countries."[84] Moreover, Hughes pointed out that, unlike many other mystic practices, *zikr* had the explicit sanction of The Prophet.[85] There were essentially two broad types of *zikr*—

namely, *zikr-i-jali* (to be performed aloud) and the *zikr-i-khafi* (performed either in a low voice or in one's mind).[86] Despite the transnational genealogy, in Bengal, the *zikr* had developed its own complex philosophical understanding. One of the first explicitly Bengali statements of the philosophy of *jikir* appeared in the writings of Baba Jan Sharif Shah Sureswari Qibla. In a treatise entitled *Nur-e-Haqq, Ganj-e-Nur* Baba Sureswari devoted a lengthy chapter to the exposition of the philosophy of *jikir.* Though the exact date of the composition is uncertain, since Baba Jan passed away in 1919 (1326 BE), therefore the composition most certainly predated that date. In *Nur-e-Haqq,* Baba Jan outlined a theory based on the existence of a set of *latifa*s within the body. He compared these *latifa*s to fakirs meditating upon Allah while residing in their fixed positions within the body. There is no need to enter into a full discussion of Baba Jan's theory here, but it should be noted that this theory is quite distinct from the one observed by Hans Harder among the Maijbhandari mystics recently. The Maijbhandaris, even while using the terminology of *latifa*s, assert that there are only six such *latifas* and identifies them with six "energy centres."[87] This suggests a connection to either Sufi notions of *muqam*s or the Tantric notions of chakras. Possibly, it draws upon both.[88] This, however, is totally distinct from Baba Jan's theory. Baba Jan spoke explicitly of ten *latifa*s, anthropomorphized them, and, far from equating them with "energy centres," he identified a heterogeneous set of entities including the heart (*qalb*), the head (*sir*), the soul (*ruh*), the lower soul (*nafs*), the four elemental atoms (water: *ab*, fire: *atash*, earth: *khak*, wind: *bad*), etc.[89] Just as Tarashankar's *Arogya Niketan* provided fictional, but reasonably reliable, testimony that the practice of *sonkirton* for healing had survived into the twentieth century, similarly, Syed Waliullah's realist novel *Lal Salu* vividly depicted and use and abuse of *jikir* as a therapeutic alternative.[90]

These major streams of aural therapy do not exhaust the entire gamut of such therapeutic alternatives. James Newell, for instance, has ethnographically documented the use of *qawwali* for healing in western India and it is more than likely that similar practices were available in Bengal as well.[91] There are also rituals, such as those involving Satya Pir or Satya Narayan, the Islamic saint who is also worshipped as a Hindu god, that are used for healing purposes where a key element is the recitation of lengthy narratives about the superhuman figure invoked.[92] There were a number of such smaller aural practices that contributed toward constituting the entire range of aural therapies. One of the earliest commentators on Bengali soteriological practices, the Reverend William Ward, had written at the dawn of the nineteenth century that "the Hindoos . . . do not depend for cures altogether upon medicine. They repeat the names of their gods . . . repeat many charms. . . . They also listen to parts of different shastras, or to forms of praise to Doorga or Rama."[93]

Aural therapies were not unique in pre-reform Ayurvedic medicine. Miri Shefer-Messensohn, in her account of Ottoman medicine in the early modern period, draws attention to the importance of "music therapy." According to Shefer-Messensohn, "music therapy was very common among Turkish peoples throughout history. It was a cure for sickness and also a method for guarding one's health as part of a preventive lifestyle."[94] Ottoman "music therapy," however, modified this. From the fifteenth century onward, Ottoman "music therapy" drew upon "music theories" that conceptualized music as the medium for putting the world into "order, harmony and rhythm." Ideas of balance became preeminent, particularly the balancing of elements and tempers. From the fifteenth century, there existed elaborate "detailed tables of musical modes appropriate for each physical condition." By the sixteenth century, physicians like Daud Al-Antaki, began to draw upon these and included chapters in medical texts on the therapeutic use of music. By the time of Sultan Mehmet II, the traveler Evliya Celebi reported a regular band of ten singers and musicians as associated with Edirne Hospital. He also reported the royal military band, the *mehterhane-i haqani*, playing at hospitals.[95]

By contrast, in the Bengali context, there seems to have been no attempt by physicians to rationalize the use of aural therapies by drawing upon musical theory. In fact, the very distinction between music, speech and sound is often blurred and little effort is made to clarify it. The therapeutic power of sounds in Bengal, irrespective of which religious tradition one looks at, seems largely to be grounded in a belief in the innate power of certain sounds, especially names or formulaic sequences, be they mantras or segments of shastras. The ideas about balance and harmony that influenced the developments of Ottoman music therapy do not seem to have mattered in Bengal.

Even more interestingly, while some efforts were made to rationalize and explain the way sounds actually worked upon the body, these efforts, so far as I have been able to discern, did not involve physicians acting in the medical realm. This is especially striking given that in certain other neighboring disciplines there had been attempts to theorize the impact of mantric sounds upon the body more rigorously. Daud Ali, for instance, has recently drawn attention to some erotological literature where the action of mantras was related to evolving ideas about the role of *nadis* within the body.[96] In the medical realm, there seemed to be a certain degree of consensus about the therapeutic potency of sounds without any attempt to explicate it.

Notwithstanding any of these traditions, survivals or consensus, modern Ayurveda from early on developed an apologetic tone toward aural therapies. Bhagvat Singh Jee's *History of Aryan Medical Science*, while admitting the presence of aural therapies, distanced itself from such beliefs by saying that such therapies were "supposed to" work and that it was "curious to note" that such therapies had actually been widespread across the globe.[97] Much later, while treating fevers, Nagendranath Sengupta described a certain type of fever called "fever caused by incantations," but avoided any mention of aural or incantation-based therapies for fever in the section on treatments, despite the fact that mantras were indeed an established part of fever therapy even within the canonical literature.[98] This gradual marginalization of aural therapies contributed to the position echoed by scholars such as Zysk when they speak of mantras as a "superstitious" residue from an earlier era of "unscientific" therapeutics.[99]

Cosmo-Therapeutics

These diverse and hesitant elisions that I have been detailing above constituted the contours of modern Ayurveda. While it remained and even developed its distinctiveness from its contemporary biomedicine, the range or distance between the two was bounded by a broad agreement on the elisions. Neither biomedicine nor modern Ayurveda accepted intangible agencies, the hyperphysical body, the ontology of superhuman health or aural therapies. This was the shared template of modern rationality that undergirded both biomedical and modern Ayurvedic therapeutic repertoires.

From a pragmatic point of view, this change can also be conceptualized as a shrinking of the canvas upon which therapeutics operated. The change, I argue, ought to be seen as a shift from a cosmo-therapeutics to a physiogramatic therapeutics. Although I borrow the term "cosmo-therapeutics" from the writings of Harish Naraindas, my usage is slightly different from his. Whereas Naraindas speaks of "cosmo-therapeutics" as contributing to the absolute Othering of East and West,[100] I do not want to imply any such Othering by my usage. Instead, I see "cosmo-therapeutics" simply as a therapeutic modality that considered the entire cosmos as the canvas upon which it acted. By contrast, the physiogramatic therapeutics was a therapeutic modality that concentrated narrowly upon specific physiograms or materialized body images. These latter were not simply somatic maps. Indeed, as I have noted in the foregoing chapters of this book, they often included "spiritual" or invisible elements in the body. There was also a lot of variation about the physiograms themselves. The body image was not a single, coherent and universal figure like an anatomical chart hung on a wall. Drawing upon Naraindas once again, I contend that, while cosmo-therapeutics blended the "cosmological" and the "physiological" in explicit ways, physiogramatic therapeutics was more exclusively "anthropological" or, perhaps more accurately, "anthropocentric."[101]

The anthropocentric perspective engendered in physiogramatic therapeutics eschewed any direct engagement with the larger cosmology and focused on the body. In so doing, it also rendered the anatomo-politics of the human body immune to irrational, and therefore unforeseeable, manipulations. By denying the hyperphysicality of the body, the body was also rendered discrete and rationalizable—namely, a space in which anatomo-politics might be engendered. Moreover, as Michel Foucault points out, anatomo-politics is merely one end of the biopolitical organization of modern power. The other end is that of "government of the population."[102] The population can only emerge once bodies are rendered identical and enumerable. One cannot speak of "populations" if bodies constituting the potential population are capable of radical self-transformation through unrationalized processes that are innately incapable of being rationalized. Such potential for radical and unpredictable transformations would make a mockery of any enumeration and undermine any attempt at government of populations. I am reminded here of Lawrence Cohen's brilliant insight that "any inscription of a proposed postmodern science must write from and through an insurrectionary abandonment rather than a pluralizing of the field of discourse."[103]

What I am arguing then is that the cosmo-therapeutics that was gradually overshadowed and exiled from modern Ayurveda was one that was deeply incommensurate with biopolitical regimes that formed the aspirational horizons of colonial governmentality.[104] Apocryphal stories of how hyperphysical bodies and their superhuman *siddhis* confounded colonial government abound. One such story is remarkable in its explicit linking of biopolitics and cosmo-therapeutics.

In 1885, one M. J. Walhouse recounted to the members of the Anthropological Institute of Great Britain and Ireland an incident that had happened more than thirty years ago. Around 1850–51, Walhouse, then an employee of the East India Company, had been posted in Mysore and cholera was raging in the area. His subordinates brought him information that a young woman of about thirty was going from village to village claiming to be the cholera goddess and willy-nilly extracting tribute from the villagers. Such living cholera goddesses were not at all novel. Arnold refers to several such in different parts of British India.[105] Eager to stamp out any threat to the

Company authority, Walhouse ordered the woman arrested. Though she offered no resistance, most of the Company's employees refused to carry out Walhouse's orders for fearing of offending her. Eventually, Walhouse sent two carefully chosen Muslim constables who "despised Hindu superstitions." The woman, now arrested, came willingly but refused to stop her peregrinations unless ordered by god himself. When Walhouse threatened imprisonment, the woman, now identified as Nagamani, willingly accepted, only stipulating that so long as she lived in prison no morsel of food would pass her mouth. Walhouse turned once again to a jailor who was a "grim and fanatical old Mussulman" to take charge of her and ensure that she eats. The "old fellow," said Walhouse, "grinned scornfully" and looked forward to his task of breaking her.

After a week, however, the scornful grin looked somewhat sheepish as he reported that despite his best efforts she had not eaten anything at all and that it was impossible to have done so without his knowing. Yet Nagamani appeared just as she was at her arrest. When another week passed with no change of circumstances, Walhouse was moved to act. He spoke to her at length. Nagamani explained that when she was but a young girl, she had met Shiva in the jungles of Seringapatnam and he had entered her body. Ever since, he abides in her and speaks and acts through her. Walhouse also found that, despite coming from an agricultural caste, even Brahmins believed in her power. It was widely believed, reported Walhouse, that by passing her hand over objects and uttering a simple mantra in praise of Shiva she was able to charge the objects with superphysical power and auspiciousness. So impressed was Walhouse at the authenticity of the case that not only did he have her released and escorted across the boundaries of the British empire, but in his reminiscences he equated her to the Sibyls.[106]

There are numerous stories of colonial officers being confounded by superhuman saints. The numerous stories about Tailanga Swami of Benares, for instance, recount tales of one hapless British officer's frustration upon another's in their efforts to lock the saint up in prison.[107] Every time they attempted it, they found him strolling outside or calling upon the magistrate for a cup of tea, even as a twenty-four-hour guard saw him sitting in his prison cell all along. But unlike those stories, Nagamani's story comes not from the hagiographic lore, but from a colonial officer himself. It also demonstrates many of the aspects I have been discussing above. There is an intangible agent of disease who abides within the apparently "human" body of Nagamani thereby transforming it into a hyperphysical body and bequeathing her with superhuman capacities as well as the knowledge of mantras with healing powers. In short, she embodied and engendered the cosmo-therapeutics that I have been recounting. It is also clear that if we can temporarily suspend our disbelief and consider the ramifications of someone like her actually existing, we immediately notice what a huge problem she would pose to the biopolitical aspirations of the colonial state. She could neither be locked up nor could her power be understood or controlled. Worst of all, she was a young peasant girl who had accidentally been transformed by a serendipitous meeting with Shiva. The peasantry after all were not supposed to be superhuman. Indeed, they were the bulk of the nameless, faceless bodies making up a body politic called "population." It was upon the calculation of the laboring power and controllability of that faceless mass that colonial power in its biopolitical formation eventually resided. If other peasant girls, even one in a million, started to be thus transformed, the "government of populations" would rapidly evaporate.

The eclipse of cosmo-therapeutics was not achieved suddenly. In fact, the braiding of traditional Ayurvedic practices with electro-magnetism might be seen to be a clear intermediary stage. As Suman Seth has argued, there was, by the end of the nineteenth century, an electro-magnetic worldview.[108] Electro-magnetism, far from being reductionist, then, was capable of sustaining an alternate cosmo-therapeutics. It was only with the further shift toward endocrinology that the wider view was eclipsed and a more anthropocentric view was articulated.

At the heart of the shift from cosmo-therapeutics to physiogramatic therapeutics is the related movement from pataphysics to metaphysics. Metaphysics attempts to explain the world and being in terms of the universal and the particular; pataphysics, a term coined by Alfred Jarry, on the other hand, seeks to extrapolate a science of the singular, the unrepeatable and the exceptional.[109] Metaphysics seeks out regularities and explanations; pataphysics seeks out exceptions and limits to explicability. Cosmo-therapeutics operated with limited generalizations. It accepted the possibility of the exceptional and the singular. In so doing, pre-modernized Ayurveda, it can be said, was anexact—that is, "deliberately and not accidentally inexact."[110] But modernized Ayurveda, in surrendering that anexactitude, disavowed, or at the very least seriously marginalized, the possibility of the unrepeatable singularity within its logical apparatus.

Authors of modern Ayurvedic works such as Gananath Sen and Ashutosh Roy repeatedly emphasized that, whereas biomedicine was merely interested in the physical, their therapeutics was equally invested in the physical and the metaphysical. But what they meant literally was that they were interested in an enhanced physicality: namely, what Alter would call a meta-materiality, not a hyperphysicality that would be singular, exceptional, quirky, mystical, and resolutely non-biopolitical. Indeed, as modern Ayurveda was gradually fitted into the governing apparatus of the modern state, it had to make its therapeutics amenable to biopolitical structures.[111] Simultaneously, the pharmaceuticalization of Ayurveda also required a plausibility of population-level drug-impact studies.[112] The pataphysics of singularities and mysticism engendered in hyperphysical bodies with superhuman capacities, intangible agencies, and the mysterious therapeutic power of sounds had to be transplanted by a metaphysics engendered in the discrete predictability of standardized human biology, the calculable rationalities of population-level drug trials and governmental policies, and the neat essentialisms of colonial identity politics.

ACKNOWLEDGMENTS

This book has a taken a long time to write. In the meanwhile, I have moved to four different countries on three continents, switched jobs several times, and finished a book I started long after this one. Along the way, I have naturally incurred a lot of debt.

My greatest debt is to my three gurus, David Arnold, Gautam Bhadra, and Majid Siddiqi. Each of them has taught me much, not only by what they have written or said, but also by who they are and how they have chosen to live their lives. I feel incredibly lucky to have had the chance to learn from them. I am equally grateful to David Hardiman for his friendship. I have learned as much from him as from anyone else. His simplicity and the clarity of his politics have been especially inspiring.

As for my teachers in India and the UK, I have profited greatly from Daud Ali, Neeladri Bhattacharya, Sabyasachi Bhattacharya, Kunal Chakrabarti, Subhas Ranjan Chakraborty, Waltraud Ernst, Ashis Nandy, K. N. Panikkar, Peter Robb, Francis Robinson, Tanika Sarkar, and Heeraman Tiwari. Besides my teachers, I have also benefited from my participation within a larger scholarly community in India and the UK. Those within this community to whom I owe a debt of gratitude include Guy Attewell, Nandini Bhattacharya, Sanjoy Bhattacharya, Ian Burney, Pratik Chakrabarti, Prasun Chatterjee, Burton Cleetus, Chris Fuller, Rajarshi Ghose, Sharmistha Gooptu, Deepak Kumar, Prabhat Kumar, John Bosco Loudusamy, John Matthew, Boria Majumdar, Harish Naraindas, Nitin Sinha, Rajat Kanti Sur, and Michael Worboys. In Bangladesh, my greatest debts are to Sirajul Islam and Sonia Nishat Amin.

Having moved to North America five years ago, I feel lucky to have discovered a warm, welcoming, and intellectually stimulating circle of interlocutors. Among these, I would especially like to acknowledge my debt to

Babak Ashrafi, Ishita Banerjee, Nicole Barnes, Rachel Berger, Lundy Braun, Thomas Broman, Dipesh Chakrabarty, Indrani Chatterjee, Lawrence Cohen, Angela Creager, Alex Csiszar, James Delbourgo, Michael Dodson, Saurabh Dube, Mats Fridlund, Yulia Frumer, Michael Gordin, Katja Guenther, Jeremy Greene, Christopher Hamlin, Marta Hanson, Prakash Kumar, Gabriella Soto Laveaga, Eugenia Lean, Rochona Majumdar, Clapperton Chakanetsa Mavhunga, Erika Milam, Suzanne Moon, Graham Mooney, Carla Nappi, Emily Lynn Osborne, Randall Packard, Ishita Pande, Amit Prasad, Joanna Radin, Arvind Rajagopal, Charles Rosenberg, Mitra Sharafi, Asif Siddiqi, Kalyanakrishnan Sivaramakrishnan, Kavita Sivaramakrishnan, Pamela Smith, Thomas Trautmann, John Harley Warner, Keith Wailoo, and David Wright. They have each in their own ways taught me much. I am particularly thankful to Douglas Haynes and Mary Fissell. Their generosity, guidance, and approachability have been a great source of strength and inspiration.

One of the few good things about the "global village" some of us inhabit today is the genuine possibility for long-distance intellectual collaborations. I have benefited immensely from these possibilities and, as a result, owe deep intellectual debts to many in continental Europe and the Antipodes. Among them, I owe the most to Warwick Anderson, Sekhar Bandyopadhyay, Jean-Paul Gaudilliere, Hans Pols, Laurent Pordie, Lissa Roberts, William Sax, Dominik Wujastyk, and Dagmar Wujastyk.

Luckily, I have always had wonderfully supportive colleagues in every institution I have worked at. At McMaster University, I would like to specially acknowledge the help, friendship, and support I received from Pamela Swett and Stephen Heathorn, though the entire History Department was enormously supportive while I was there. At the University of Pennsylvania, I would like to particularly thank Daud Ali, Anthea Butler, Shampa Chatterjee, Jody Chavez, Jamal Elias, Srilata Gangulee, Toorjo Ghose, Suvir Kaul, Justin McDaniel, Michael Meister, Rahul Mukherjee, Ania Loomba, Deven Patel, Annette Reed, Terenjit Sevea, and Pushkar Sohoni. In my home department, History and Sociology of Science, the entire cohort has been enormously supportive, but I would particularly like to thank Robby Aronowitz for his "bro-ship," John Tresch for his friendship, Harun Kucuk for his *Abi*-ship, Susan Lindee for her mentorship, Beth Linker for her guruship, and Heidi Voskuhl for the shoulder bumps. While David Barnes, Etienne Benson, Ann Greene, Andi Johnson, and Jonathan Moreno have also been wonderfully supportive colleagues, it has been a great honor to learn from Steve Feierman, Rob Kohler, the late Riki Kuklick, Ruth Schwartz Cowan, and Nathan Sivin. Finally, a big thanks to Pat Johnson.

Among my students, I would particularly like to thank Baishakh Chakrabarti, Allegra Giovine, and Kaushik Ramu.

I must also add a special word of thanks for Gautam Bhadra, Christopher Hamlin, Marta Hanson, and Mick Worboys, along with the two anonymous referees who read and commented on draft chapters of the book. Their comments have greatly enriched the final form this book has taken. I also owe a special thanks to my colleague John Tresch. Conversations with him have always opened up exciting new areas of exploration, while the breadth of his erudition has been a constant source of inspiration.

This book would not have been possible without the various archives in India and the UK that I have drawn on or the old booksellers in Calcutta through whom I have found many of my sources. I am particularly grateful to Abhijit Bhattacharya, Subhashis Bhattacharya, the late Indranath Majumdar, Tushar Majumdar, and Ashim Mukhopadhyay; as well as the staff at the Asian and African Studies, and the Rare Books and Music Reading Rooms of the British Library, the staff at the Wellcome Library, the Duke Humphreys Reading Room at the Bodleian Library, and the Inter Library Loan section of the University of Pennsylvania, and Pushkar Sohoni, the South Asia bibliographer at the University of Pennsylvania.

I have benefited much from the friendship of Santanu Banerjee, Vivek Boray, Ashok Kesari, Prabat Kumar, Boria Majumdar, Saurabh Mishra, and Anshuman Rane. The warm familiarity and generous friendship of Pushkar, Dipeshda/Rochonadi, Waltraud/Mike, Srilatapishi, Shampadi, Suvir/Ania, Daud/Sugra, and Robby/Jane have also gone a long way in ameliorating the sense of exile that attends the nomadic life of the twenty-first century academic.

I would like to thank Karen Merikangas Darling at the University of Chicago Press for the many discussions that helped shape this book, Susan Karani for her sensitive editing, and Leonard Rosenbaum for the indexing. Though unfortunately none of my grandparents lived to see me finish school, their memory has been a constant source of inspiration. It was from my grandparents that I inherited a love of books, an abiding interest in the past, and above all the courage to be critical of that past while still respecting it. The legacies and memories of my late father and both my parents-in-law, being too close in time, are more difficult to label. Yet I feel sure that their legacies too have shaped this book as it has developed gradually over the last decade and a half. To my mother, I owe a debt that is beyond words. It is because of her that I am what I am. Like always, despite her occasional frustrations at "the book" still not being done, she has looked the other way and allowed me to do my thing.

Unlike my mother, Siraj and Mohanlal have resolutely refused to look the other way and tried every trick in their feline repertoires to make me stop writing. Luckily for my sanity, sometimes they succeeded and reminded me of the wonderful world beyond the computer screen. Without them, this book might have been finished a few weeks earlier, but it would decidedly have been a poorer and mirthless one.

Finally, above all, my deepest, truest and most abiding debt is to my beloved wife, Manjita. As this book has dragged on through the years and eaten into larger and larger chunks of our family time, Manjita has borne it all with remarkable fortitude and has continued to encourage me to push forward. Without her, neither the book nor I would have made it through these last few years. I hope this book has been worthy of at least some of the sacrifices she has made for it.

NOTES

INTRODUCTION

1. Vivek Bald's fascinating new research has revealed a long-forgotten world of Bengali Muslim peddlers—mostly from Hooghly district—selling embroideries, curios, and other "fantasies of the East" to a white-collar American lower middle class in the 1880s and 1890s in the New York–New Jersey–Philadelphia region and beyond. It is highly likely that the bottle dug up at Pittsburgh came through one of these peddlers from Hughli. Whatever the route by which it came to the US, it is intriguing to wonder through what frameworks of meaning the producer and consumer of this bottle of medicine made sense of the substance that connected them. How did the lower middle class American man or woman who bought the medicine understand an Ayurvedic remedy? Did his understanding match that of the Hooghly peddler? These are questions I cannot fully pursue here. Vivek Bald, *Bengali Harlem and the Lost Histories of South Asian America* (Cambridge, MA: Harvard University Press, 2013).
2. On Ayurveda and Hindi, Hindu, and Punjabi nationalisms, see Kavita Sivaramakrishnan, *Old Potions, New Bottles: Recasting Indigenous Medicine in Colonial Punjab, 1850–1945* (Hyderabad: Orient Longman, 2006). On Unani Tibb and Muslim nationalism, see Guy Attewell, *Refiguring Unani Tibb: Plural Healing in Late Colonial India* (Hyderabad: Orient Longman, 2007); and Neshat Quaiser, "Unani Medical Culture: Memory, Representation, and the Literate Critical Anticolonial Public Sphere," in *Contesting Colonial Authority: Medicine and Indigenous Responses in 19th and 20th Century India*, ed. Poonam Bala (Maryland: Lexington Books, 2012), 115–36. On Siddha and Tamil nationalism, see Richard S. Weiss, *Recipes of Immortality: Healing, Religion and Community in South India* (Oxford: Oxford University Press, 2009).
3. For a thorough discussion of the case against seeing modern Ayurveda simply as an invented tradition, see David Hardiman, "Indian Medical Indigeneity: From Nationalist Assertion to the Global Market," *Social History* 34, no. 3 (2009): 263–83.
4. Bhushan Patwardhan, "Let's Plan for National Health," *Journal of Ayurveda and Integrated Medicine* 2, no. 3 (2011): 103–4.
5. Poonam Bala, *Imperialism and Medicine in Bengal: A Socio-Historical Perspective* (Delhi: Sage, 1991); Gananath Obeyesekere, "Impact of Ayurvedic Ideas on the Culture and the Individual in Sri Lanka," in *Asian Medical Systems: A Comparative Study*, ed. Charles Leslie (Berkeley: University of California Press, 1976), 201–26; Charles Leslie, "The Ambiguities of Medical Revivalism in Modern India," in *Asian Medical Systems*, ed. Charles

Leslie, 356–67; Brahmananda Gupta, "Indigenous Medicine in Nineteenth and Twentieth Century Bengal," in *Asian Medical Systems*, ed. Charles Leslie, 368–82; K. N. Panikkar, "Indigenous Medicine and Cultural Hegemony: A Study of the Revitalization Movement of Kerala," *Studies in History* 8, no. 2 (1992): 283–308.

6. I see Sivaramakrishnan's work as the first purely historical work because Bala's earlier work was both self-confessedly "sociohistorical" and almost completely eschewed any engagement with Ayurvedic writing as such, relying almost entirely on external, colonial reporting of it. Sivaramakrishnan, *Old Potions, New Bottles*.
7. Dagmar Wujastyk and Frederick M. Smith, eds., *Modern and Global Ayurveda: Pluralism and Paradigms* (Albany, NY: SUNY Press, 2008).
8. Madhulika Banerjee, *Power, Knowledge, Medicine: Ayurvedic Pharmaceuticals at Home and in the World* (Hyderabad: Orient Blackswan, 2009).
9. Rachel Berger, *Ayurveda Made Modern: Political Histories of Indigenous Medicine in North India, 1900–1955* (Basingstoke: Palgrave Macmillan, 2013).
10. Burton Cleetus, "Indigenous Traditions and Practices in Medicine and the Impact of Colonialism in Kerala, 1900–1950" (PhD diss., Jawaharlal Nehru University, 2007).
11. Jean M. Langford, *Fluent Bodies: Ayurvedic Remedies for Postcolonial Imbalance* (Durham, NC: Duke University Press, 2002); Maarten Bode, *Taking Traditional Knowledge to the Market: The Modern Image of the Ayurvedic and Unani Industry, 1980–2000* (Hyderabad: Orient Longman, 2008).
12. Attewell, *Refiguring Unani Tibb*; Seema Alavi, *Islam and Healing: Loss and Recovery of an Indo-Muslim Medical Tradition, 1600–1800* (Basingstoke: Palgrave Macmillan, 2008); Claudia Liebeskind, "Arguing Science: Unani Tibb, Hakims and Biomedicine in India," in *Plural Medicine: Tradition and Modernity, 1800–2000*, ed. Waltraud Ernst (London: Routledge, 2002); Markus Daechsel, *The Politics of Self-Expression: The Urdu Middle-class Milieu in Mid-Twentieth Century India and Pakistan* (London: Routledge, 2006).
13. E. Valentine Daniel, "The Pulse as an Icon in Siddha Medicine," *Contributions to Asian Studies* 18 (1984): 115–26; Richard Weiss, *Recipes of Immortality: Medicine, Religion and Community in South India* (Oxford: Oxford University Press, 2009); Gary J. Hausman, "Siddhars, Alchemy and the Abyss of Tradition: "Traditional" Tamil Medical Knowledge in "Modern" Practice" (PhD diss., University of Michigan, 2006).
14. Laurent Pordie, ed., *Tibetan Medicine in the Contemporary World: Global Politics of Medical Knowledge and Practice* (London: Routledge, 2008).
15. Some anthropological works, such as that of Jean Langford's, are an exception to this trend and do indeed attempt to explore the cognitive and practical content of the medical tradition. Unfortunately, however, the lack of historical depth in these scholars renders the inquiries rather flat and historically unenlightening.
16. Hardiman, "Indian Medical Indigeneity."
17. Nathan Sivin, *Traditional Medicine in Contemporary China* (Ann Arbor: University of Michigan Press, 1987).
18. Volker Scheid, *Chinese Medicine in Contemporary China: Plurality and Synthesis* (Durham, NC: Duke University Press, 2002); Sean Hsiang-lin Lei, *Neither Donkey nor Horse: Medicine in the Struggle over China's Modernity* (Chicago: University of Chicago Press, 2014).
19. Vincanne Adams, Mona Schrempf and Sienna R. Craig, eds., *Medicine between Science and Religion: Explorations on Tibetan Grounds* (Oxford: Berghahn, 2011).
20. Steven Feierman, "Explanation and Uncertainty in the Medical World of the Ghaambo," *Bulletin of the History of Medicine* 74, no. 2 (2000): 317–44; Julie Livingston,

"Productive Misunderstandings and the Dynamism of Plural Medicine in Mid-Century Bechuanaland," *Journal of Southern African Studies* 33, no. 4 (2007): 801–10; Stacey A. Langwick, *Bodies, Politics and African Healing: The Matter of Maladies in Tanzania* (Bloomington: Indiana University Press, 2011).

21. Charles Rosenberg, *Explaining Epidemics and Other Studies in the History of Medicine* (Cambridge: Cambridge University Press, 1992), 12.
22. Ibid., 23.
23. Poonam Bala, *Medicine and Medical Policies in India: Social and Historical Perspectives* (Lanham, MD: Lexingtons, 2007), 25.
24. Jayanta Bhattacharya, "The First Dissection Controversy: Introduction to Anatomical Education in Bengal and British India," *Current Science* 101, no. 9 (2011): 1231.
25. Ibid.
26. Shighehisa Kuriyama, *The Expressiveness of the Body and the Divergence of Greek and Chinese Medicine* (New York: Zone, 1999), 153.
27. Ruth Richardson, *Death, Dissection and the Destitute* (Chicago: University of Chicago Press, 1987); Michael Sappol, *A Traffic of Dead Bodies: Anatomy and Embodied Social Identity in Nineteenth Century America* (Princeton, NJ: Princeton University Press, 2002); Carin Berkowitz, "The Beauty of Anatomy: Visual Displays and Surgical Education in Nineteenth-Century London," *Bulletin of the History of Medicine* 85, no. 2 (2011): 248–78.
28. Nelly Oudshoorn, *Beyond the Natural Body: An Archaeology of Sex Hormones* (London: Routledge, 1994), 4.
29. On the basic chronology of Ayurvedic education and professionalization see Brahmananda Gupta, "Indigenous Medicine."
30. Dominik Wujastyk "Interpreting the Image of the Human Body in Premodern India," *International Journal of Hindu Studies* 13, no. 2 (2009): 189–228; David Gordon White, "On the Magnitude of the Yogic Body," in *Yogi Heroes and Poets: Histories and Legends of the Naths*, ed. David Lorenzen and Adrian Muñoz (Delhi: Oxford University Press, 2011), 80.
31. Kapil Raj, "The Historical Anatomy of a Contact Zone: Calcutta in the Eighteenth Century" *Indian Economic and Social History Review* 48, no. 1 (2011): 78.
32. John Tresch, "Cosmologies Materialized: History of Science and the History of Ideas," in *Rethinking Modern European Intellectual History*, ed.Darrin McMahon and Sam Moyn (New York: Oxford University Press, 2014), 153–72.
33. Robert E Kohler, "A Generalist's View," *Isis* 96, no. 2 (2005): 224–29; James A. Secord, "The Big Picture," *British Journal for the History of Science* 26 (1993): 387–483.
34. Feierman, "Explanation and Uncertainty."
35. Wujastyk "Image of the Human Body"; Murphy Halliburton, "Rethinking Anthropological Studies of the Body: Manas and Bodham in Kerala," *American Anthropologist* 104, no. 4 (2002): 1123–34; Harish Naraindas, "Of Relics, Body Parts and Laser Beams: The German Heilpraktiker and his Ayurvedic Spa," *Anthropology & Medicine* 18, no. 1 (2011): 67–86; Helen Lambert, "The Cultural Logic of Indian Medicine: Prognosis and Aetiology in Rajasthani Popular Therapeutics," *Social Science & Medicine* 34, no. 10 (1992): 1069–76; Robert Desjarlais, *Body and Emotions: The Aesthetics of Illness and Healing in the Nepal Himalayas* (Philadelphia: University of Pennsylvania Press, 1992).
36. Joseph Alter, *Yoga in Modern India: The Body between Science and Philosophy* (Princeton, NJ: Princeton University Press, 2004); Langford, *Fluent Bodies*.
37. Langford, *Fluent Bodies*, 141–47.

38. Annemarie Mol, *The Body Multiple: Ontology in Medical Practice* (Durham, NC: Duke University Press, 2002). For an excellent review of some of the latest anthropological investigations into the many bodies of biomedicine, see Margaret Lock and Vinh-Kim Nguyen, *An Anthropology of Biomedicine* (Willey-Blackwell: Chichester, 2010).
39. Scheid, *Chinese Medicine*, 12.
40. Sandra Harding, *Sciences from Below: Feminisms, Postcolonialities, and Modernities* (Durham, NC: Duke University Press, 2008), 145.
41. Alter, *Yoga in Modern India*, 8.
42. Langford, *Fluent Bodies*, 12.
43. Alter, *Yoga in Modern India*, 30.
44. Martin Heidegger, "The Question Concerning Technology," in *The Question Concerning Technology and Other Essays*, trans. William Lovitt (New York: Garland, 1977), 9.
45. Bimal Krishna Matilal, "Causality in the Nyaya-Vaisesika School," *Philosophy East and West* 25, no. 1 (1975): 41–48.
46. Douglas E. Haynes et al., eds., *Towards a History of Consumption in South Asia* (New Delhi: Oxford University Press, 2009).
47. Sanjay Joshi, ed., *Middle Class in Colonial India* (New Delhi: Oxford University Press, 2010).
48. Douglas Haynes, "Making the Ideal Home? Advertising of Electrical Appliances and the Education of the Middle Class Consumer in Bombay, 1925–40," unpublished paper.
49. Partha Chatterjee, *Our Modernity* (Rotterdam/Dakar: SEPHIS CODESRIA, 1997); Sudipta Kaviraj, "Modernity and Politics in India," *Daedalus* 129, no. 1 (2000): 137–62.
50. Warwick Anderson, "Making Global Health History: The Postcolonial Worldliness of Biomedicine," *Social History of Medicine* 27, no. 2 (2014): 372–84.
51. Dipesh Chakrabarty, "The Muddle of Modernity," *American Historical Review* 116, no. 3 (2011): 663–75.
52. Lynn Hunt, "Modernity: Are Modern Times Different?," *Historia Critica* 54 (2014): 107–24.
53. Stanley Joel Reiser, *Medicine and the Reign of Technology* (Cambridge: Cambridge University Press, 1978), x.
54. Joel D. Howell, *Technology in the Hospital: Transforming Patient Care in the Early Twentieth Century* (Baltimore: Johns Hopkins University Press, 1996).
55. Margarete Sandelowski, *Devices and Desires: Gender, Technology, and American Nursing* (Chapel Hill: University of North Carolina Press, 2000).
56. Graham Mooney, "The Material Consumptive: Domesticating the Tuberculosis Patient in Edwardian England," *Journal of Historical Geography* 42 (2013): 152–66.
57. Alter, *Yoga in Modern India*; Guy Attewell, "Haptic Iterability and Credibility: Extra-Diagnostic X-rays, Body Parts and Distributed Agency—with a 'Bonesetting' Clinic in Hyderabad City," unpublished paper; Servando Z. Hinojosa, "Bonesetting and Radiography in the Southern Maya Highlands," *Medical Anthropology* 23, no. 4 (2004): 263–93.
58. Jaminibhushan Ray, "Baidya Sommelone Sobhapotir Obhibhashon," *Ayurveda* 1, no. 5 (1916 [1323 BE]): 189–90.
59. Marie-Noelle Bourguet, Christian Liccope, and H. Otto Sibum, eds., *Instruments, Travel and Science: Itineraries of Precision from the Seventeenth to the Twentieth Century* (New York: Routledge, 2014), 7.
60. Bourguet et al., 11.
61. For the usefulness of "circulation" as a paradigm that undermines the binarism of the "global" and the "local," see Kapil Raj, "Beyond Postcolonialism . . . and

Post-positivism: Circulation and the Global History of Science," *Isis* 104, no. 2 (2013): 337–47.

62. David Arnold, *Everyday Technologies: Machines and the Making of India's Modernity* (Chicago: Chicago University Press, 2013), 5.
63. Gyan Prakash, "Review of David Arnold's Everyday Technologies," *Isis* 105, no. 2 (2014): 445.
64. Wiebe E. Bijker, *Of Bicycles, Bakelite and Bulbs: Towards a Theory of Sociotechnical Change* (Cambridge, MA: MIT Press, 1997); Frank Dikotter, *Things Modern: Material Culture and Everyday Life in China* (London: C. Hurst, 2006); David Edgerton, "Creole Technologies and Global Histories: Rethinking How Things Travel in Time and Space," *Journal of History of Science and Technology* 1 (2007): 75–112; Ruth Schwartz Cowan, *More Work for Mother* (New York: Basic Books, 1983); Dipesh Chakrabarty, *Provincializing Europe: Postcolonial Thought and Historical Difference* (Princeton, NJ: Princeton University Press, 2000).
65. Langdon Winner, "Do Artifacts Have Politics?," *Daedalus* 109, no. 1 (1980): 121–36.
66. Bijker, *Of Bicycles, Bakelite and Bulbs*, 280.
67. Aditya Bharadwaj, *Local Cells, Global Science: The Rise of Embryonic Stem Cells Research in India* (London: Routledge, 2009), 40.
68. Lissa Roberts and Simon Schaffer, preface to *The Mindful Hand: Inquiry and Invention from the Late Renaissance to Early Industrialisation*, ed.Lissa Roberts, Simon Schaffer and Peter Dear (Amsterdam: Koninklijke Nederlandse Akademie van Wetenschappen, 2007), xxi.
69. Ian Inkster, "The West Had Science and the Rest Had Not? Queries of a Mindful Hand," *History of Technology* 29 (2009): 205.
70. Andrew Pickering, *The Mangle of Practice: Time, Agency and Science* (Chicago: University of Chicago Press, 1995), 20.
71. Raj, "Historical Anatomy of a Contact Zone."
72. George Basalla, "The Spread of Western Science," *Science* 156, no. 3775 (1967): 611–22.
73. David Wright, Nathan Flis, and Mona Gupta, "The "Brain-Drain" of Physicians: Historical Antecedents to an Ethical Debate, c. 1960–79," *Philosophy, Ethics and Humanities in Medicine* 3, no. 24 (2008), e-pub.
74. Jan Golinski, "Is It Time to Forget Science? Reflections on a Singular Science and Its History," *Osiris* 27, no. 1 (2012): 19–36.
75. Bruno Latour, *Science in Action: How to Follow Scientists and Engineers through Society* (Cambridge, MA: Harvard University Press, 1987).
76. For a fuller discussion, see Projit Bihari Mukharji, "Cultures of Fear: Technonationalism and the Postcolonial Responsibilities of STS," *East Asian Science, Technology and Society* 6 (2012): 267–74.
77. Helen Tilley, "Global Histories, Vernacular Science, and African Genealogies; Or Is the History of Science Really for the World?," *Isis* 101, no. 1 (2010): 110–19.
78. Kohler, "A Generalist's View."
79. David Pingree, *Census of the Exact Sciences in Sanskrit* (Philadelphia: American Philosophical Society, 1970, vols. 1–5); Kim Plofker, *Mathematics in India* (Princeton, NJ: Princeton University Press, 2009); S. R. Sarma, *The Archaic and the Exotic: Studies in the History of Indian Astronomical Instruments* (Delhi: Manohar, 2008); J. A. McHugh, "Blattes de Byzance in India: Mollusk Opercula and the History of Perfumery," *Journal of the Royal Asiatic Society* 23, no. 1 (2013): 53–67; David Gordon White, *The Alchemical Body: The Siddha Tradition in Medieval India* (Chicago: University of Chicago Press, 1996); G. J. Meulenbeld, *A History of Indian Medical Literature* (Groningen:

E. Forsten, 1999, vols. 1–3). For medicine see also, Kenneth Zysk, *Asceticism and Healing in Ancient India: Medicine in the Buddhist Monastery* (Delhi: Motilal Banarsidass, 1998); and Dominik Wujastyk, ed. and trans., *The Roots of Ayurveda: Selections from Sanskrit Medical Writings* (New Delhi: Penguin, 1998); Daud Ali, "Garden Automata and the Production of Wonder Across the Indian Ocean" (paper presented at the History and Sociology of Science Workshop, University of Pennsylvania, Philadelphia, February 4, 2013).

80. Dominik Wujastyk, "Change and Creativity in Early Modern Indian Medical Thought," *Journal of Indian Philosophy* 33, no. 1 (2005): 95–118.
81. Simon Schaffer et al., eds., *The Brokered World: Go-Betweens and Global Intelligence, 1770–1820* (Sagamore Beach, MA: Science History Publications, 2009).
82. Marwa Elshakry, *Reading Darwin in Arabic, 1860–1950* (Chicago: University of Chicago Press, 2013). See also, Daniel A Stolz, "'By Virtues of Your Knowledge': Scientific Materialism and the Fatwas of Rashid Rida," *Bulletin of the School of Oriental and African Studies* 75, no. 2 (2012): 223–47.
83. Simon Schaffer, "The Asiatic Enlightenments of British Astronomy," in *Brokered World*, ed.Schaffer et al., 51.
84. Nicholas Thomas, *Entangled Objects: Exchange, Material Culture, and Colonialism in the Pacific* (Cambridge, MA: Harvard University Press, 1991).
85. Scheid, *Chinese Medicine*.
86. Nancy Rose Hunt, *A Colonial Lexicon: Of Birth Ritual, Medicalization, and Mobility in the Congo* (Durham, NC: Duke University Press, 1999); Langwick, Bodies, Politics, Healing.
87. Hunt, *Colonial Lexicon*, 8, 12.
88. Monica Juneja, "Objects, Frames, Practices: A Post Script on Agency and Braided Histories of Art," *Medieval History Journal* 15, no. 2 (2012): 418.
89. Juneja, "Braided Histories," 419. The exact quote is attributed to Sheldon Pollock by Finbarr Barry Flood from whom Juneja borrows it. For "braiding" see also, Ranajit Guha, "Some Aspects of the Historiography of Colonial India" in Ranajit Guha and Gayatri Chakravorty, eds. *Selected Subaltern Studies* (New York: Oxford University Press, 1988): 42 and Gautam Bhadra, *Iman O Nishan: Unish Shotoker Banglaye Krishok Choitonyer EK Odhyay, c. 1800–1850* (Calcutta: Subarnareka, 1988): 318–319.
90. Michael Werner and Benedicte Zimmermann, "Beyond Comparison: Histoire Croisee and the Challenge of Reflexivity," *History & Theory* 45, no. 1 (2006): 37–38.
91. Christopher Pinney, "Creole Europe: A Reflection of a Reflection," *Journal of New Zealand Literature* 20 (2003): 125–61; Jane Bennett, *Vibrant Matter: A Political Ecology of Things* (Durham, NC: Duke University Press, 2010).
92. George Sarton, "Second Preface to Volume XXIII: History of Science versus the History of Medicine," *Isis* 23, no. 2 (1935): 313–20; Henry E. Sigerist, "The History of Medicine *and* the History of Science: An Open Letter to George Sarton, Editor of *Isis*," *Bulletin of the Institute of the History of Medicine* 4 (1936): 1–13.
93. Bhagavat Singh Jee, *A Short History of Aryan Medical Science* (London: Macmillan, 1896).
94. Gyan Prakash, *Another Reason: Science and the Imagination of Modern India* (Princeton, NJ: Princeton University Press, 1999).
95. Chandak Sengoopta, *The Most Secret Quintessence of Life: Sex, Glands and Hormones, 1850–1950* (Chicago: University of Chicago Press, 2006), 2.
96. John Harley Warner, "Science in Medicine," *Osiris* 1 (1985): 37.
97. Marwa Elshakry, "When Science Became Western?: Historiographical Reflections," *Isis* 101, no. 1 (2010): 146–58.

98. Tresch, "Cosmologies Materialized," 164.
99. See, for instance, Meera Nanda's almost paranoid denunciation of anyone who so much as questions the totalizing universalism of a singular "Western science." *Prophets Facing Backwards: Postmodern Critiques of Science and Hindu Nationalism in India* (New Brunswick, NJ: Rutgers University Press, 2003). While I often find her arguments intellectually indefensible and colored by political paranoia, I sympathize with the political anxieties that drive her to make such arguments.
100. See, for instance, Steven Shapin and Simon Schaffer, *The Leviathan and the Air-Pump: Hobbes, Boyle and the Experimental Life* (Princeton, NJ: Princeton University Press, 2011). See also, Simon Schaffer, "Godly Men and Mechanical Philosophers: Souls and Spirits in Restoration Natural Philosophy," *Science in Context* 1, no. 1 (1987): 53–85.
101. Peter van der Veer, *Imperial Encounter: Religion and Modernity in India and Britain* (Princeton, NJ: Princeton University Press, 2001).
102. See, for instance, Meera Nanda, *The God Market: How Globalization Is Making India More Hindu* (New York: NYU Press, 2011).
103. On the political nature of the label "pseudo-science," see Michael Gordin, "How Lysenkoism Became Pseudoscience: Dobzhansky to Velikovsky," *Journal of the History of Biology* 45, no. 1 (2012): 443–68. In South Asian scholarship, Joseph Alter is one of the few people to recognize and flag the slipperiness of the distinction between "science" and "pseudo-science"; see chap. 2 in *Yoga in Modern India*.
104. Newspaper baron Mrinalkanti Ghosh described Hemchandra Sen's involvement in Theosophy in his spiritual memoir. Mrinalkanti Ghosh, *Poroloker Kotha* (n.p., 1933), 57–60. Another eminent Ayurvedic author who also dabbled in Theosophy was Shyamacharan Sengupta, who authored a medical dictionary titled, *Ayurvedarthochondrika* (Kashipur: Pallibikashini Yantra, 1887 [1294 BE]). His theosophical connections are mention in Surendramohan Mukhopadhyay, *Jonmantor Rohosyo* (n.p., 1899 [1306 BE]), 215–24.
105. Lawrence Cohen, "Whodunit?," *Configurations* 2, no. 2 (1994): 343–47.
106. Langford, *Fluent Bodies.*
107. Lawrence Cohen, "The Epistemic Carnival: Meditations on Disciplinary Intentionality and Ayurveda," in *Knowledge and the Scholarly Medical Tradition*, ed. Don Bates (Cambridge: Cambridge University Press, 1995), 320–43; Harish Naraindas, "Of Spineless Babies and Folic Acid: Evidence and Efficacy in Biomedicine and Ayurvedic Medicine," *Social Science & Medicine* 62, no. 11 (2006): 268–69.
108. Wujastyk, "Image of the Human Body."
109. Ranajit Guha, *Dominance without Hegemony: History and Power in Colonial India* (Cambridge, MA: Harvard University Press, 1997).
110. Naraindas, "Spineless Babies"; see also, Murphy Halliburton, "Resistance or Inaction? Protecting Ayurvedic Medical Knowledge and Problems of Agency," *American Ethnologist* 38, no. 1 (2011): 86–101; Jean-Paul Gaudilliere, "An Indian Path to Biocapital? The Traditional Knowledge Digital Library, Drug Patents, and the Reformulation Regime of Contemporary Ayurveda," *East Asian Science, Technology and Society: An International Journal* 8, no. 4 (2014): 391–415.
111. Naraindas, "Spineless Babies."
112. Jean Baudrillard and Marc Guillaume, *Radical Alterity* (Los Angeles: Semiotexte/Smart Art, 2008).
113. Partha Chatterjee, *The Nation and Its Fragments: Colonial and Postcolonial Histories* (Princeton, NJ: Princeton University Press, 1993).
114. Here I am deeply influenced by the writings of Ashis Nandy and Shiv Visvanathan.

Particularly, see Ashis Nandy and Shiv Visvanathan, "Modern Medicine and Its Non-Modern Critics: A Study in Discourse," in *Dominating Knowledge: Development, Culture and Resistance*, ed. Frederique Apffel Marglin and Stephen A Marglin (Oxford: Clarendon, 1990). See also, Walter Mignolo, "The Many Faces of Cosmo-Polis: Border Thinking and Critical Cosmopolitanism," *Public Culture* 12, no. 3 (2000): 721–48.

115. Homi Bhabha, *The Location of Culture* (London: Routledge, 1994), 86.
116. Cohen, "Whodunit?," 344.
117. Berger, *Ayurveda Made Modern*.
118. On pharmaceuticalization, see Madhulika Banerjee, "Ayurveda in Modern India: Standardization and Pharmaceuticalization," in *Modern and Global Ayurveda*, ed. Wujastyk and Smith, 201–14.

CHAPTER ONE

1. See, for instance, Bala, *Imperialism and Medicine in Bengal*, 124, 138–39.
2. Ibid., 29.
3. Pierre Bourdeiu, "The Social Space and the Genesis of Groups," *Theory & Society* 14, no .6 (1985): 723–44.
4. Sivaramakrishnan, *Old Potions, New Bottles*, 14–32.
5. Marshall Sahlins, *Islands of History* (Chicago: University of Chicago Press, 1987), xiv.
6. Steven Feierman, *Peasant Intellectuals: Anthropology and History in Tanzania* (Madison: University of Wisconsin Press, 1990), 13.
7. On the impact of Gangadhar Ray and the modernization process in Bengal upon the rest of the subcontinent, see Priyavrat Sharma, *Ayurved ka Vaigyanik Itihas* (Benares: Chaukhamba Orientalia, 1975), 221–22, 542.
8. Originally published in *Chikitsa Sammilani* 1, no. 9 (1884): 76. Reproduced in Subrata Pahari, *Unish Shotoker Banglaye Sonatoni Chikitsa Byabosthar Sworup* (Calcutta: Progressive Publishers, 2001), 89.
9. Gupta, "Indigenous Medicine."
10. The theatrical reference is made in the course of a dialogue in a popular Bengali satirical play on quacks. Ami, *Thengapathik Bhuinphod Daktar* (Calcutta: Manimohan Rakshit, 1886 [1293 BE]).
11. See Gupta, "Indigenous Medicine"; Pahari, *Unish Shotoker.*
12. Ronald B. Inden and Ralph W. Nicholas, *Kinship in Bengali Culture* (Chicago: Chicago University Press, 1977). Though this discussion is in the context of a slightly different sort of *guru*, the terminology still works with Baidya *gurus* and I have heard it used as such.
13. See Shiv Sharma, *Ayurvedic Medicine: Past and Present* (Calcutta: Dabur, 1975), 15.
14. Sivaramakrishnan, *Old Potions, New Bottles*, 46–52.
15. Gopalchandra Sengupta, "Grahokdiger Nam" (List of Subscribers), *Ayurveda Sarsamgraha*, vol. 2 (Calcutta: Columbian Press, 1871 [1278 BE]), unnumbered page.
16. Originally published in *Chikitsa Sammilani* 1, no. 9 (1884): 76. Reproduced in Subrata Pahari, *Unish Shotoker*, 89.
17. Gananath Sen, "Bharotborshe Ayurveder Jagriti," *Ayurveda Bikash* 1, no. 12 (1913 [1320 BE]): 361.
18. HFJT Maguire, *Report on the Census of Calcutta Taken on the 26th February 1891* (Calcutta: Bengal Secretariat Press, 1891), cxvii.
19. See James Wise, *Notes on the Races, Castes and Trades of Eastern Bengal* (London: Harrison and Sons, 1883).
20. Brahmananda Gupta, *Sutanati Parishad Smarak Grantha* (n.p., n.d.).

21. Madhusudan Sensharma, "Baidya-Brahmin Somiti Bhabanipur Kendrer Barshik Biboron," *Baidya Hitoishi* 3, no. 4 (1926 [1333 BE]): 121–24.
22. Anon., "Bhabanipur Kendrer Prothom Barshik Sadharon Sobha," *Baidya Hitoishi* 3, no. 4 (1926 [1333 BE]): 148–51.
23. Sivaramakrishnan, *Old Potions, New Bottles*, 20–21.
24. William Adam, *Second Report on the State of Education in Bengal: District of Rajshahi* (Calcutta: Bengal Military Orphan Press, 1836), 74.
25. Adam, *Second Report*, 74.
26. Ibid., 75–79.
27. Pahari, *Unish Shotoker*, 82–87.
28. Bala, Imperialism and Medicine in Bengal, 29.
29. Dines Chandra Sircar, "The Ambashtha Jati," *Journal of the United Provinces Historical Society* 17, no. 1 (1944): 148–61.
30. Diane P. Mines, *Caste in India*, Key Issues in Asian Studies 3 (Association of Asian Studies, 2009).
31. The differences between *varna* and *jati* are central to how "caste" operates as a form of hierarchic social organization. Whereas the actual group affiliations which determine everyday sociality and commensality is the localized *jati*, each *jati* attempts to raise itself up the hierarchy or is pushed down it by neighboring *jatis* with reference to the *jati*'s derivation from this or that *varna*. Thus, those seeking to raise a *jati* within this system will claim that it is closer in ritual practice or descent to a higher *varna*—for instance, Kshatriyas rather than Vaishyas. Those seeking to push it down or deny it upward mobility will make the obverse claim arguing that, ritually or ancestrally, the *jati* is closer to Vaishyas than Kshatriyas. For a general discussion of *varna* and *jati*, see Diane P. Mines, *Caste in India*. For a recent and engaging discussion of how these categories continue to be renegotiated within an increasingly complex political milieu, see Ishita Banerjee-Dube, "Caste, Race and Difference: The Limits of Knowledge and Resistance," *Current Sociology* 42, no. 4 (2014): 512–30.
32. Hitesranjan Sanyal, "Continuities of Social Mobility in Traditional and Modern Society in India," *Journal of Asian Studies* 30, no. 2 (1971): 315–39.
33. William Ward, *Account of the Writings, Religion and Manners of the Hindoos*, vol. 4 (Serampore: Mission Press, 1811), 72. For an internalist view of the events, see also Rasiklal Gupta, Moharaj Rajballabh Sen o Totsomokalborti Banglar Itihaser Sthul Sthul Biboron (Calcutta: Sakhi Press, n.d.), 95–112.
34. S. N. Mukherjee, "Caste, Class and Politics in Calcutta, 1815–38," in *Elites in South Asia*, ed.Edmund Leach and S. N. Mukherjee (Cambridge: Cambridge University Press, 1970), 59.
35. Nagendranath Basu, "Gangadhar Kobiraj," *Bishwokosh*, vol. 5 (Calcutta: Sri Rakhalchandra Mitra, 1894 [1301 BE]),157.
36. Srikanta Ray, "Gangadhar Sen," in *Bengal Celebrities* (Calcutta: City Book Society, 1906), 27.
37. Anon., *Ambastha-Kulochondrika Orthat Baidyajatir Chokshudan* (Calcutta: Rakhaldas Borat, 1892 [1299 BE]), 1–2.
38. Tryambakeshwar Ray, "Acharjyo Gangadharer Jiboni," *Ayurveda Bikash* 2, no. 1 (1914 [1321 BE]): 24–25.
39. Bharatamallick, "Obotoronika" in *Ratnaprabha: Radhiya Baidyakuloponjika*, ed. Binodlal Sen (Calcutta, 1891 [1298 BE]).
40. Kumkum Chatterjee, "King of Controversy: History and Nation-making in Late Colonial India," *American Historical Review* 110, no. 5 (2005): 1459.

41. Bharatmullick, *Chandraprabha: Baidyakulapanjika*, ed. Binodlal Sen (Calcutta, 1892 [1299 BE]): back cover.
42. P. K. Sen, advertisement, *Baidya Protibha* 2, no. 1 (1925 [1332 BE]): back cover.
43. P. K. Sen, advertisements, *Baidya Protibha* 2, no. 2 (1925 [1332 BE]): unnumbered back page and cover; P. K. Sen, advertisement, *Baidya Protibha* 2, no. 3 (1925 [1332 BE]): unnumbered back page and cover.
44. Nagendranath Sen & Co. Ltd, Advertisement, "Baidyabritti o Baidyok Grontho," *Baidya Hitoishi* 3, no. 7 (1927 [1334 BE]): 226.
45. M. N. Srinivas, "A Note on Sanskritization and Westernization," *Far Eastern Quarterly* 15, no. 4 (1956): 481–96.
46. See Sekhar Bandyopadhyay, *Caste, Culture and Hegemony: Social Dominance in Colonial Bengal* (New Delhi: Sage, 2004).
47. Wise, Races, *Castes and Trades of Eastern Bengal*, 201.
48. E. A. Gait, *Report of the Census of India*, vol. 1, pt. 1 (Calcutta: Government Printing Press, 1903), 351–52.
49. Gananath Sen, "Baidyer Jatiyo Obonotir Karon (Oitihasik Tottwo)," *Baidya Hitoishi* 4, no. 11 (1928 [1335 BE]): 368–72.
50. Khagendranath Choubey, "'Ambashtanang' Chikitsitam," *Baidya Hitoishi* 4, no. 6 (1926 [1333 BE]): 192.
51. For the long-term association between Sanskrit learning and Baidya *jati* mobility, see Pascale Haag, "I Wanna Be a Brahmin Too: Grammar, Traditions and Mythology as Means for Social Legitimisation among the Vaidyas in Bengal," in *Samskrta-sadhuta "Goodness of Sanskrit": Studies in Honour of Professor Ashok N. Aklujkar*, ed.Chikafumi Watanabe, Michele Marie Desmarais, and Yoshichika Honda (New Delhi: D. K. Printworld, 2012), 226–49.
52. Chaitanya Mahaprabhu was born in a Brahmin family in Nabadwip in the Nadia district of modern-day West Bengal. He was initially known mainly as an erudite young scholar, but after meeting his spiritual preceptor while on a trip to Gaya to perform his father's funerary rights he began to change. Eventually he preached a simple creed whereby merely the singing of Krishna's name in public and with all honesty was enough to achieve salvation. He often disregarded orthodox caste and ritual codes and thus helped create a massive congregation throughout eastern South Asia. He worked within the longer South Asian tradition of Bhakti that preaches a personalized, mystic form of religion. Chaitanya came to be seen as a god even while he was still alive. For a fuller discussion of Chaitanya's biography as well as the complexities of the biographical representations see Tony K. Stewart, *The Final Word: The Caitanya Caritamrita and the Grammar of Religious Tradition* (Oxford University Press, 2010).
53. For an excellent discussion of the history of Gaudiya Vaishnavism, see Ramakanta Chakrabarty, *Bonge Vaishnava Dharma: Ekti Oitihasik Ebom Samajtattwik Odhyayon* (Calcutta: Ananda Publishers Pvt. Ltd., 1996).
54. Ibid., 131.
55. Ibid.
56. Wise, *Races, Castes and Trades of Eastern Bengal*, 201.
57. Umeshchandra Gupta, *Jati-tottwo Baridhi* (Calcutta: Majumdar Library, 1902), 358–64.
58. Ibid., 371–81.
59. Report of the Commission Appointed by the Govt. of Bengal to Enquire into the Excise of Country Spirit in Bengal, 1883–84, vol. 2 (Calcutta: Bengal Secretariat Press, 1884), 250. The word chakra literally means "circle"; in this context it means a "tantric

circle of ritual participants." It seems to mainly be associated with folk Tantrism in Bengal. For a gloss on the term, see June McDaniel, *Offering Flowers, Feeding Skulls: Popular Goddess Worship in Bengal* (Oxford: Oxford University Press, 2004), 10.

60. J. T. O'Connell, "Chaitanya Vaishnava Devotion (bhakti) and Ethics as Socially Integrative in Sultanate Bengal," *Bangladesh eJournal of Sociology* 8, no. 1 (2011): 58.
61. Basu, "Gangadhar Kobiraj," 157.
62. Anon., "Kobiraj Surendranath Goswami," *Janmabhumi* 27, no. 6 (1922): 204.
63. Anon., "Astanga Ayurveda o Astanga Ayurveda Bidyaloy," *Ayurveda* 1, no. 1 (1916 [1323 BE]): 32.
64. Anon., "Kobiraj Surendranath Goswami," *Janmabhumi* 27, no. 8 (1922): 251–52.
65. Surendranath Goswami, *Ayurved o Malaria-jwor* (Calcutta: Janmabhumi Press, 1914 [1321 BE?]), 12.
66. Ibid., 13.
67. Anon., "Samkshipto Jiboni," *Ayurveda Bikash* 2, nos. 11–12 (1914–15): 281.
68. For a much fuller discussion of this point, see David Gilmartin and Bruce Lawrence, eds., *Beyond Turk and Hindu: Rethinking Identities in Islamicate South Asia* (Gainesville, FL: University Press Florida, 2000).
69. Srinivas, "Sanskritization and Westernization," 481–96.
70. H. H. Risley, *The Tribes and Castes of Bengal: Ethnographic Glossary*, vol. 1 (Calcutta: Bengal Secretariat Press, 1892), 49.
71. E. A. Gait, *Census of India, 1901*. Vol. 6. *The Lower Provinces of Bengal and Their Feudatories*. Part 1 (Calcutta: Bengal Secretariat Press, 1902), 486.
72. Michael S. Dodson, *Orientalism, Empire and National Culture: India, 1770–1880* (Basingstoke: Palgrave Macmillan, 2007), 53.
73. Haag, "I Wanna Be a Brahmin."
74. H. H. Wilson, "Medical and Surgical Sciences of the Hindus," in *Works of the Late Horace Hayman Wilson*, vol. 3 (London: Trubner, 1864); John Forbes Royle, *An Essay on the Antiquity of Hindoo Medicine* (London: W. H. Allen, 1837); F. J. Mouat, "Hindu Medicine," in *Calcutta Review*, vol. 8 (Calcutta: Sanders, Cones and Co., 1847), 379–433; T. A. Wise, *Review of the History of Medicine*, vols. 1–2 (London: J. Churchhill, 1867); Allan Webb, *The Historical Relations of Ancient Hindu with Greek Medicine* (Calcutta: Military Orphan Press, 1850).
75. Wise, *Review*, 2:457.
76. Tony Ballantyne, *Orientalism and Race: Aryanism in the British Empire* (Basingstoke: Palgrave Macmillan, 2002), 54.
77. Punjabi Vaids, too, were influenced by broadly the same set of orientalist authors, see Sivaramakrishnan, *Old Potions, New Bottles*, 132.
78. Gananath Sen, *Ayurveda Samhita* (Calcutta: Gobardhan Press, 1924 [1331 BE]), 13.
79. Brajaballabh Ray, "Ayurveda," *Ayurveda* 1, no. 1 (1916 [1323 BE]): 5.
80. Thomas R. Trautmann, *Aryans and British India* (Berkeley: University of California Press, 1997), 3.
81. Binodlal Sen, *Ayurveda Bigyan*, vol. 1 (Calcutta: Adi Ayurveda Machina Yantra, 1930 [1337 BE]), 1.
82. In the novel, Jibon Moshai learns "Western" medicine from a brilliant autodidact, Rangalal *daktar*.
83. Wise, *Review*, 1:lxxxiv–lxxxviii.
84. Mukharji, *Nationalizing the Body*, 5; Bala, *Imperialism and Medicine*, table A-16.

85. Bhattacharya, "The First Dissection Controversy: Introduction to Anatomical Education in Bengal and British India," *Current Science* 101, no. 9 (2011): 1231.
86. Kuriyama, Expressiveness of the Body.
87. Letter from J. Donald, Secretary to the Government of Bengal, dated 18 December 1916, Financial Dept., Medical No. 3340 Medl. File 3M-5/19, West Bengal State Archives.
88. The request was from a college called the Jatiyo Ayurbigyan Bidyaloy. File Medl. 10–44, Proceedings 64–66, Local Self-Government Dept., Medical Branch Proceedings, July 1922, West Bengal State Archives.
89. File Medl. 3M-11, Proceedings B 307 and 308, Local Self-Government Dept., Medical Branch Proceedings, February 1928, West Bengal State Archives.
90. See Sappol, *Traffic of Dead Bodies*; Berkowitz, "The Beauty of Anatomy."
91. The term *daktar* is a vernacularization of the term *doctor* and refers to the Bengali practitioners of "Western" medicine. The medicine they practiced was itself a vernacularized form of "Western" medicine called *daktari*. See Mukharji, *Nationalizing the Body*.
92. Tarashankar Bandyopadhyay, *Arogya Niketan* (Calcutta: Prakash Bhavan, 1958).
93. John Harley Warner and Lawrence J. Rizzolo, "Anatomical Instruction and Training for Professionalism: From the 19th to the 21st Centuries," *Clinical Anatomy* 19, no. 5 (2006): 403–14.
94. Carol A. Breckenridge, ed., *Consuming Modernity: Public Culture in a South Asian World* (Minneapolis: University of Minnesota Press, 1995); Christophe Jaffrelot and Peter van der Veer eds., *Patterns of Middle Class Consumption in India and China* (New Delhi: Sage, 2008); Mark Leichty, *Suitably Modern: Making Middle Class Culture in a New Consumer Society* (Princeton, NJ: Princeton University Press, 2003); Daechsel, *Politics of Self-Expression*; Joshi, *Fractured Modernity*; Haynes et al., *History of Consumption in South Asia*; Joanne Punzo Waghorne, *Diaspora of the Gods: Modern Hindu Temples in an Urban Middle Class World* (Oxford: Oxford University Press, 2004).
95. Jaffrelot and van der Veer, *Middle Class Consumption*, 11.
96. Ibid., 12.
97. Joshi, *Fractured Modernity*, 2.
98. Daechsel, *Politics of Self-Expression*, 93.
99. Sen, *Ayurveda Bigyan*, 1:1.
100. Satyacharan Sengupta, "*Onukorone Amader Obostha*," *Ayurveda* 1, no. 1 (1916 [1323 BE]): 428.
101. Surendralal Sensharma, "*Dasham Graha*," *Baidya Protibha* 1, no. 1 (1925 [1332 BE]): 62.
102. Advertisement, "Deccan Watch Co.," *Journal of Ayurveda* 4, no. 8 (1928): xii. On the same page, see also the advertisements of the Lord Wellington Watch Co.
103. Advertisement, "Ik-Mik Cooker," *Baidya Hitoishi* 4, no. 8 (1928 [1335 BE]): 290; 4, no. 11 (1928): 370.
104. Anon. "Baidya Sommelone Sobhapotir Obhibhashon," *Ayurveda* 1, no. 5 (1916 [1323 BE]): 189.
105. Christopher Pinney, "The Parallel Histories of Anthropology and Photography," in *Anthropology and Photography, 1860–1920*, ed. Elizabeth Edwards (New Haven, CT: Yale University Press, 1994), 91.
106. I was able to retrieve it from a family website maintained by the descendants of Goswami. The photograph on the site seems to be a scan of published photograph. http://

www.thakurkanaifamily.com/thakurkanai/index.php/en/shrimat-surendranath-goswami, accessed on July 25, 2014.

107. Surendranath Goswami, "Bhut-bigyan," *Janmabhumi* 18, no. 1 (1910 [1317 BE]): 2–7.
108. Sanjit Narwekar, *Films Division and the Indian Documentary* (New Delhi: Publications Division, Ministry of Information and Broadcasting, Government of India, 1992), 11.
109. Charles Leslie, "Interpretations of Illness: Syncretism in Modern Ayurveda," in *Paths to Asian Medical Knowledge*, ed.Charles Leslie and Allan Young (Berkeley: University California Press, 1992), 177–208.
110. Sivaramakrishnan, *Old Potions, New Bottles*, 158–83.
111. Berger, *Ayurveda Made Modern*.
112. Brahmananda Gupta, *Sutanati Porishod* (n.p., n.d.), 18; Gupta, "Indigenous Medicine," 376.
113. Local Self Govt. Dept., Proceedings, 16–21, 1921, File: Medl. 2R-12 (1), West Bengal State Archives.
114. Ibid.
115. Anon., "Kalikata Baidya Brahmin Samitir Dwitiya Barshik Adhibeshan," *Baidya Hitoishi* 3, no. 1 (1333 BE), 17–18.
116. Local Self Govt. Dept., Proceedings, 215, 1922, File: Medl. 2R-17, West Bengal State Archives.
117. Local Self Govt. Dept., Proceedings, 7, 1921, File Medl. 2R-28(1), West Bengal State Archives.
118. Girindranath Mukhopadhyay, *History of Indian Medicine*, vol. 2 (New Delhi: Oriental Books Reprint Corp., 1974), 51.
119. Rathindra Nath Sen, *Life and Times of Deshbandhu Chittaranjan Das* (New Delhi: Northern Book Centre, 1989), 60.
120. Mukhopadhyay, *History of Indian Medicine*, 2:53.
121. K. S. Roy in fact was to go on to later formulate the plan for an independent Bengal, to avert the division of the province in 1947 along with C. R. Das's young protégé, H. S. Suhrawardy and Subhash Bose's elder brother, Sarat Bose. Cf. Salahuddin Ahmed, *Bangladesh: Past and Present* (New Delhi: APH Publishing Corporation, 2004), 290. Sundari Mohan Das, on the other hand, had been close to the radical Congress leader, Bepin Chandra Pal, and had worked alongside K. S. Roy in his professional life. He was also the first superintendent of the hospital established in Calcutta in the memory of C. R. Das upon the latter's death. http://sundarimohansevabhawan.com/sundarimohan.htm, accessed June 30, 2010.
122. Mukhopadhyay, *History of Indian Medicine*, 2:58.
123. M. K. Gandhi, "Speech at Astanga Ayurveda Vidyalaya," in *The Collected Works of Mahatma Gandhi*, vol. 31 (Delhi: Publications Division, Ministry of Information and Broadcasting, Government of India, 1969), 280–83; Gandhi, "Ayurvedic System" in *Collected Works*, 31:460–62.
124. Personal interview with Brahmananda Gupta, Calcutta, January 21, 2006.

CHAPTER TWO

1. Bandyopadhyay, *Arogya Niketan*, 125–26.
2. Berger, *Ayurveda Made Modern*.
3. Reiser, *Reign of Technology*, 96.
4. Ibid.

5. T. Mouat, "Epidemic Diseases which occurred at Bangalore during the Year 1833," *Transactions of the Medical and Physical Society of Calcutta: Volume the Seventh* (Calcutta: Baptist Mission Press, 1835) 306, 312, 318.
6. Duncan Stewart, "Observations on the Fever which Prevailed at Howrah," *Transactions of the Medical and Physical Society of Calcutta: Volume the Seventh* (Calcutta: Baptist Mission Press, 1835), 367–69.
7. Langford, *Fluent Bodies*.
8. Lawrence Cohen, *No Aging in India: Alzheimer's, the Bad Family, and Other Modern Things* (Berkeley: University of California Press, 1998), 128.
9. Ahsan Jan Quaiser, *The Indian Response to European Technology and Culture, AD 1498–1707* (New Delhi: Oxford University Press, 1982), 64–68.
10. Chatterjee, *Nation and Its Fragments*, 76–115; U. Kalpagam, "Temporalities, Histories and Routines of Colonial Rule in India," *Time & Society* 8, no. 1 (1999): 141–59.
11. See, for instance, Nishimoto Ikuko, "The 'Civilization' of Time: Japan and the Adoption of the Western Time System," *Time & Society* 6, no. 2/3 (1997): 237–59; Atilla Bir, Sinasi Acar, and Mustafa Kacar, "The Clockmaker Family Meyer and Their Watch Keeping the *Alla Turca* Time" in *Science between Europe and Asia: Historical Studies on the Transmission, Adoption and Adaptation of Knowledge*, ed.Feza Gunergun and Dhruv Raina (Dordrecht: Springer, 2011); Takehiko Hashimoto, "The Adoption and Adaptation of Mechanical Clocks in Japan," in *Science between Europe and Asia*, ed. Gunergun and Raina; Daniel Stolz, *The Lighthouse and the Observatory: Islam, Authority, and Cultures of Astronomy in Late Ottoman Egypt* (PhD diss., Princeton University, 2013).
12. Sankar Sen Gupta, *Folklorists of Bengal* (Calcutta: Indian Publications, 1965), 92.
13. Pahari, *Unish Shotoker*, 84, 92.
14. Nilambor's son and disciple was the famous Kobiraj Gangaprasad Sen. Numerous eminent later Kobirajes who played a central role in refiguring Ayurveda, such as the redoubtable Gananath Sen, traced their intellectual genealogy back to Nilambor and Gangaprasad. Pahari, *Unish Shotoker*, 84. See also chap. 1 of this book.
15. Langford, *Fluent Bodies*, 190–91.
16. Leslie, "The Ambiguities of Medical Revivalism in Modern India," 356–57.
17. Sarva Dev Upadhyay, *Nadi Vijnana: Ancient Pulse Science* (Benares: Chaukhamba Vidyabhawan, 1986).
18. Jacob Rosenbloom, "The History of Pulse Timing with Some Remarks on Sir John Floyer and His Physicians' Pulse Watch" in *Annals of Medical History*, ed. Francis Randolph Packard, vol. 4 (New York: Paul B. Hoeber, 1922). See also Reiser, *Reign of Technology*, 96–98.
19. Roberta Bivins, *Alternate Medicine? A History* (Oxford: Oxford University Press, 2007), 29–30.
20. Harold J. Cook, "Conveying Chinese Medicine to Seventeenth Century Europe," in *Science between Europe and Asia*, ed. Gunergun and Raina, 209–33.
21. Silas Weir Mitchell, *Doctor and Patient* (Philadelphia: JB Lippincott Company, 1888), 34–35.
22. Malcolm Nicholson, "The Art of Diagnosis: Medicine and the Five Senses," in *Companion Encyclopaedia of the History of Medicine*, ed.W. F. Bynum and Roy Porter, vol. 2 (London: Routledge, 1993), 806.
23. These graphical machines includes gadgets such as Julius Hérisson's sphygmometer, Jean Léonard Marie Poiseuille's hemodynamometer, Carl Ludwig's kymographion and Karl Vierordt's sphygmograph to name only a few. Reiser, *Reign of Technology*, 98–106.

24. Sumit Sarkar, *Beyond Nationalist Frames: Postmodernism, Hindu Fundamentalism, History* (Delhi: Permanent Black, 2002); see also Sumit Sarkar, *Writing Social History* (Oxford: Oxford University Press, 1997).
25. Anon., "The Great Anarchy," *Calcutta Review*, vol. 107 (Calcutta: City Press, 1899), 216.
26. "David Hare," in *Calcutta, Old and New: A Historical and Descriptive Handbook to the City*, ed. H. E. A. Cotton (Calcutta: W. Newman, 1907), 422–23.
27. British Library, Indian Office Records and Private Papers, IOR/Z/E/4/37/H149.
28. A. Upjohn, "Calcutta in the Olden Time—Its Localities," *Calcutta Review* 18 (1852): 292.
29. Obituary notice for Mrs. Henrietta Eliza Peters, *Calcutta Monthly Journal and General Register* (Calcutta: Samuel Smith, 1840), 111.
30. Anon., "Residents in Calcutta," *Bengal and Agra Annual Guide and Gazetteer*, vol. 1 (Calcutta: William Rushton, 1842), 189.
31. "Hamilton & Co., Jewellers and Silversmiths," "Calcutta Review Advertiser," *Calcutta Review* 74 (1861): 2–3.
32. For the new temporal regimes, see Sumit Sarkar, "Colonial Times: Clocks and Kali-Yuga," in *Beyond Nationalist Frames*, 10–37.
33. Anon., *Report of the Salaries Commission* (Calcutta: Bengal Secretariat Press, 1886), 203.
34. H. F. J. T. Maguire, *Report on the Census Taken in the City of Calcutta* (Calcutta: Bengal Secretariat Press, 1891), cv-cvi.
35. Quaisar, *Indian Response to European Technology*, 68.
36. Markus Vink, *Dutch Sources on South Asia c. 1600–1825*, vol. 4 (Delhi: Manohar Publishers, 2012), 42.
37. H. F. J. T. Maguire, *Report on the Census of Calcutta Taken on the 26th February 1891* (Calcutta: Bengal Secretariat Press, 1891), cvi–cvii.
38. Nirmalshib Bandopadhyay, "Ghorioala," *Birbhumi Nabaparjaya* 3, no. 2 (1913 [1320 BE]): 120–26.
39. Bandopadhyay, *Arogya Niketan*, 36.
40. Hindu Woman with Watch, Chitrashala Press, Pune, 1882. Print no. 18, Wellcome Trust Collection, record no. 583013i.
41. Gopalchandra Sengupta, *Ayurveda Sarsamgraha*, vol. 1 (Calcutta: Columbian Press, 1871 [1278 BE]), 12–15.
42. Sengupta, *Ayurveda Sarsamgraha*, 1: 17.
43. Sengupta, *Ayurveda Sarsamgraha*, 1: 15–16.
44. White's critique of the macrocosm/microcosm formulation is twofold. First, White points out that the macro/micro entails a miniaturization (i.e., the macrocosm exists in the microcosm as a miniature version). Yet, in the Yogic or Tantric texts he addresses, he does not see any mention of miniaturization. The individual body is not a scale model of the universe. It *is* the universe. White's second critique is that the non-miniaturization points to a very different cosmophysiological model where the *yogin* through the correct steps can actually absorb the entire universe into his body. The model here is Shiva, from whom the universe emerged and into whom it is will withdraw. This is what White calls "macranthropy"—a term originally coined by White's teacher, Mircea Eliade, to designate the psychophysiological experience of temporary oneness with the world. But White's usage is slightly different. He avoids Eliade's phenomenolization of macranthropy, pointing instead to its ontic reality within the Yoga-Tantric tradition. David Gordon White, "On the Magnitude of the Yogic Body,"

in *Yogi Heroes and Poets: Histories and Legends of the Naths*, ed. David N. Lorenzen and Adrian Munoz (Albany: SUNY Press, 2011).

45. Faizullah, *Goroksho Bijoy*, ed. Abdul Karim (Calcutta: Bangiya Sahitya Parisat, 1917 [1324 BE]).
46. Srisa Chandra Vasu, trans., *Siva Samhita* (Allahabad: Indian Press, 1914).
47. Upendranath Bhattacharya, *Banglar Baul o Baul Gan* (Calcutta: Orient Book Company, 2001 [1957]).
48. Sudhir Chakraborty, *Bangla Dehotottwer Gan* (Calcutta: Pustak Bipani, 1990).
49. G. T. Bettany and rev. Richard Hankins, "Guy, William Augustus (bap. 1810, d. 1885)," *Oxford Dictionary of National Biography* (Oxford University Press, 2004 [online May 2009]), accessed May 28, 2015, http://dx.doi.org/10.1093/ref:odnb/11801.
50. See, for instance, W. A. Guy, "On the Effects Produced on the Pulse by Change of Posture," *Eclectic Journal of Medicine* 3, no. 11 (1839): 403–6.
51. William A Guy and John Harley, *Hooper's Vade Mecum* (London: Henry Renshaw, 1869).
52. Karl Magee, "Graves, Robert James (1796–1853)," *Oxford Dictionary of National Biography* (Oxford University Press, 2004), http://dx.doi.org/10.1093/ref:odnb/11317, accessed on May 28, 2015.
53. Sengupta, *Ayurveda Sarsamgraha*, 1:50.
54. Ibid., 1:49–57.
55. Ibid., 1:21.
56. Martha Ann Selby, "Narratives of Conception, Gestation and Labour in Sanskrit Ayurvedic Texts," *Asian Medicine: Tradition and Modernity* 1, no. 2 (2005): 254–75.
57. Sengupta, *Ayurveda Sarsamgraha*, 1:53.
58. Jashodanandan Sarkar, trans., *Susruta Samhita (Mul o Bonganubad)* (Calcutta: Bangabasi Steam Machine Press, 1894).
59. Nagendranath Sengupta, *Sochitro Kobiraji Siksha*, vol. 1 (Calcutta: Nagendra Steam Printing Works, 1930), 21.
60. While the immense success of Nagendranath's text clearly points to the dominance of his model, one should be careful about expecting a perfect and complete linear transition from Gopalchandra's model to Nagendranath's. Even in 1909, for instance, one Dwarakanath Bidyaratna published a book called *Chikitsaratna*, wherein a robustly transmaterial Kobiraji *nadipariksha* is placed side-by-side with what he calls "*Nadipariksha* by the watch," without any attempt to reconcile the two. Dwarakanath Bidyaranta, *Chikitsaratna* (Calcutta: Sanskrita Jantra, 1909 [1316 BE]), 31–32.
61. S. V. Gupta, *Units of Measurement: Past, Present and Future. International System of Units* (Dordrecht: Springer, 2010), 3.
62. Edward Balfour, a Scottish surgeon in British India and a contemporary of Haralal, represented a different relationship between the minute and the *pols.* He stated that the *pol* was "a measure of time, a moment, or a minute, of which there are 60 in a *ghori.*" Edward Balfour, *Cyclopaedia of India and of Eastern and Southern Asia*, 3:82. This, however, is almost certainly a mistake. It was the kind of "rough translation"—which made partial similarities look absolute—that Dipesh Chakrabarty has described as a crucial technology of colonial domination. Chakrabarty, *Provincializing Europe*, 17.
63. Abul Fazl-i-Allami, *Ain-i-Akbari*, vol. 3, trans. Colonel H. S. Jarrett (Calcutta: Baptist Mission Press, 1894), 15–16.
64. Ibid., 16.
65. Ibid.

66. Fleet, "The Ancient Indian Water-Clock," *Journal of the Royal Asiatic Society of Great Britain and Ireland* 47, no. 2 (1915): 213–30.
67. Ibid., 214.
68. Quaisar, *Indian Response to European Technology*, 64–65.
69. Avner Wishnitzer, "The Transformation of Ottoman Temporal Culture During the Long Nineteenth Century" (PhD diss., Tel Aviv University, 2009).
70. Kumudnath Mullick, *Nadia Kahini* (Ranaghat: n.p., 1910), 155.
71. Sen, *Ayurveda Bigyan*, 1:190.
72. Debendranath Sengupta and Upendranath Sengupta, *Ayurved Samgraha*, vol. 1 (Calcutta: Dipayan, n.d. [reprint 1892]), 361.
73. Jarrett, *Ain-i-Akbari*, 3:17.
74. Rabindranath Tagore, *Farewell Song*, trans. Radha Chakravarty (New Delhi: Penguin, 2011), 36.
75. While the issue of cyclical time and linear time has been much debated in general South Asian history, medical temporalities have received little historical attention. In neighboring China, Marta Hanson points out that there were two distinct schools of medical temporality. Marta E. Hanson, *Speaking of Epidemics in Chinese Medicine: Disease and the Geographic Imagination in Late Imperial China* (Abingdon: Routledge, 2011), 12. For South Asian debates on cyclical time, see Romila Thapar, *Time as a Metaphor of History: Early India* (Delhi: Oxford University Press, 1996).
76. Pitambar Sen, *Nadiprokash* (Calcutta: Chaitanyachandrodaya Jantra, 1865), 21–23.
77. Amritalal Gupta, *Nadigyan Rohosyo* (Calcutta: Bandemataram Oushodhaloy, 1916 [1323 BE]), 53–55.
78. Anandachandra Barman, *Sarkaumudi* (Calcutta: Sudhasindhu Yantra, 1867 [1274 BE]), 22.
79. The partial mutability of gender difference is a characteristic also seen in earlier epochs of Chinese medicine. See Charlotte Furth, "Androgynous Males and Deficient Females: Biology and Gender Boundaries in Sixteenth- and Seventeenth-Century China," *Late Imperial China* 9, no. 2 (1988): 1–31.
80. Pitambar Sen, *Nadiprokash*, 3–4.
81. On the rise of biochemistry, see Robert E. Kohler, *From Medical Chemistry to Biochemistry: The Making of a Biomedical Discipline* (New York: Cambridge University Press, 1982). On the routinization of laboratory testing and how it transformed medicine in general and hospitals in particular, see Reiser, *Reign of Technology* and Howell, *Technology in the Hospital*.
82. Christopher Crenner, "Race and Laboratory Norms: The Critical Insights of Julian Herman Lewis (1891–1989)," *Isis* 105, no. 3 (2014): 477–507.
83. William H. Howell, *Textbook of Physiology for Medical Students and Physicians* (Philadelphia, 1907), 576–77.
84. Sengupta, *Ayurveda Sarsamgraha*, 1:50–52.
85. Haralal Gupta, *Nadi-siksha* (Calcutta: Rajendranath Sengupta, 1911 [1318 BE]), 17–20.
86. Sengupta, *Kobiraji Siksha*, 1:21.
87. Basantakumar Pramanik, "Nadigyan Lobhopay," *Chikitsa Sammilani* 12, nos. 2–3 (1907): 149–51.
88. Howell, *Textbook of Physiology*, 576–77.
89. Salimuddin Ahmad Bidyabinod, *Padye Nadi-gyan* (Calcutta: Altafi Press, 1913 [1320BE]), 24.
90. Sengupta, *Ayurveda Sarasamgraha*, 1:51.
91. Ibid., 361.

92. Nagendranath Sengupta, *Kobiraji Siksha*, 1:21.
93. Pramanik, "Nadigyan Lobhopaye," 150.
94. Wujastyk, *Roots of Ayurveda*, 43–44.
95. Satishchandra Sharma, trans., *Charaka Samhita* (Calcutta: Bhaisajya Steam Machine Jantra, 1904 [1311 BE]), 63–64.
96. Upadhyay, *Nadi Vijnana*.
97. Gopalchandra Sengupta, *Ayurveda Sarsamgraha*, vol. 1 (Calcutta: Columbian Press, 1871 [1278 BE]), 53.
98. Wujastyk, *Roots of Ayurveda*, 250.
99. Severine Pilloud and Micheline Louis-Courvoisier, "The Intimate Experience of the Body in the Eighteenth Century: Between Interiority and Exteriority," *Medical History* 47, no. 4 (2003): 462.
100. Philip van der Eijk, "Medicine and Health in the Graeco-Roman World," in *The Oxford Handbook of the History of Medicine*, ed. Mark Jackson (Oxford: Oxford University Press, 2011), 30.
101. Pilloud and Louis-Courvoisier, "The Intimate Experience."
102. Roy Porter, *Blood and Guts: A Short History of Medicine* (New York: Norton, 2004), 27–29.
103. John C. Murray, "On the Nervous, Bilious, Lymphatic and Sanguine Temperaments: Their Connection with Races in England, and their Relative Longevity," *Anthropological Review* 8 (1870): 14–15.
104. Gyan Prakash, *Another Reason: Science and the Imagination of Modern India* (Princeton, NJ: Princeton University Press, 1999), 145–46.
105. Shruti Kapila, "Race Matters: Orientalism and Religion, India and Beyond, c. 1770–1880," *Modern Asian Studies* 41, no. 3 (2007): 471–513.
106. K. P. Mukherjee, preface in Russick Lall Gupta, *Science of Sphygmica or Sage Kanad on Pulse* (Calcutta: S. C. Addy, 1891), v.
107. T. A. Wise, *Commentary on the Hindu System of Medicine: New Issue* (London: Trubner, 1860), 76–79; Russick Lal Gupta, *Hindu Anatomy, Physiology, Therapeutics, History of Medicine and Practice of Physic* (Calcutta: S. C. Addy, 1892), 46–49.
108. Sarkar, *Susruta Samhita*, 212–19.
109. T. A. Wise, *Hindu System of Medicine*, 76–77.
110. Murray, "Temperaments," 16–17.
111. Gupta, *Hindu Anatomy*, 46–47.
112. Dagmar Wujastyk, *Well-Mannered Medicine: Medical Ethics and Etiquette in Classical Ayurveda* (New York: Oxford University Press, 2012), 129.
113. Sarkar, *Susruta Samhita*, i–ii.
114. Francis Zimmermann, *The Jungle and the Aroma of the Meats* (Delhi: Motilal Banarsidass, 2011).
115. Murray, "Temperaments," 17.
116. Mark Harrison, "'The Tender Frame of Man': Disease, Climate and Racial Difference in India and the West Indies, 1760–1860," *Bulletin of the History of Medicine* 70, no. 1 (1996): 90.
117. Mitchell, *Doctor and Patient*, 35.
118. Jimena Canales, *A Tenth of a Second: A History* (Chicago: University of Chicago Press, 2009), 31.
119. Charles J Cullingworth, *The Nurse's Companion: A Manual of General and Monthly Nursing* (London: J & A Churchill, 1876), 51.
120. Attewell, *Refiguring Unani Tibb*.

121. Sengupta and Sengupta, *Ayurveda Samgraha*, 1:360.
122. Sandelowski, *Devices and Desires*; Christopher Hamlin, *More than Hot: A Short History of Fever* (Baltimore: Johns Hopkins University Press, 2014).
123. On Bengali opposition to hospitalization even in the case of serious infectious diseases see Mukharji, *Nationalizing the Body*, 165.
124. Mridula Ramanna, *Health Care in Bombay Presidency, 1896–1930* (Delhi: Primus, 2012), 76–109.
125. Anon., "Medical Women in India and Africa," *Physician & Surgeon* 7 (1885): 213.
126. Anon., *"Phuler Saji," Bangadarshan* 9, no. 10 (1882 [1289 BE]): 518–19.
127. Gupta, "Indigenous Medicine," 372–73.
128. Parasuram, *"Chikitsa Sonkot,"* in *Goddalika* (Calcutta: M. C. Sarkar & Sons Ltd., 1951 [1358 BE]), 45–46.
129. Daniel, "The Pulse as an Icon in Siddha Medicine," 120.
130. Elisabeth Hsu, *Pulse Diagnosis in Early Chinese Medicine: The Telling Touch* (Cambridge: Cambridge University Press, 2010).
131. See, for instance, Eeva Sointu, *Theorizing Complementary and Alternative Medicines: Wellbeing, Self, Gender, Class* (Hampshire: Palgrave Macmillan, 2012).
132. See, for instance, David Frawley, *Ayurveda, Nature's Medicine* (Delhi: Motilal Banarsidass, 2004), v.
133. T. N. Ganguly, *Svarnalata or Scenes from Hindu Village Life in Bengal*, trans. D. C. Roy (Calcutta: Sanyal, 1906), 295.
134. Sheryle J. Whitcher and Jeffrey D. Fisher, "Multidimensional Reaction to Therapeutic Touch in a Hospital Setting," *Journal of Personality and Social Psychology* 37, no. 1 (1979): 87–96; J. D. Fisher, M. Rytting, and R. Heslin, "Hands Touching Hands: Affective and Evaluative Effects of an Interpersonal Touch," *Sociometry* 39, no. 4 (1976): 416–21.
135. See Wujastyk, "Image of the Human Body," 205–11; Wujastyk, *Roots of Ayurveda*, 31.
136. Abinashchandra Kobirotno, "Jadihasti tadanyatra jannastih na tatkkachit," *Chikitsa Sammilani* 9, no. 1 (1893): 30.
137. Nagendra Nath Sen Gupta, "Jib Yantre Ayurveder Bisheshotyo," *Ayurveda Bikash* 3, no. 4 (1915–16): 67.
138. Sen Gupta, "Jib Yantre," 3, no. 5 (1915–16): 98.
139. Gananath Sen, "Sharir Bidya," *Ayurveda* 4, no. 5 (1919 [1326 BE]): 193.
140. Elshakry, *Reading Darwin*, 141. See also, Stolz, "'By Virtues of Your Knowledge.'"
141. Haramohan Majumdar, *Ayurveda Gurhotottwo Prokash* (Calcutta: Bangabasi Ltd., n.d.), 3–5.
142. Descartes, *Oeuvres*, 9:322, cited in Laurens Laudan's "The Clock Metaphor and Probabilism: The Impact of Descartes on English Methodological Thought, 1650–65," *Annals of Science* 22, no. 2 (1966): 78.
143. Laudan, "The Clock Metaphor," 81.
144. On Shiv Sharma's version of Ayurveda and his clash with Sen, see Charles Leslie, "Interpretations of Illness."
145. Shiv Sharma, *The System of Ayurveda* (Bombay: Khemraj Shrikrishnadas Shri Venkateshwar Steam Press, 1929), 184.
146. Rivka Feldhay, "Religion," in *The Cambridge History of Science: Volume 3; Early Modern Science*, ed. Katharine Park and Lorraine Daston (Cambridge: Cambridge University Press, 2006), 749.
147. John Tresch, *The Romantic Machine: Utopian Science and Technology after Napoleon* (Chicago: University of Chicago Press, 2012), 12.

CHAPTER THREE

1. Bandyopadhyay, *Arogya Niketan*, 341.
2. James Long, *A Descriptive Catalogue of Bengali Works* (Calcutta: Sanders, Cones, and Co., 1855), 35.
3. J. M. S. Pearce, "A Brief History of the Clinical Thermometer," *Quarterly Journal of Medicine* 95 (2002): 251–52.
4. Reiser, *Reign of Technology*, 114–15.
5. Reiser, *Reign of Technology*, 111–13.
6. Hasok Chang, *Inventing Temperature: Measurement and Scientific Progress* (New York: Oxford University Press, 2004), 40–43.
7. Heinz Otto Sibum, "Reworking the Mechanical Value of Heat: Instruments of Precision and Gestures of Accuracy in Early Victorian England," *Studies in the History & Philosophy of Science* 26, no. 1 (1995): 73–106.
8. Reiser, *Reign of Technology*, 111–13.
9. S. T. Anning, "Clifford Albutt and the Clinical Thermometer," *Practitioner* 197, no. 182 (1966): 820.
10. Quoted in ibid., 7.
11. Ibid., 8.
12. Anon., "Review of Clinical Lectures on Dengue by T. Edmonston Charles," *Calcutta Review* 55 (Calcutta: City Press, 1872): vi–viii.
13. Edward Lawrie, "On the Temperature in Health," *Indian Medical Gazette* 7, no. 10 (1873): 253–56.
14. Volker Hess, "Standardizing Body Temperature: Quantification in Hospitals and Daily Life, 1850–1900," in *Body Counts: Medical Quantification in Historical and Sociological Perspective*, ed. Gerard Jorland, A. Opinel, and G. Weisz (Montreal: McGill-Queen's University Press, 2005), 109–26.
15. Hamlin, *More than Hot*, 1–16.
16. Christiane Sinding, "Les multiples usages de la quantification en medicine: Le cas du diabète sucré," in *Body Count*, ed. Jorland et al., 127–44.
17. Hamlin, *More than Hot*; Reiser, *Reign of Technology*, 117.
18. Sandelowski, *Devices and Desires*, 73.
19. Arnold, Everyday Technologies, 6.
20. Rosenberg, *Explaining Epidemics*, 144.
21. *Indian Medical Gazette*, 1 January 1868, 7.
22. Sarkar, *Susruta Samhita*, unnumbered introductory page.
23. Hamlin, *More than Hot*, 255.
24. Robert A Aronowitz, "When Do Symptoms Become a Disease?," *Annals of Internal Medicine* 134, no. 9/2 (2001): 803–8.
25. Sengupta, *Kobiraji Siksha*, 1:23–25.
26. Bidyaratna, *Chikitsaratna* , 32–36.
27. "fever, n.1," *OED Online* (Oxford: Oxford University Press, March 2015, accessed May 30, 2015), http://www.oed.com/view/Entry/69682?rskey=b6KQiB&result=1.
28. Hamlin, *More than Hot*, 253.
29. I am indebted to Christopher Hamlin for pointing this out.
30. Reiser, *Reign of Technology*, 114.
31. For a discussion of the biography of Madhavakara and the history of the *Nidan* see G. J. Meulenbeld, introduction to *The Madhavanidana and Its Chief Commentaries: Chapters 1—10* (Leiden: Brill, 1974).
32. Ibid., 87.

33. Ibid., 87–88.
34. Ibid., 89.
35. Udaychand Dutta, *Nidan* (Calcutta: Ayurveda Yantra, 1880), 16–34.
36. Meulenbeld, *Madhavanidana*, 89.
37. Nicholson, "The Art of Diagnosis," 806.
38. Meulenbeld, *Madhavanidana*, 92.
39. Hamlin, *More than Hot*, 253.
40. Frederique Apffel-Marglin, *Smallpox in Two Systems of Knowledge* (Helsinki: WIDER Publications, 1987), 17.
41. Cohen, No Aging in India, 155.
42. Projit Bihari Mukharji, "In-disciplining Jwarasur: The Folk/Classical Divide and the Transmateriality of Fevers in Colonial Bengal," *Indian Economic & Social History Review* 50, no. 3 (2013): 261–88.
43. For Mullick's founding of the free college, see Gupta, "Indigenous Medicine," 376.
44. Ramchandra Mullick, *Bishwo Chikitsa* (Calcutta: Ramayana Press, 1889 [1296 BE]), 27.
45. Mullick, *Bishwo Chikitsa*, 28.
46. Mullick, *Bishwo Chikitsa*, 29.
47. On the restructuring of medical and nursing care through the thermometer, see Hamlin, *More than Hot*, 250–80.
48. See, for instance, Panchanan Mandal, *Chithi Potre Somaj Chitro*, vol. 2 (Santiniketan: Vishwabharati, 1952 [1359 BE]), 66–80.
49. Mandal, *Chithi Potre*, 2:71.
50. Ibid., 2:72.
51. Rabindranath Tagore, *Chithi Potro*, vol. 1 (Santiniketan: Vishwabharati Granthabibhag, 1993 [1942]), 35.
52. Tagore, *Chithi Potro*, 1:39.
53. Ibid., 1:100.
54. Kishorimohan Chattopadhyay, *Swapnatattwa* (Calcutta: Metcalf Press, n.d.), 233.
55. Rosenberg, "Therapeutic Revolution."
56. Mullick, *Bishwo Chikitsa*, 26.
57. Markus Daechsel, "The Civilizational Obsessions of Ghulam Jilani Barq," in *Colonialism as Civilizing Mission: Cultural Ideology in British India*, ed. Harald Fischer-Tine and Michael Mann (London: Anthem, 2004), 283.
58. Douglas Haynes, "Masculinity, Advertising and the Reproduction of the Middle Class Family in Western India, 1918–1940," in *Being Middle Class in India: A Way of Life*, ed. Henrike Donner (Abingdon: Routledge, 2011), 23–46.
59. Bruce Clarke, *Energy Forms: Allegory and Science in the Era of Classical Thermodynamics* (Ann Arbor: University of Michigan Press, 2001), 2.
60. Sen, *Ayurveda Bigyan*, 2:9.
61. Binodbihari Ray, "Jwor," *Chikitsak* 1, no. 1 (1889 [1296 BE]): 14.
62. Ibid., 16–17.
63. Ray, "Vayu-Pitta-Kapha," *Chikitsak* 1, no. 3 (1889 [1296 BE]): 52–53.
64. Ibid., 78.
65. Ibid., 8.
66. Binodbihari Ray, *Podye Ayurveda Siksha* (Calcutta: Lila Printing Works Yantra, 1908 [1315 BE]), 8.
67. Surendranath Goswami, "Vayur Boigyanik Byakhya," *Janmabhumi* 24, no. 1 (1916 [1323 BE]): 22.

68. Surendranath Goswami, "Vayur Boigyanik Byakhya," *Janmabhumi* 24, no. 2 (1916 [1323 BE]): 50.
69. Surendranath Goswami, "Vayur Boigyanik Byakhya," *Janmabhumi* 24, no. 4 (1916 [1323 BE]): 132.
70. Ibid., 51.
71. Anon., "Ayurbigyane Vayu Pitta Kapha," *Ayurveda Hitoishini* 2, no. 4 (1912 [1319]): 121–22.
72. Anon., "Ayurbigyane Vayu Pitta Kapha," 122–25.
73. Amritalal Gupta, "Tej o Pitta," *Ayurveda Bikash* 3, nos. 2–3 (1915–16 [1322 BE]): 34.
74. Gupta, *Nadigyan Rohosyo*, 11.
75. Sen, *Ayurveda Bigyan*, 2:10.
76. Ray, *Podye Ayurveda Siksha*, 5.
77. Rosenberg, *Explaining Epidemics*, 23–24.
78. Rosenberg, *Explaining Epidemics*, 27–29.
79. John Harley Warner, "'The Nature-Trusting Heresy': American Physicians and the Concept of the Healing Power of Nature in the 1850s and 1860s," *Perspectives in American History* 11 (1977–78): 291–324.
80. Neshat Quaiser, "Politics, Culture, and Colonialism: Unani's Debate with Doctory," in *Health, Medicine, and Empire: Perspectives on Colonial India*, ed. Biswamoy Pati and Mark Harrison (Hyderabad: Orient Longman, 2001), 317–55.
81. Mukharji, *Nationalizing the Body*, 158–63.
82. Gupta, *Nadigyan Rohosyo*, 13.
83. Gupta, *Nadigyan Rohosyo*, 15.
84. Gupta uses the words "*pran*, *jib*, *jibatma*, and *atma* in closely overlapping ways to designate a sentient being that is "life."
85. Gupta, *Nadigyan Rohosyo*, 13.
86. Bruce Clarke, *Energy Forms: Allegory and Science in the Era of Classical Thermodynamics* (Ann Arbor: University of Michigan Press, 2001), 72.
87. Wujastyk, *Roots of Ayurveda*, 31.
88. Sen, *Ayurveda Bigyan*, 2:9.
89. Ibid.
90. Gupta, *Nadigyan Rohosyo*, 12.
91. Ibid., 25.
92. Gupta, "Tej o Pitta," *Ayurveda Bikash* 3, nos. 2–3 (1915–16 [1322 BE]): 35.
93. Anson Rabinbach, *The Human Motor: Energy, Fatigue, and the Origins of Modernity* (Berkeley: University of California Press, 1992), 2.
94. Cited in ibid., 124.
95. Cited in Frank A. J. L. James, "Thermodynamics and the Sources of Solar Heat, 1846–1862," *British Journal of the History of Science* 15, no. 2 (1982): 157.
96. Cited in ibid., 157.
97. Ibid., 155–81.
98. Anon., "Ayurbigyane Vayu Pitta Kapha," *Ayurveda Hitoishini* 2, no. 4 (1912 [1319 BE]): 121.
99. Clarke, *Energy Forms*, 66.
100. Goswami, "Vayur Boigyanik Byakhya," *Janmabhumi* 24, no. 1 (1916 [1323 BE]): 25.
101. T. Henry Green, *Pathology and Morbid Anatomy* (Philadelphia: Lea Brothers, 1900), 18.
102. Binodbihari Ray, *Podye Ayurveda Siksha*, 7.
103. Wujastyk, "Early Modern Indian Medicine," 13.
104. Ibid., 14.

105. Anandachandra Barman, *Sarkaumudi* (Calcutta: Sudhasindhu Jantra, 1867 [1274 BE]), 2.
106. Sarkar, *Susruta Samhita*, 36.
107. Amritalal Gupta, "Agni," *Ayurveda* 1, no. 2 (1916 [1323 BE]): 86–87.
108. Gupta, "Tej o Pitta," *Ayurveda Bikash* 3, no. 1 (1915–16 [1322 BE]): 15.
109. Graham Maurice, "Round the World—And Some Gas Works," *Journal of Gas Lighting, Water Supply Etc.* 103 (1908): 26.
110. Anon., "Humours Vs. Hormones," *Journal of Ayurveda* 4, no. 8 (1928): 281.
111. Sarachchandra Sengupta, "Pitta," *Ayurveda* 8, no. 3 (1923 [1330 BE]): 70.
112. Udoy Chand Dutt, preface to first edition in *Nidana: A Sanskrit System of Pathology* (Calcutta: Ayurveda Yantra, 1880 [2nd edn.]).
113. Information on what books Bengali Kobirajes actually studied was provided by William Ward, Francis Buchanan, and William Adam. See William Ward, *A View of the History, Literature and Mythologies of the Hindoos*, vol. 4 (London: Black, Kingsbury, Parbury & Allen, 1820 [3rd edn.]), 341–42; Francis Buchanan, *An Account of the District of Purnea* (Patna: Bihar & Orissa Research Society, 1928), 185; William Adam, *Report on the State of Education in Bengal* (Calcutta: Bengal Military Orphan Press, 1835), 59–60.
114. Dutt, *Nidan*, 3.
115. Bhagvat Singh Jee, *A Short History of Aryan Medical Science* (London: Macmillan, 1896), 85.
116. Wujastyk, *Roots of Ayurveda*, 8.
117. Anthony Cerulli, *Somatic Lessons: Narrating Patienthood and Illness in Indian Medical Literature* (Albany: SUNY Press, 2014), 24.
118. Cerulli, *Somatic Lessons*, 25.
119. Wujastyk, *Roots of Ayurveda*, 31.
120. Poonam Bala, *Medicine and Medical Policies in India: Social and Historical Perspectives* (Plymouth: Lexington, 2007), 64.
121. Wujastyk, *Roots of Ayurveda*, 31.
122. Dutt, *Nidan*, 16.
123. Sengupta, *Kobiraji Siksha*, 1:36.
124. Sanyal, "Dhatu—Allopathy-mote," *Chikitsa Sammilani* 5, nos. 10–12 (1888 [1295 BE]): 198.
125. Prasannakumar Maitra, "Dhatu-byakhya," *Chikitsa Sammilani* 6 (1889 [1296 BE]): 161.
126. It is worth noting here that Sanyal confessed to a certain unease in translating *snayus* as "nerves." He added a footnote saying that he was following the contemporary Bengali convention on the matter, but Susruta describes "nerves" and "ligaments" both as *snayus*. Pulinbihari Sanyal, "Dhatu: Allopathy-mote," *Chikitsa Sammilani* 5 (1888 [1295 BE]): 205.
127. Binodbihari Ray, "Vayu Pitta Kapha," *Chikitsak* 1, no. 4 (1889–90 [1296–97 BE]): 78.
128. Binodbihari Ray, "Vayu, Pitta, Kapha," *Chikitsak* 1, no. 2 (1889–90 [1296–97 BE]): 33.
129. Laura Otis, *Networking: Communicating between Bodies and Machines in the Nineteenth Century* (Ann Arbor: University of Michigan Press, 2001), 16.
130. Iwan Rhys Morus, *Shocking Bodies: Life, Death and Electricity in Victorian England* (Stroud: History Press, 2011).
131. Thomas Stretch Dowse, *The Brain and the Nerves* (London: Bailliere, Tyndall & Cox, 1884), 9.

132. Langwick, *Bodies, Politics and African Healing*, 152–53.
133. Morus, *Shocking Bodies*.
134. Edward Brown, "Neurology and Spiritualism in the 1870s," *Bulletin of the History of Medicine* 57 (1983): 563–77; Aura Satz, "'The Conviction of Its Existence': Silas Wier Mitchell, Phantom Limbs and Phantom Bodies in Neurology and Spiritualism," in *Neurology and Modernity: A Cultural History of Nervous Systems 1800–1950*, ed. Laura Salisbury and Andrew Shail (Basingstoke: Palgrave Macmillan, 2010), 113–29.
135. Morus, *Shocking Bodies*, 159–82.
136. Ibid., 39–48, 70–80, 112–23.
137. Ibid., 113.
138. See, for instance, Ernst Haeckel, *Gesammelte populare Vortrage aus dem Gebiete der Entwicklungslehre* (Bonn: Emil Strauss, 1878), 149. Reproduced in Katja Guenther, *A Body Made of Nerves: Reflexes, Body Maps and the Limits of the Self in Modern German Medicine* (PhD diss., Harvard University Press, 2009), 55.
139. The only inscriptions were two words—i.e., *nadi* and *dhomoni*—unattached to any arrows. It is not clear whether the words are meant to designate particular organs or spots on the body. Both words are known to be synonyms of the word *snayu* under certain conditions as well.
140. Sanyal, "Dhatu," 206.
141. Ray, "Vayu, Pitta, Kapha," *Chikitsak* 1, no. 1 (1889–1890 [1296–97 BE]): 9.
142. Ray, *Podye Ayurveda Siksha*, 7–8.
143. On the telegraph system in colonial India see Deep Kanta Lahiri Choudhury, *Telegraphic Imperialism: Crisis and Panic in the Indian Empire, c. 1830–1920* (Basingstoke: Macmillan, 2010).
144. Otis, *Networking*, 12.
145. Guenther, *Body Made of Nerves*, 54.
146. Sengupta, *Kobiraji Siksha*, 2:1618–19.
147. Ibid., 2:1641–46.

CHAPTER FOUR

1. Anon., "Chikitsa," *Anubikshan* 1, no. 1 (1875 [1282 BE]): 3–13.
2. On Victorian notions regimes of vision, see Srdjan Smajic, *Ghost-Seers, Detectives and Spiritualists: Theories of Vision in Victorian Literature and Science* (Cambridge: Cambridge University Press, 2010), 20–33.
3. Gananath Sen, "Sharir-bidya," *Ayurved* 4, no. 5 (1919 [1326 BE]): 194.
4. Majumdar, *Ayurveda Gurhotottwo*, 10.
5. D. N. Ray, "Modern Problems in Ayurveda," *Journal of Ayurveda* 7, no. 4 (1930): 125.
6. T. G. Ramamurthi Iyer, "A Scientific Study of Septic Cases Treated in Indigenous System," *Journal of Ayurveda* 7, no. 4 (1930): 130.
7. Brajaballabh Ray, "Jwor," *Ayurveda* 1, no. 6 (1916 [1323 BE]): 255.
8. Majumdar, *Ayurveda Gurhototwo*, 28.
9. Bhagabat Kumar Goswami, foreword in *Ayurveda Ratnakarah*, by Jogendranath Darshanshastri (Calcutta: Jyotirindra Bhattacharya, 1936 [1343 BE]): unnumbered page.
10. Chakrabarty, *Provincializing Europe*, 175–78.
11. Smajic, *Ghost-Seers, Detectives and Spiritualists*, 34–44.
12. Chakrabarty, *Provincializing Europe*, 175–78.
13. Nicholson, "Art of Diagnosis," 803.
14. Sarkar, *Susruta Samhita*, 21.

15. Ahmed Sharif, *Banglar Sufi Sahityo* (Dhaka: Samay Prakashan, 1969), 88.
16. Chris Otter, *The Victorian Eye: A Political History of Light and Vision in Britain, 1800–1910* (Chicago: University of Chicago Press, 2008).
17. Capt. Everest, "On the Compensation Measuring Apparatus of the Great Trigonometrical Survey of India," *Asiatic Researches* 18 (1833), 203–5.
18. Allan Webb, *The Historical Relations of Ancient Hindu with Greek Medicine* (Calcutta: Military Orphan Press, 1850), 20.
19. John Harley Warner, "'Exploring the Labyrinths of Creation': Popular Microscopy in Nineteenth Century America," *Journal of the History of Medicine and Allied Sciences* 37, no. 1 (1982): 7–33.
20. Webb, *Historical Relations*, 6.
21. Allan Webb, *Elephentiasis Orientalis, and Especially Elephentiasis Genitalis in Bengal*(Calcutta: Bengal Military Orphan Press, 1855).
22. Sengupta, *Ayurveda Sarsamgraha*, 1:71.
23. Howell, *Technology in the Hospital*, 70–71.
24. Bhagwan Dash and Lalitesh Kashyap, *Basic Principles of Ayurveda* (New Delhi: Concept Publishing, 2003), 42–45.
25. Guy Attewell, "Islamic Medicine: Perspectives on the Greek Legacy in the History of Islamic Medical Traditions in West Asia," in *Medicine across Cultures: History and Practice of Medicine in Non-Western Cultures*, ed. Helen Selin (New York: Kluwer, 2003), 326.
26. Projit Mukharji, "Lokman, Chholeman and Manik Pir: Multiple Frames for Institutionalising Islamic Medicine in Modern Bengal," *Social History of Medicine* 24, no. 3 (2011): 726.
27. Wujastyk, "Early Modern Indian Medical Thought," 97.
28. Sengupta, *Ayurveda Sarsamgraha*, 69–70.
29. Ibid., 71–72.
30. Ibid., 67–72.
31. Ibid., 70–71.
32. Howell, *Technology in the Hospital*, 81.
33. Jonathan Winawer et al., "Russian Blues Reveal Effect of Language on Color Discrimination," *Proceedings of the National Academy of the Sciences* 104, no. 19 (2007): 7780.
34. Ibid.
35. Otter, *The Victorian Eye*, 182–96.
36. Sengupta, *Ayurveda Sarsamgraha*, 68.
37. Ibid., 69.
38. Mol, *The Body Multiple*.
39. Siddheswar Ray, *Mutro Tottwo*, 221.
40. Ibid., 223.
41. Ibid., 224.
42. Ibid., 261.
43. Ibid., 260–86.
44. Sarkar, "Bigyapon" in *Susruta Samhita*, i.
45. Nancy J. Tomes and John Harley Warner, "Introduction to the Special Issue on Rethinking the Reception of the Germ Theory of Disease: Comparative Perspectives," *Journal of the History of Medicine and Allied Sciences* 52 (1997): 10.
46. David Barnes, *The Great Stink of Paris and Nineteenth-Century Struggle against Filth and Germs* (Baltimore: Johns Hopkins University Press, 2006), 107–8.

47. Michael Worboys, "Was There a Bacteriological Revolution in Late Nineteenth-Century Medicine," *Studies in History and Philosophy of Biological and Biomedical Sciences* 38 (2007): 24.
48. Michael Worboys, "The Emergence of Tropical Medicine: A Study in the Establishment of a Scientific Speciality," in *Perspectives on the Emergence of Scientific Disciplines*, ed. G. Lemaine, R. Macleod, M. Mulkay, and P. Weingart (Hague: Mouton, 1976), 75–98.
49. For the decline in medical geography in Europe, see Barnes, *The Great Stink of Paris*, 108–20. For the ecological paradigm in Tropical Medicine, see Megan Vaughan, *Curing Their Ills: Colonial Power and African Illness* (Stanford, CA: Stanford University Press, 1991), 29–54.
50. Ramsahay Kabyatirtha, "Arya Rishira Jibanutottwo Janiten Ki Na?," *Ayurveda* 2, no. 7 (1917 [1324 BE]): 281–85.
51. Michael Worboys, *Spreading Germs: Disease Theories and Medical Practice in Britain, 1865–1900* (Cambridge: Cambridge University Press, 2000), 86.
52. Pratik Chakrabarti, "'Living Versus Dead': The Pasteurian Paradigm and Imperial Vaccine Research," *Bulletin of the History of Medicine* 84, no. 3 (2010): 387–423.
53. Jadabeshwar Tarkaratna, "Prachin Bharote Kitanu Tottwo," *Ayurveda* 3, no. 12 (1919 [1326 BE]): 442–45.
54. Saradacharan Sen, "Hookworm ba Krimi," *Ayurveda* 3, no. 10 (1919 [1326]): 376–78.
55. Meulenbeld, *Madhavanidana and Its Chief Commentary*, 285–95.
56. Gobindadas, *Bhaisajyaratnabali*, trans. Binodlal Sen (Calcutta: Adi Ayurveda Machine Press, 1929 [1876]), 251–58.
57. Sen's position is reminiscent of Julie Livingston's observation of how Tswana healers later in the twentieth century vernacularized TB bacilli as "worms." Livingston, "Productive Misunderstandings," 808.
58. Sean Hsian-lin Lei, *Neither Donkey nor Horse: Medicine and the Struggle over China's Modernity* (Chicago: Chicago University Press, 2014), 167–92.
59. Lei, *Neither Donkey nor Horse*, 21–44.
60. Bridie Andrews, "Tuberculosis and the Assimilation of Germ Theory in China, 1895–1937," *Journal of the History of Medicine and Allied Sciences* 52, no. 1 (1997): 114–57.
61. Attewell, Refiguring Unani Tibb, 92.
62. Girindranath Mukhopadhyaya, *History of Indian Medicine: Containing Notices, Biographical, of the Ayurvedic Physicians and Their Works on Medicine, from the Earliest Ages to the Present Time* (New Delhi: Munshiram Manoharlal Publishers Private Limited, 1994); Girindranath Mukhopadhyaya, *Ancient Hindu Surgery: Surgical Instruments of the Hindus; With a Comparative Study of the Surgical Instruments of the Greek, Roman, Arab, and the Modern European Surgeons* (New Delhi: Cosmo, 1994).
63. Girindra Nath Mukerjee, "Human Parasites in the Atharva Vedas," *Journal of Ayurveda* 4, no. 5 (1927): 173, 180.
64. Joseph Dumit, *Picturing Personhood: Brain Scans and Biomedical Identity* (Princeton, NJ: Princeton University Press, 2003), 6–7.
65. Warner, "Exploring the Labyrinths of Creation," 11.
66. J. Andrew Mendelsohn, "Lives of the Cell," *Journal of the History of Biology* 36 (2003): 4.
67. Hemchandra Sen, "Nadichakra," *Chikitsa Sammilani* 11, nos. 11–12 (1907): 41.
68. Sen, "Nadichakra," 42.
69. Haramohan Majumdar, "Sharir Vayu," *Ayurveda* 1, no. 9 (1917 [1324 BE]): 403.
70. Personal interview with Brahmananda Gupta, Calcutta, January 21, 2006.
71. Bandyopadhyay, *Arogya Niketan*, 353.

72. Guenther, "Body Made of Nerves," 36.
73. Robert Brain, "Protoplasmania: Huxley, Haeckel, and the Vibratory Organism in Late Nineteenth-Century Art and Science," in *The Art of Evolution: Darwin, Darwinisms and Visual Culture*, ed. B. Larson and E. Brauer (Hanover: University Press of New England, 2009), 101–4.
74. Wujastyk, "Image of the Human Body."
75. Cerulli, *Somatic Lessons*, 24.
76. Surendranath Goswami, "Bhut Bigyan," *Janmabhumi* 18, no. 3 (1911): 126–27.
77. Emily Martin, "Towards an Anthropology of Immunology: The Body as Nation State," *Medical Anthropology Quarterly* 4, no. 4 (1990): 410–26 (quote on 417).
78. Tony K. Stewart, "Replicating Vaisnava Worlds: Organizing Devotional Space through the Architectonics of the Mandala," *South Asian History and Culture* 2, no. 2 (2011): 300–336.
79. Ashutosh Roy, *Pulse in Ayurveda* (Calcutta: Journal of Ayurveda, 1929): 6.
80. Roy, *Pulse in Ayurveda*, 6.
81. Ibid., 7.
82. Goswami, *Ayurveda o Malaria-jwor*, 7.
83. Goswami's positive deployment of the Puranas is both innovative and striking. Prakash has pointed out how the scientific claims of neo-Hindu propagandists set up the Vedic era in opposition to the Puranic one. Prakash's neo-Hindu Arya Samajists consistently disparaged the Puranas as a body of fables and myths that had obscured and undermined the original scientific temper of the Vedic Hindus. See Prakash, *Another Reason*, 107–13.
84. Goswami, *Ayurveda o Malaria-jwor*, 16.
85. Ibid., 25.
86. Ibid., 17.
87. Ibid., 12–13.
88. Ray, "Vayu-Pitta-Kapha," *Chikitsak* 1, no. 4 (1890 [1297 BE]): 79.
89. Scheid, *Chinese Medicine in Contemporary China*, 152–53.
90. Sen, "Nadichakra," 41–42.
91. Brain, "Protoplasmania."
92. Anon., "Sharirotpotti," *Chikitsa Sammilani* 12, nos. 2–3 (1907): 152–56.
93. Goswami, "Bhut Bigyan," 124–25.
94. Surendranath Goswami, "Vayur Boigyanik Byakhya," *Janmabhumi* 24, no. 2 (1916 [1323 BE]): 51–53.
95. Gananath Sen, Hindu Medicine: Address Delivered at Ceremony for Foundation of Benares Hindu University (Calcutta: Sushilkumar Sen, 1916), 13.
96. Roy, *Pulse in Ayurveda*, 8.
97. Ibid., 9.
98. Ibid., 26.
99. Anon., "Humour vs. Hormone in Ayurveda," *Journal of Ayurveda* 4, no. 8 (1928): 281–82.

CHAPTER FIVE

1. Amulyakumar Dasgupta ("Sambuddha"), *"Torunayon"* in *Dialektiks* (Calcutta: Ranjan Publishing House, 1946 [1353BE]), 33–64.
2. Chandak Sengoopta, "'Dr Steinach Coming to Make Old Young!': Sex Glands, Vasectomy and the Quest for Rejuvenation in the Roaring Twenties," *Endeavour* 27, no. 3 (2003): 122.

3. Nelly Oudshoorn, *Beyond the Natural Body: An Archeology of Sex Hormones* (London: Routledge, 1994), 16–17.
4. See Michael J. Aminoff, *Brown-Sequard: An Improbable Genius Who Transformed Medicine* (New York: Oxford University Press, 2011).
5. Anon., "Humour vs. Hormone in Ayurveda," 281.
6. The two pamphlets were, respectively, *Original Researches in the Treatment of Tropical Diseases* (Calcutta: Record Press, 1902), and *A Thesis on Tropical Abscess of the Liver* (Calcutta: Record Press, 1902). Both were reviewed in an American medical journal. The reviewer mentioned that the first pamphlet "consists largely of reprints of articles published elsewhere." The reviewer was clearly taken by Sen's argument and also noted that "it is curious in reading these pamphlets to remark how much the ancient Hindoos really anticipated many of our modern ideas of physiology and therapeutics." F. R. P., "Reviews," *American Journal of the Medical Sciences* 125, no. 2 (1903): 342.
7. Mukharji, *Nationalizing the Body*, 158–63.
8. Hemchandra Sen, "Swabhabik Rogbadhok Shokti," *Chikitsa Sammilani* 12, nos. 2 and 3 (1907): 137–38.
9. Sen, "Swabhabik Rogbadhok Shokti," 138.
10. The time lag is in keeping with European developments. A number of authors have pointed out that despite its origins in the 1880s, the endocrinal model only became dominant by the 1920s and 30s. See Chandak Sengoopta, *The Most Secret Quintessence of Life: Sex, Glands and Hormones, 1850–1950* (Chicago: University of Chicago Press, 2006); Oudshoorn, *Beyond the Natural Body*.
11. Roy, *Pulse in Ayurveda*, 3.
12. Ibid., 6.
13. Ibid., 7.
14. Shiv Sharma, *The System of Ayurveda* (Bombay: Khemraj Shrikrishnadas Shri Venkateshwar Steam Press, 1929), 177.
15. Sharma, *System of Ayurveda*, 177–78.
16. Ibid., 188.
17. Ibid., 190.
18. Langford, *Fluent Bodies*, 150–51.
19. In recent years, a number of scholarly works have appeared on the history of medicinal advertisements in colonial India. Most of these works have used the advertisements to explore the construction of ideals, identities, and selfhood. Charu Gupta, for instance, uses them to understand the constructions of masculinity. Markus Dachsel sees them as sites of middle-class self-expression. Douglas Haynes interrogates them for their constructions of ideals of conjugality. Rachel Berger uses them as windows into the constructions of indigeneity. Along a somewhat different and interesting track, Madhuri Sharma explores medical advertising as a crucial component of the medical market. Relatively little account, however, has been taken of the medical information contained in these advertisements. Most of the extant studies, for instance, disregard differences in the type of medicines being sold. Thus, hormonal products, tonics, patent medicines, vitamins, etc., are often treated together without any attempt to distinguish the pharmaceutical registers these products lie in. Charu Gupta, *Sexuality, Obscenity, Community: Women, Muslims and the Hindu Public in Colonial India* (New York: Palgrave, 2002); Daechsel, *Politics of Self-Expression*; Douglas Haynes, "Selling Masculinity: Advertisements for Sex Tonics and the Making of Modern Conjugality

in Western India, 1900–1945," *South Asia: Journal of South Asian Studies* 35, no. 4 (2012): 787–831; Berger, *Ayurveda Made Modern*; Madhuri Sharma, *Indigenous and Western Medicine in Colonial India* (New Delhi: Foundation Books, 2012).

20. Advertisement, "Colwell's Hormones," *Journal of Ayurveda* 4, no. 2 (1927): ii.
21. Advertisement, "Elements of Endocrinology," *Journal of Ayurveda* 4, no. 1 (1927): iv.
22. Advertisement, "Hormotone," *Journal of Ayurveda* 4, no. 3 (1927): vii.
23. Advertisement, "Viriligen," *Journal of Ayurveda* 4, no. 4 (1927): vii.
24. Advertisement, "Testacoids," *Journal of Ayurveda* 5, no. 1 (1928): vi.
25. Advertisement, "Protonuclein," *Journal of Ayurveda* 5, no. 3 (1928): vi.
26. Haynes, "Selling Masculinity," 788.
27. The historiography of hormonal pharmaceuticals in the period has mainly looked at European firms. The advertisements of the *Journal of Ayurveda* suggest that relatively lesser-known American firms too were important global players. For the European hormonal pharmaceuticals, see Oudshoorn, *Beyond the Natural Body*, 81–110; Jean-Paul Gaudilliere, "The Visible Industrialist: Standards and the Manufacture of Sex Hormones," in *Evaluating and Standardizing Therapeutic Agents, 1890–1950*, ed. Jonathan Simon and Christoph Gradmann (Basingstoke: Palgrave Macmillan, 2010), 174–201.
28. Haynes, "Selling Masculinity," 806–10.
29. Daechsel, however, feels that the language of hormones and glands is unusual for the (Urdu?) medical advertisements of the time. Daechsel, *Politics of Self-Expression*, 179.
30. Cohen, *No Aging in India*, 127.
31. Cohen, *No Aging in India*, 127–33.
32. For details of what Kobirajes actually read, see footnote 113 in chap. 3.
33. Judith Farquhar, *Appetites: Food and Sex in Post-Socialist China* (Chapel Hill, NC: Duke University Press, 2002).
34. Cohen, *No Aging in India*, 125.
35. Chakrapanidatta, *Charakchaturanan Srimach Chakrapanidatta Pranita Chikitshasamgraha*, ed. Pyarimohan Sengupta (Calcutta: Bidyaratna Jantra, 1888 [1295 BE]): 697.
36. Sengupta, ed., *Chakrapanidatta*, 695.
37. Ibid., 696.
38. Interestingly, one does not find in Chakrapanidatta the distinction between *medha* and *buddhi* that Cohen has reported elsewhere.
39. Sengupta, ed., *Chakrapanidatta*, 697.
40. Ibid., 698.
41. Ibid., 699.
42. Gobindadas, *Bhaisajyaratnabali*, 908–9.
43. Ibid., 908, 909, 916, 917.
44. Sengoopta, "'Dr Steinach Coming to Make Old Young!,'" 122–26.
45. Advertisement, "Chyavanaprash," *Journal of Ayurveda* 4, no. 1 (1927): xv.
46. Cohen, No Aging in India, 132.
47. Cited in Aminoff, *Brown-Sequard*, 242–43.
48. Sen, "Swabhabik Rogbadhok Shokti," 138.
49. It is worth noting that originally Brown-Sequard himself had initially experimented upon himself by injecting an emulsion of either guinea pig or canine testes. Aminoff, *Brown-Sequard*, 205.
50. Gaudilliere, "The Visible Industrialist."

51. John Anderson, *A Guide to the Calcutta Zoological Gardens* (Calcutta: City Press, 1883), 21–22.; Ram Brahma Sanyal, *A Handbook of the Management of Animals in Captivity in Lower Bengal* (Calcutta: Bengal Secretariat Press, 1892), 124–25.
52. Anderson, *Guide to the Calcutta Zoological Gardens*, 25.
53. Wujastyk, *Roots of Ayurveda*, 5–6.
54. White, *Alchemical Body*, 1996. See particularly p. 188.
55. Sharif, *Banglar Sufi Sahityo*; Chakraborty, *Dehotottwer Gan*.
56. See, for instance, Roy, *Pulse in Ayurveda*, 28–29.
57. Oudshoorn, *Beyond the Natural Body*, 9–13.
58. Hemchandra Sen, "Pitter Byabohar," *Chikitsha Sammilani* 11, no. 10 (1906): 17–21.
59. Wujastyk, "Image of the Human Body," 199–200.
60. Ray, "Vayu Pitta Kapha," 1:2, 33.
61. Sengupta, *Ayurveda Sarsamgraha*, 17.
62. Alter, *Yoga in Modern India*, 62.
63. Brain, "Protoplasmania."
64. Sen, "Nadichakra," 40.
65. Ibid., 43.
66. Ibid., 41–43.
67. Ibid., 36.
68. Bhattacharya, *Banglar Baul o Baul Gan*, 438.
69. Ibid., 438–39.
70. On Woodroffe, see Kathleen Taylor, Sir John Woodroffe, *Tantra and Bengal: An Indian Soul in a European Body?* (London: Routledge, 2001).
71. Roy, *Pulse in Ayurveda*, 22.
72. Ibid.
73. Ibid.
74. Ibid.
75. Ibid., 21.
76. Ashutosh Roy, "The Nervous System of the Ancient Hindus," *Journal of Ayurveda* 6, no. 9 (1930): 330.
77. Ibid., vol. 6, no. 8 p. 298.
78. Ibid., vol. 6, no. 10, p. 369.
79. Ibid., vol. 6, no. 8, p. 297.
80. Ibid., vol. 6, no. 9, p. 330.
81. Vasant G. Rele, *The Mysterious Kundalini: The Physical Basis of the "Kundalini (Hatha) Yoga" in terms of Western Anatomy and Physiology* (Bombay: D. B. Taraporevala Sons, 1931), 2.
82. Rele, *Mysterious Kundalini*, 1.
83. Probir K. Bondyopadhyay, "Sir J. C. Bose's Diode Detector Received Marconi's First Transatlantic Wireless Signal of December 1901 (The "Italian Navy Coherer" Scandal Revisited)," *Proc. IEEE* 86, no. 1 (1998): 259–85.
84. Partha Sarathi Gupta, *Radio and the Raj, 1921–47* (Calcutta: K. P. Bagchi, 1995), 4–5.
85. Alasdair Pinkerton, "Radio and the Raj: Broadcasting in British India (1920–1940)," *Journal of the Royal Asiatic Society of Great Britain & Ireland* 18, no. 2 (2008): 171.
86. Sengoopta, *Most Secret Quintessence of Life*, 33–67.
87. Roy, "Nervous System," vol. 6, no. 10, p. 372.
88. Ibid., vol. 6, no. 10, p. 374.
89. Ibid.
90. Ibid., vol. 6, no. 10, p. 375.

91. Alter, *Yoga in Modern India*, 95, 128.
92. Langford, *Fluent Bodies*, 141.
93. Ibid., 147.
94. Richardson, *Death, Dissection and the Destitute*, 34, 48.
95. David Arnold, *Colonizing the Body: State Medicine and Epidemic Disease in Nineteenth-Century India* (Berkeley: University of California Press, 1993).
96. Sappol, *Traffic of Dead Bodies*.
97. Richardson, *Death, Dissection and the Destitute*, 31.
98. Langford, *Fluent Bodies*, 143–44.
99. Wujastyk, "Image of the Human Body."
100. Chintaharan Chakravarti, *Tantras: Studies on Their Religion and Literature* (Calcutta: Punthi Pustak, 1963), 66. Also, Sures Banerji, *Tantra in Bengal: A Study in Its Origin, Development, and Influence* (Calcutta: Naya Prokash, 1977), xxxii–xxxiii, 111–12.
101. David Edwin Pingree, *Census of Exact Sciences in Sanskrit*, ser. A, vol. 3 (Montpelier, VT: Beckman and Beckman, 1992), 137.
102. Rajendralal Mitra, *Notices of Sanskrit MSS*, vol. 1 (Calcutta: Baptist Mission Press, 1871), 276. See also, Pingree, *Census of Exact Sciences*, 120.
103. Sen, "Nadichakra," 33–37.
104. Ibid., 38.
105. Vasu, *Siva Samhita*, 16.
106. Chakrabarty, *Dehotottwer Gan*, 29–30.
107. Roy, *Pulse in Ayurveda*, 25.
108. Ibid.
109. Ibid.
110. Ashutosh Roy, "Astrology in Hindu Medicine," *Journal of Ayurveda* 7, no. 4 (1930): 139–47.
111. Ashutosh Roy, "Astrology in Hindu Medicine," *Journal of Ayurveda* 7, no. 3 (1930): 114–15.
112. Sen, "Nadichakra," 41.
113. Roy, "Nervous System," vol. 6, no. 12, p. 446.

CHAPTER SIX

1. Sengupta, *Ayurveda Sarsamgraha*, 1:7.
2. Sen, *Ayurveda Bigyan*, 1:191.
3. Sengupta and Sengupta, *Ayurveda Samgraha*, 1:360.
4. Dhurmo Dass Sen Gupta, *Nari Vijnana, Or An Exposition of the Pulse* (Calcutta: Union Printing Works, 1893), 15.
5. See, for instance, *Samsad English-Bengali Dictionary*, 692.
6. Prayagchandra Joshi, *Shrikanadmaharishipranit Nadivigyanam* (Benares: Chowkhamba Sanskrit Series Office, 1959), 5.
7. Sarkar, *Susruta Samhita*, 8–9.
8. Wilson, "Medical and Surgical Sciences," 382.
9. Royle, Antiquity of Hindoo Medicine, 50.
10. N. H. Keswani, "Medical Education in India Since Ancient Times," in *History of Medical Education: A Symposium*, ed. Charles Donald O'Malley (Berkeley: University of California Press, 1970), 339.
11. Roberts and Schaffer, preface in *Mindful Hand*, xiii.
12. Pamela Smith, *The Body of the Artisan: Art and Experience in the Scientific Revolution* (Chicago: University of Chicago Press, 2004), 17–18.

13. Sarkar, *Susruta Samhita*, 9.
14. Yulia Frumer, "Translating Time: Habits of Western-Style Timekeeping in Late Edo Japan," *Technology & Culture* 55, no. 4 (2014): 787.
15. Sheldon Pollock, "The Theory of Practice and the Practice of Theory in Indian Intellectual History," *Journal of the American Oriental Society* 105, no. 3 (1985): 519.
16. Marta Hanson, "Hand Mnemonics in Classical Chinese Medicine: Texts, Earliest Images and Arts of Memory," *Asia Major*, ser. 3, vol. 21, no. 1 (2008): 328–29.
17. The familiar mode of explanation for this growth of emphasis on ritual purity has been to simply assert that "Hindus" regressed from their ancient Golden Age of reason and rationality into a postclassical era of superstition and unreason promoted by wily Brahmins seeking to consolidate their social authority. See, for instance, Bala, *Medicine and Medical Policies in India.*
18. Wujastyk, *Roots of Ayurveda*, 246.
19. Langwick, *Bodies, Politics and African Healing*, 113.
20. Sen, *Bhaisajyaratnabali*, 5.
21. I am reminded here of Lissa Roberts's account of the partial and protracted "death of the sensuous chemist" in the nineteenth century. See Lissa Roberts, "The Death of the Sensuous Chemist: The "New" Chemistry and the Transformations of Sensuous Technology," *Studies in the History & Philosophy of Science* 26, no. 4 (1995): 503–29.
22. Sen, *Bhaisajyaratnabali*, 6.
23. Sengupta, *Kobiraji Siksha*, 1:12–13.
24. Ibid., 1:25.
25. Satyacharan Sengupta, "Chikitsoker Kotha," *Ayurveda* 8, no. 4 (1923–24 [1330 BE]): 91.
26. Ibid., 92.
27. It is cogent to point out here that I do not use the word *biomoral* in the sense that Rachel Berger has recently used it, that is, as a self-consciously mobilized notion deployed by orientalists, colonial officials, and eventually Ayurvedic propagandists to designate "Ayurveda's moral features outstripping its biological components." Berger, *Ayurveda Made Modern*, 42–49. Instead, my usage is indebted to Joseph Alter's usage of the term as a heuristic term used to signify the much more thoroughgoing interpenetration of the biological and the moral. In Alter's usage there is no question of the "moral" outpacing the "biological," but rather a fundamental redefinition of the term *biological* itself in a way so as to make inseparable from the "moral." For Alter's definition of biomorality, see Joseph Alter, *Gandhi's Body: Sex, Diet and the Politics of Nationalism* (Philadelphia: University of Pennsylvania Press, 2000), 155. For the difference in Alter's usage and the kind of self-conscious deployment that Berger is interested in, see Alter, *Yoga in Modern India.*
28. Berger, *Ayurveda Made Modern*, 173.
29. Anon., "Baidya Chikitsa," *Ayurveda* 7, no. 7 (1922–23 [1329 BE]): 203.
30. Amritalal Gupta, "Baidya Britti," *Ayurveda* 1, no. 12 (1916–17 [1323 BE]): 541.
31. Jogendrakishore Loh, "Chikitsoker Kortobyo," *Ayurveda* 2, no. 12 (1917–18 [1324 BE]): 492–96.
32. Anon., "Baidya Chikitsa," 205.
33. Sengupta, "Chikitsaker Katha," 90.
34. Indubhushan Sengupta, "Ayurveder Bonoushodhi," *Ayurveda* 8, nos. 10–11 (1923–24 [1330 BE]): 251.
35. Sarangadhar, *Sarangadhar Samhita*, ed. Peary Mohan Sengupta (Calcutta: Benimadhab Dey, 1889), 20.
36. Binodlal Sen, *Arya Grihachikitsa* (Calcutta: n.p., 1879 [1285 BE]), 5.

37. Sailendra Biswas, *Samsad Bengali—English Dictionary* (Calcutta: Sahitya Samsad, 2000), 633.
38. Rev. Lal Behari Dey, *Govinda Samanta* (London: Macmillan, 1874), 250.
39. Dey, *Govinda Samanta*, 249.
40. Risley, *Tribes and Castes of Bengal*, 1:364–65.
41. Nagendranath Sengupta, *Pachon o Mushtijog* (Calcutta: Nagendra Steam Printing Works, 1911), 2.
42. Nagendranath Sengupta, *Pachon o Mushtijog*, 4th ed. (Calcutta: Nagendranath Sengupta, 1920).
43. Ashwinikumar Chattopadhyay, *Pachon o Tahar Byabohar Siksha* (Calcutta: AK Chatterjee, 1927).
44. Advertisement, "Bharot Ayurveda Oushodhaloy," *Amrita Bazar Patrika* (January 8, 1893): 3.
45. Advertisement, "Bijoy Ratna Sen Kobironjon's Ayurvedic Oushodhaloy," *Amrita Bazar Patrika* (June 29, 1894): 8.
46. Advertisement, "Peacock Chemical Works," *Amrita Bazar Patrika* (December 19, 1904): 9.
47. Advertisement, "Adi Ayurveda Oushodhaloy," *Bankura Darpan* (November 8, 1903): 1.
48. Debendranath Sengupta and Upendranath Sengupta, *Pachon Samgraha* (Calcutta: n.p., 1895).
49. Kaliprasanna Bidyaratna, *Pachon Samgraha o Kobiraji Siksha* (Calcutta: n.p., 1906).
50. Sengupta, "Chikitsoker Kotha," 91.
51. Pramathanath Tarkabhushan, "Ayurveder Unnoti ki Obonoti?," *Ayurveda* 1, no. 11 (1916–17 [1323 BE]): 476.
52. William Ward, *History, Literature and Mythologies of the Hindoos*, 4:341.
53. Shib Chunder Basu, *The Hindoos as They Are* (Calcutta: Thacker, Spink & Co., 1883), 217.
54. A fascinating recent book on the advertising culture of Calcutta also strongly suggests that the first bottled perfumes, hair oils, and medicines begun to appear by the late 1880s. Krishnapriya Dasgupta, *Botoler Nanaprokar Koutuk ba Purono Kolkatar Dokandari ar Ponyo Pasarer Bigyapon: Oushudh-Ator-Toiladi* (Calcutta: Gangchil, 2012).
55. Advertisement, "GN Roy's Kalpatarusuda," *Amrita Bazar Patrika* (July 7, 1887): 3.
56. Advertisement, "C. K. Sen & Co.," *Amrita Bazar Patrika* (August 18, 1887): 3.
57. Advertisement, "B. Basu & Co. and Roy & Co.," *Amrita Bazar Patrika* (February 12, 1893): 3.
58. Manu Goswami, "From Swadeshi to Swaraj: Nation, Economy and Territory in Colonial South Asia," *Comparative Studies in Society and History* 40, no. 4 (1998): 609–36.
59. Anon., "News of the Day," *Amrita Bazar Patrika* (October 10, 1911): 9.
60. Advertisement, "The Calcutta Glass & Silicate Company, Limited," *Amrita Bazar Patrika* (May 28, 1920): 4.
61. Advertiesment, "Support Home Industry," *Amrita Bazar Patrika* (January 13, 1922): 1.
62. Advertisement, "Essence of Chiretta," *Amrita Bazar Patrika* (September 27, 1911): 12.
63. Advertisement, "Pushpasar," *Amrita Bazar Patrika* (May 4, 1910): 1.
64. Advertisement, "Three Sovereign Remedies," *Amrita Bazar Patrika* (July 22, 1922): 1.
65. Gupta, "Indigenous Medicine," 374.
66. Stuart Anderson, "Pharmacy and Empire: The 'British Pharmacopoeia' as an Instrument of Imperialism, 1864–1932," *Pharmacy in History* 52, nos. 3–4 (2010): 112–21.
67. Jean-Paul Gaudilliere, "Biologics in the Colonies: Emile Perrot, Kola Nuts and the Industrial Reordering of Pharmacy," in *Biologics: A History of Agents Made from Living*

Organisms in the Twentieth Century, ed. Alexander von Schwerin, Heiko Stoff, and Bettina Wahrig (London: Pickering & Chatto, 2003), 47.

68. Gaudilliere, "Biologics in the Colonies," 62.
69. Ibid., 63.
70. Harry Collins, *Tacit and Explicit Knowledge* (Chicago: University of Chicago Press, 2010), 4.
71. Sengupta, *Pachon o Mushtijog*, 15.
72. Chattopadhyay, *Pachon o Tahar Byabohar*, 3.
73. Loh, "Chikitsoker Kortobyo," 493–94.
74. Ibid., 494.
75. Salimuddin Ahmad Bidyabinod, *Podye Nadigyan* (Calcutta: Altafi Press, 1923), 7.
76. Upadhyay, *Nadi Vijnana*, 58.
77. Ibid., 62–63.
78. Ibid., 64.
79. Gupta, *Nadigyan Siksha*, 15.
80. Ray, *Podye Ayurveda Siksha*, 10.
81. Upadhyay, *Nadi Vijnana*, 63.
82. Ray, *Podye Ayurveda Siksha*, 10.
83. Sitalchandra Chattopadhyay, "Ayurveder Koshay Mahatmyo," *Ayurveda* 1, no. 9 (1916–17 [1323 BE]): 409.
84. Sengupta, Pachon o Mushtijog, 1.
85. Anon., "Prostabona," *Ayurveda Hitoishini* 2, no. 4 (1912 [1319 BE]): 152.
86. Nabinchandra Dey, "Drobyo Porichoy," *Ayurveda Bikash* 2, no. 2 (1914–15 [1321 BE]): 61.
87. Dey, "Drobyo Porichoy," 61.
88. Anon., "Nobabishkrito Desiyo Bheshoj o Tahar Proyog," *Ayurveda Bikash* 3, no. 1 (1915–16 [1322 BE]): 9–12.
89. Sengupta, Pachon o Mushtijog, 1.
90. It is interesting to contrast this mode of vernacular ethnology with the more overt colonial union on ethnography, botany, and pharmacy that Gaudilliere describes as enabling the production of "colonial biologics." Gaudilliere, "Biologics in the Colonies."
91. Anon., "Prostabona," 151.
92. Kshitishchandra Lahiri, "Prachin Chikitsoker Totka o Mushtijog," *Ayurveda* 4, no. 3 (1919–20 [1326 BE]): 141–42.
93. Indubhushan Sengupta, "Prachin Chikitsoker Porikshito Mushtijog," *Ayurveda* 6, no. 11 (1921–22 [1327 BE]): 407–8.
94. Sengupta, "Chikitsoker Kotha," 92.
95. Alter, *Yoga in Modern India*, 142–80.
96. By speaking of "Baidya parochialism," I do not mean to insinuate any negative judgment against it, which would then reflect positively on the Hindu nationalism it interrupted. My usage is informed by Uday Mehta's comment that parochialism "is just a way to characterize what for other peoples are the conditions of a meaningful life—the modes of experience by which things hang together." Uday Singh Mehta, *Liberalism and Empire: A Study in Nineteenth-Century British Liberal Thought* (Chicago: University of Chicago Press, 1999), 27.
97. Rosenberg, "Therapeutic Revolution," 9.
98. Dipesh Chakrabarty, *Habitations of Modernity: Essays in the wake of Subaltern Studies* (Chicago: University of Chicago Press, 2002), 51–64.

99. For an excellent discussion on Sen's romanticism and politics, see Dipesh Chakrabarty, "Romantic Archives: Literature and the Politics of Identity in Bengal," *Critical Inquiry* 30, no. 3 (2004): 654–83.
100. Dineshchandra Sen, "Therapeutics," *Ayurveda* 6, no. 2 (1921–22 [1328 BE]): 57–58.

CONCLUSION

1. Mrinalkanti Ghosh, *Poroloker Kotha* (Calcutta: n.p., 1933), 57–60.
2. Sen, *Bhaisajyaratnabali,* 50.
3. Ibid., 51.
4. Mukhopadhyay, *Jonmantor Rohosyo,* 215–21.
5. Sarkar, *Susruta Samhita,* 488–89.
6. Priya vrat Sharma trans., *Cakradatta* (Delhi: Chaukhamba Orientalia, 2007), 557–66.
7. Sen, *Bhaisajyaratnabali,* 872–74.
8. Zysk, *Asceticism and Healing in Ancient India.*
9. Michael Meister, "Regional Variations in Matrika Conventions," *Artibus Asia* 47, nos. 3–4 (1986): 238–39.
10. Wujastyk, *Roots of Ayurveda,* 105.
11. Meister, "Regional Variations," 233.
12. Ibid., 240, n. 27.
13. Apffel-Marglin, *Smallpox in Two Systems of Knowledge*; Susan S. Wadley, "Sitala: The Cool One," *Asian Folklore Studies* 39, no. 1 (1980): 33–62; Lauren Nauta Minsky, "Pursuing Protection from Disease: The Making of Smallpox Prophylactic Practice in Colonial Punjab," *Bulletin of the History of Medicine* 83, no. 1 (2009): 163–89; Harish Naraindas, "Preparing for the Pox: A Theory of Smallpox in Bengal and Britain," *Asian Journal of Social Science* 31, no. 2 (2003): 304–39.
14. On Sitala's subaltern affiliations, see Arnold, *Colonizing the Body,* 116–58. On Sitala's presence in the textual tradition of Ayurvedic scholarship, see Meulenbeld, *History of Indian Medical Literature,* 2B:265, n. 76.
15. Mukharji, "In-Disciplining Jwarasur."
16. Roy, *Pulse in Ayurveda,* 3.
17. Premavati Tiwari, *Kasyapa-samhita Or Vriddhajivakiya Tantra* (Benares: Caukhamba Visvabharati, 1996), 99.
18. Sen, Bhaisajyaratnabali, 121.
19. Amaresh Datta, *Encyclopaedia of Indian Literature,* vol. 1 (New Delhi: Sahitya Akademi, 1987), 24.
20. Panchanan Mandal, *Punthi Porichoy,* vol. 3 (Santiniketan: Vishwabharati, 1963), 165–74.
21. Atulchandra Datta, *Gupto Mantra* (Calcutta, 1907 [1314 BE]).
22. Gupta Prakashak, *Bongiyo Bishwasyo Mantraboli* (Calcutta?: Prakrita Jantralaya?, 1863?).
23. Shri Dere Babaji, *Indrajal* (Calcutta: Basak Press, n.d.), 414.
24. Dipesh Chakrabarty, "Community, State and the Body: Epidemics and Popular Culture in Colonial India," in *Medical Marginality in South Asia: Situating Subaltern Therapeutics,* ed. David Hardiman and Projit Bihari Mukharji (Abingdon: Routledge, 2012), 46.
25. Rabinbach, *Human Motor.*
26. Peter van der Veer, *Imperial Encounters,* 55–82.
27. See Mukharji, "In-disciplining Jwarasur," 280–84.
28. Roy, *Pulse in Ayurveda,* 3.

29. Ruy Blanes and Diana Espirito Santo, eds., *The Social Life of Spirits* (Chicago: University of Chicago Press, 2013), 17.
30. Blanes et al., *Social Life of Spirits*, 136–37.
31. Gopinath Kobiraj, "The Conception of Physical and Superphysical Organism in Sanskrit Literature," *Notes on Religion and Philosophy* (Benares: Sampurnanand Sanskrit University, 1987, 212).
32. Noted Indologist Frits Staal points out that "the Indian classical doctrine [on the body], accepted in both Hinduism and Buddhism, is a plurality of onionskin-like layers, 'Sheaths' (*kosa*) or 'joints' (*skandha*)." This doctrine was first formulated in the *Taittiriya Upanishad* and has generally come to speak of five sheaths or *kosas* comprising progressively of the *annam* (food), *prana* (life/breath), *manas* (mind), *vijnana* (knowledge), and *ananda* (bliss) *kosas*. Frits Staal, "Indian Bodies," in *Self as Body in Asian Theory and Practice*, ed. Thomas Kasulis, Roger T. Amis, and Wimal Dissanayake (Albany: SUNY Press, 1993), 59–102.
33. Kobiraj, *Notes on Religion and Philosophy*, 214.
34. Mel Y. Chen, *Animacies: Biopolitics, Racial Mattering and Queer Affect* (Durham, NC: Duke University Press, 2012), 3.
35. Haradhan Sharma, *Ayurvedokto Nuton Chikitsa Darshan* (Calcutta: Bishwambhar Laha, 1883 [1290 BE]), 3.
36. Gopinath Kobiraj, *Aspects of Indian Thought* (Burdwan: University of Burdwan, 1966), 232.
37. Gavin Flood, *The Tantric Body: The Secret Tradition of Hindu Religion* (London: I. B. Tauris, 2006), 156.
38. Shashibhusan Dasgupta, *Obscure Religious Cults* (Calcutta: University of Calcutta Press, 1946), 353.
39. Dasgupta, *Obscure Religious Cults*, 350–51.
40. Ibid., 354–55.
41. David Kinsley, *Tantric Visions of the Divine Feminine: The Ten Mahavidyas* (Berkeley: University of California Press, 1997), 28.
42. Sharif, *Banglar Sufi Sahityo*, 84.
43. Joseph Westermeyer, "Folk Medicine in Laos: A Comparison between two Ethnic Groups," *Social Science & Medicine* 27, no. 8 (1988): 769–78.
44. Jean Mottin, "A Hmong Shaman's Séance," in *Shamans in Asia*, ed. Clark Chilson and Peter Knecht (London: Routledge/Curzon, 2003), 89.
45. Kristofer Marinus Schipper, *The Taoist Body*, trans. Karen C Duval (Berkeley: University of California Press, 1993), 103.
46. Kristofer Marinus Schipper, "The Taoist Body," *History of Religion* 17, nos. 3–4 (1978): 368.
47. Dwarkanath Datta, *Chikitsa Chakrasar* (Calcutta: Jyotishprakash Jantra, 1882 [1289 BE]), 10.
48. For the exemplary statements referring to the sun and the moon residing in specific *nadis*, see Datta, *Chikitsa Chakrasar*, 10; Gupta, *Nadigyan Rohosyo*, 39.
49. Jayakanta Mishra, "Kaviraj Ji as I Knew Him" in *Gopinath Kobiraja Commemoration Volume* (Allahabad: Ganganatha Jha Kendriya Sanskrit Vidyapeetha, 1976), iv.
50. Joseph Alter, "Heaps of Health, Metaphysical Fitness: Ayurveda and the Ontology of Good Health in Medical Anthropology," *Current Anthropology* 40, no. S1 (1999): S43-S66.
51. Datta, *Chikitsa Chakrasar*, 24–25.
52. Sharma, *Cakradatta*, 593. (Sharma's translations.)

53. Sen, *Bhaisajyaratnabali*, 982.
54. Ibid., 901.
55. Kalyani Mullick, *Nath Somprodayer Itihas, Darshan o Sadhonpronali* (Calcutta: Calcutta University Press, 1946), 292.
56. Mullick, *Nath Somprodayer Itihas*, 511.
57. Ibid., 546.
58. White, *Alchemical Body*, 2.
59. Mullick, *Nath Somprodayer Itihas*, 293.
60. Ibid., 543.
61. Kobiraj, *Notes on Religions and Philosophy*, 217–19.
62. Bandyopadhyay, *Arogya Niketan*, 136.
63. For a detailed discussion of the nature of this *siddhodeho*, see Kalyani Mullick, *Nath Somprodayer Itihas, Darshan o Sadhonpronali* (Calcutta: Calcutta University Press, 1946), 390–91.
64. Munshi Faizullah, *Goroksho Bijoy*, ed. Abdul Karim (Calcutta: Bangiya Sahitya Parishat, 1917 [1324 BE]).
65. Gananath Sen, *Ayurveda Porichoy* (Calcutta: Vishwabharati, 1944).
66. Kenneth G. Zysk, "Mantra in Ayurveda: A Study of the Use of Magico-Religious Speech in Ancient Indian Medicine," in *Understanding Mantras*, ed. Harvey P Alper (Albany: State University of New York Press, 1989), 126.
67. Sarkar, *Susruta Samhita*, 488–89.
68. Sharma, *Cakradatta*, 557–66; Sen, *Bhaisajyaratnabali*, 872–74.
69. Sarkar, *Susruta Samhita*, 420.
70. Sharma, *Cakradatta*, 34–35; Sen, *Bhaisajyaratnabali*, 121.
71. Mandal, *Punthi Porichoy*, 3:165–74.
72. Sivaramakrishnan, *Old Potions, New Bottles*, 145.
73. Hitesranjan Sanyal, *Bangla Kirtoner Itihas* (Calcutta: K. P. Bagchi, 1989), 16–19.
74. Jadunath Sarkar, *Chaitanya's Life and Teachings: From His Contemporary Bengali Biography the Chaitanya-Charit-Amrita* (Calcutta: M. C. Sarkar & Sons, 1922), 294–95. See also, Krishnadas Kobiraj, *Chaitanya-charitamrita*, ed. Sukumar Sen (New Delhi: Sahitya Akademi, 1936), 613.
75. Lochan Das, *Chaitanyamangal* (Rangpur: Radharaman Jantra, 1892 [1299 BE]), 218–20.
76. Kobiraj, *Chaitanya-charitamrita*, 84–86.
77. Bandopadhyaya, *Arogya Niketan*, 220.
78. Mukharji, *Nationalizing the Body*, 161–62.
79. Rachel Fell McDermott, *Singing to the Goddess: Poems to Kali and Uma from Bengal* (Oxford: Oxford University Press, 2001), 11.
80. Ibid., 96.
81. Kalidas Bidyabhushan, *Chikitsatottwo o Bhaboushodh* (Calcutta: Ayurveda Bidya Mandir, 1912 [1319 BE]).
82. Bidyabhushan, *Chikitsatottwo o Bhaboushodh*, 7–8.
83. Jagadish Narayan Sarkar, *Thoughts on Trends of Cultural Contact in Medieval India* (Calcutta: OPS Publishers, 1984), 140.
84. T. P. Hughes, *Notes on Muhammadanism* (London: William H. Allen, 1875), 149.
85. Ibid., 161.
86. Ibid., 149.
87. Hans Harder, *Sufism and Saint Veneration in Contemporary Bangladesh: The Maijbhandaris of Chittagong* (Abingdon: Routledge, 2011), 51–53.

88. For the fascinating history of the complex correlation of Sufi *muqams* with Tantric chakras, see Shaman Hatley, "Mapping the Esoteric Body in the Islamic Yoga of Bengal," *History of Religions* 46 (2007): 351–68.
89. Baba Jan Sharif Sureswari Qibla, *Nur-e-Haqq, Ganj-e-Nur* (Shariatpur, Bangladesh: Syed Nur-e-Iqbal Shah Sharif al Sureswari, 2002), 67–69.
90. Syed Waliullah, *Lal Salu* (Dhaka: Nouroj Kitabistan, 2006).
91. James Newell, "Unseen Power: Aesthetic Dimensions of Symbolic Healing in *Qawwali*," *Muslim World* 97, no. 4 (2007): 640–56.
92. See Tony K. Stewart, "Satya Pir: Muslim Holy Man and Hindu God," in *Religions of India in Practice*, ed. Donald S. Lopez Jr. (Princeton, NJ: Princeton University Press, 1995), 578–98.
93. Ward, History, *Literature and Mythology of the Hindoos*, 340.
94. Miri Shefer-Mossensohn, *Ottoman Medicine: Healing and Medical Institutions 1500–1700* (Albany: SUNY Press, 2009), 69.
95. Shefer-Mossensohn, *Ottoman Medicine*, 72.
96. Daud Ali, "Padmasri's Nagarasarvasva and the World of Medieval Kamasastra," *Journal of Indian Philosophy* 39, no. 1 (2011): 41–62.
97. Singh Jee, *Short History*, 49–50.
98. Nagendranath Sengupta, *The Ayurvedic System of Medicine*, vol. 1 (Delhi: Neeraj, 1984 [1901]), 59, 69–79.
99. Zysk, "Mantras in Ayurveda," 133–35.
100. There is some ambiguity about whether Naraindas intends this critique of what he terms "cosmo-therapeutics" to apply only to the specific way in which Frederique Appfel Marglin engendered it, or more generally. Unfortunately, he does not define this enigmatic and extremely evocative term explicitly at any point. See Naraindas, "A Theory of Smallpox in Bengal and in Britain," 308. See also, Harish Naraindas, "Care, Welfare and Treason: The Advent of Vaccination in the 19th Century," *Contributions to Indian Sociology* 32, no. 1 (1998): 71, n. 4.
101. See Harish Naraindas, "Poisons, Putrescence and the Weather: A Genealogy of the Advent of Tropical Medicine," *Contributions to Indian Sociology* 30, no. 1 (1996): 31, n. 19.
102. Alessandro Fontana and Mauro Bertani, "Situating the Lectures," in Michel Foucault, *"Society Must Be Defended": Lectures at the College de France, 1975–76*, ed. Alessandro Fontana and Mauro Bertani, trans. David Macey (New York: Picador, 1997), 278.
103. Cohen, "Whodunit?," 345.
104. On the relationship between colonial governmentality and "Western" medicine, see Arnold, *Colonizing the Body.*
105. Arnold, *Colonizing the Body*, 173.
106. M. J. Walhouse, "A Hindu Prophetess," *Journal of the Anthropological Institute of Great Britain and Ireland* 14 (1885): 187–92.
107. Atulbihari Gupta, *Mrityur Pare o Punorjonmobad* (Benares: Sailendranath Gupta, 1941 [1348 BE]): 117–24. See also Umacharan Mukherjee, *Tailanga Swamir Jibonchorit o Tottwopodesh* (Calcutta: Jogendranath Mukhopadhyay, 1918 [1325 BE]), 31–33. Incidentally, Mukherjee also lists a large number of miraculous cures affected by Tailanga Swami.
108. Suman Seth, "Quantam Theory and the Electromagnetic World-View," *Historical Studies in the Physical and Biological Sciences* 35, no. 1 (2004): 67–93. It is worth noting that not everybody agrees with Seth on this matter. For an opposing view, see Shaul

Katzir, "On 'the Electromagnetic World-View': A Comment on an article by Suman Seth," *Historical Studies in the Physical and Biological Sciences* 36, no. 1 (2005): 189–92.

109. Alfred Jarry, *Exploits and Opinions of Dr Faustroll Pataphysician: A Neoscientific Novel*, trans. Simon Watson Taylor (Boston: Exact Change, 1996 [1911]).
110. Gilles Deleuze and Felix Guattari, *A Thousand Plateaus*, trans. Brian Massumi (Minneapolis: University of Minnesota Press, 1987).
111. For the insertion of Ayurveda into the modern governmental apparatus, see Berger, *Ayurveda Made Modern*.
112. For the pharmaceuticalization of Ayurveda, see Banerjee, *Power, Knowledge, Medicine*; see also, Bode, *Modern Image of the Ayurvedic and Unani Industry*.

BIBLIOGRAPHY

Adam, William. *Report on the State of Education in Bengal*. Calcutta: Bengal Military Orphan Press, 1835.

Adam, William. *Second Report on the State of Education in Bengal: District of Rajshahi*. Calcutta: Bengal Military Orphan Press, 1836.

Adams, Vincanne, Mona Schrempf and Sienna R. Craig, eds. *Medicine between Science and Religion: Explorations on Tibetan Grounds*. Oxford: Berghahn, 2011.

Advertisement. "Support Home Industry." *Amrita Bazar Patrika* (January 13, 1922): 1.

Advertisement. "Adi Ayurveda Oushodhaloy." *Bankura Darpan* (November 8, 1903): 1.

Advertisement. "B. Basu & Co. and Roy & Co." *Amrita Bazar Patrika* (February 12, 1893): 3.

Advertisement. "Bharot Ayurveda Oushodhaloy." *Amrita Bazar Patrika* (January 8, 1893): 3.

Advertisement. "Bijoy Ratna Sen Kobironjon's Ayurvedic Oushodhaloy." *Amrita Bazar Patrika* (June 29, 1894): 8.

Advertisement. "Chyavanaprash." *Journal of Ayurveda* 4:1 (1927): xv.

Advertisement. "CK Sen & Co." *Amrita Bazar Patrika* (August 18, 1887): 3.

Advertisement. "Colwell's Hormones." *Journal of Ayurveda* 4:2 (1927): ii.

Advertisement. "Deccan Watch Co." *Journal of Ayurveda* 4:8 (1928): xii.

Advertisement. "Elements of Endocrinology." *Journal of Ayurveda* 4:1 (1927): iv.

Advertisement. "Essence of Chiretta." *Amrita Bazar Patrika* (September 27, 1911): 12.

Advertisement. "GN Roy's Kalpatarusuda." *Amrita Bazar Patrika* (July 7, 1887): 3.

Advertisement. "Hamilton & Co., Jewellers and Silversmiths." Calcutta Review Advertiser. *Calcutta Review* 74 (1861): 2–3.

Advertisement. "Hormotone." *Journal of Ayurveda* 4:3 (1927): vii.

Advertisement. "Ik-Mik Cooker." *Baidya Hitoishi* 4:11 (1928 [1335 BE]): 370.

Advertisement. "Ik-Mik Cooker." *Baidya Hitoishi* 4:8 (1928 [1335 BE]): 290;

Advertisement. "Peacock Chemical Works." *Amrita Bazar Patrika* (December 19, 1904): 9.

Advertisement. "Protonuclein." *Journal of Ayurveda* 5:3 (1928): vi.

Advertisement. "Pushpasar." *Amrita Bazar Patrika* (May 4, 1910): 1.

Advertisement. "Testacoids." *Journal of Ayurveda* 5:1 (1928): vi.

Advertisement. "The Calcutta Glass & Silicate Company, Limited." *Amrita Bazar Patrika* (May 28, 1920): 4.

Advertisement. "Three Sovereign Remedies." *Amrita Bazar Patrika* (July 22, 1922): 1.

Advertisement. "Viriligen." *Journal of Ayurveda* 4:4 (1927): vii.

Advertisement. Nagendranath Sen & Co. Ltd. "Baidyabritti o Baidyak Grontho." *Baidya Hitoishi* 3:7 (1927 [1334 BE]): 226.

Ahmed, Salahuddin. *Bangladesh: Past and Present.* New Delhi: APH Publishing Corporation, 2004.

Alavi, Seema. *Islam and Healing: Loss and Recovery of an Indo-Muslim Medical Tradition, 1600–1800.* Basingstoke, UK: Palgrave Macmillan, 2008.

Ali, Daud. "Garden Automata and the Production of Wonder across the Indian Ocean." Paper presented at the History & Sociology of Science Workshop, University of Pennsylvania, Philadelphia, February 4, 2013.

———. "Padmasri's Nagarasarvasva and the World of Medieval Kamasastra." *Journal of Indian Philosophy* 39:1 (2011): 41–62.

Alper, Harvey P., ed. *Understanding Mantras.* Albany: SUNY Press, 1989.

Alter, Joseph. "Heaps of Health, Metaphysical Fitness: Ayurveda and the Ontology of Good Health in Medical Anthropology." *Current Anthropology* 40:1 (Spring 1999): 43–66.

———. *Gandhi's Body: Sex, Diet and the Politics of Nationalism.* Philadelphia: University of Pennsylvania Press, 2000.

———. *Yoga in Modern India: The Body between Science and Philosophy.* Princeton, NJ: Princeton University Press, 2004.

Ami. *Thengapathik Bhuinphod Daktar.* Calcutta: Manimohan Rakshit, 1886 (1293 BE).

Aminoff, Michael J. *Brown-Sequard: An Improbable Genius Who Transformed Medicine.* New York: Oxford University Press, 2011.

Anderson, John. *A Guide to the Calcutta Zoological Gardens.* Calcutta: City Press, 1883.

Anderson, Stuart. "Pharmacy and Empire: The 'British Pharmacopoeia' as an Instrument of Imperialism, 1864–1932." *Pharmacy in History* 52:3/4 (2010): 112–21.

Anderson, Warwick. "Making Global Health History: The Postcolonial Worldliness of Biomedicine." *Social History of Medicine* 27:2 (2014): 372–84.

Andrews, Bridie. "Tuberculosis and the Assimilation of Germ Theory in China, 1895–1937." *Journal of the History of Medicine and Allied Sciences* 52:1 (1997): 114–57.

Anning, S. T. "Clifford Albutt and the Clinical Thermometer." *Practitioner* 197:182 (1966): 818–23.

Anon. Ambastha-Kulochondrika Orthat Baidyajatir Chokshudan. Calcutta: Rakhaldas Borat, 1892 (1299 BE).

Anon. "Astanga Ayurveda o Astanga Ayurveda Bidyaloy." *Ayurveda* 1:1 (1916 [1323 BE]): 29–36.

Anon. "Ayurbigyane Vayu Pitta Kapha." *Ayurveda Hitoishini* 2:4 (1912 [1319]): 121–28.

Anon. "Baidya Chikitsa." *Ayurveda* 7:7 (1922–23 [1329 BE]): 202–6.

Anon. "Baidya Sommelone Sobhapotir Obhibhashon." Ayurveda 1:5 (1916 [1323 BE]): 185–92.

Anon. "Bhabanipur Kendrer Prothom Barshik Sadharon Sobha." Baidya Hitoishi 3:4 (1926 [1333 BE]): 148–51.

Anon. "Chikitsa." *Anubikshan* 1:1 (1875 [1282 BE]): 3–13.

Anon. "David Hare." In *Calcutta, Old and New: A Historical & Descriptive Handbook to the City*, edited by H. E. A. Cotton, 422–23. Calcutta: W. Newman, 1907.

Anon. "Fever, n.1." *OED Online.* Oxford: Oxford University Press, March 2015. Accessed on May 30, 2015. http://www.oed.com/view/Entry/69682?rskey=b6KQiB&result=1.

Anon. "The Great Anarchy." *The Calcutta Review*, vol. 107, 216–35. Calcutta: City Press, 1899.

Anon. "Humour vs. Hormone in Ayurveda." *Journal of Ayurveda* 4:8 (1928): 281–83.

Anon. "Kobiraj Surendranath Goswami." *Janmabhumi* 27:6 (1922): 191–208.

Anon. "Kobiraj Surendranath Goswami." *Janmabhumi* 27:8 (1922): 249–52.

Anon. "Medical Women in India and Africa." *Physician & Surgeon* 7 (1885): 213–14.
Anon. "News of the Day." *Amrita Bazar Patrika* (October 10, 1911): 9.
Anon. "Nobabishkrito Desiyo Bheshoj o Tahar Proyog." *Ayurveda Bikash* 3:1 (1915–16 [1322 BE]): 9–12.
Anon. *"Phuler Saji." Bangadarshan* 9:10 (1882 [1289 BE]): 517–19.
Anon. "Prostabona." *Ayurveda Hitoishini* 2:4 (1912 [1319 BE]): 151–53.
Anon. "Residents in Calcutta." In "Calcutta Guide and Directory." *The Bengal and Agra Annual Guide and Gazetteer for 1842*, vol. 1, 179–203. Calcutta: William Rushton, 1842.
Anon. "Review of *Clinical Lectures on Dengue* by T. Edmonston Charles." *The Calcutta Review* 55 (Calcutta: City Press, 1872): vi–viii.
Anon. "Samkshipto Jiboni." *Ayurveda Bikash* 2:11/12 (1914–15): 281–85.
Anon. "Sharirotpotti." *Chikitsa Sammilani* 12:2–3 (1907): 152–56.
Anon. *Report of the Salaries Commission*. Calcutta: Bengal Secretariat Press, 1886.
Anon. "Kalikata Baidya Brahmin Samitir Dwitiya Barshik Adhibeshan." *Baidya Hitoishi* 3:1 (1333 BE), 17–18.
Apffel Marglin, Frederique, and Stephen A. Marglin, eds. *Dominating Knowledge: Development, Culture and Resistance*. Oxford: Clarendon, 1990.
Appfel-Marglin, Frederique. *Smallpox in Two Systems of Knowledge*. Helsinki: WIDER, 1987.
Arnold, David. *Colonizing the Body: State Medicine and Epidemic Disease in Nineteenth-Century India*. Berkeley: University of California Press, 1993.
Arnold, David. *Everyday Technology: Machines and the Making of India's Modernity*. Chicago: University of Chicago Press, 2013.
Aronowitz, Robert A. "When Do Symptoms Become a Disease?" *Annals of Internal Medicine* 134:9/2 (2001): 803–8.
Attewell, Guy. "Haptic Iterability and Credibility: Extra-Diagnostic X-Rays, Body Parts and Distributed Agency—with a 'Bonesetting' Clinic in Hyderabad City." Unpublished paper.
———. "Islamic Medicine: Perspectives on the Greek Legacy in the History of Islamic Medical Traditions in West Asia." In *Medicine across Cultures: History and Practice of Medicine in Non-Western Cultures*, edited by Helen Selin, 325–50. New York: Kluwer, 2003.
———. *Refiguring Unani Tibb: Plural Healing in Late Colonial India*. Hyderabad: Orient Longman, 2007.
Babaji, Shri Dere. *Indrajal*. Calcutta: Basak Press, n.d.
Bala, Poonam. *Contesting Colonial Authority: Medicine and Indigenous Responses in 19th and 20th Century India*. Lanham, MD: Lexington, 2012.
———. *Imperialism and Medicine in Bengal: A Socio-Historical Perspective*. Delhi: Sage, 1991.
———. *Medicine and Medical Policies in India: Social and Historical Perspectives*. Lanham, MD: Lexington, 2007.
Bald, Vivek. *Bengali Harlem and the Lost Histories of South Asian America*. Cambridge, MA: Harvard University Press, 2013.
Balfour, Edward. *The Cyclopaedia of India and of Eastern and Southern Asia*. Vol. 3. London: Bernard Quaritch, 1885.
Ballantyne, Tony. *Orientalism and Race: Aryanism in the British Empire*. Basingstoke, UK: Palgrave Macmillan, 2002.
Bandopadhyay, Nirmalshib. "Ghorioala." *Birbhumi Nabaparjaya* 3:2 (1913 [1320 BE]): 120–26.
Bandyopadhyay, Sekhar. Caste, Culture and Hegemony: Social Dominance in Colonial Bengal. New Delhi: Sage, 2004.

Bandyopadhyay, Tarashankar. *Arogya Niketan*. Calcutta: Prakash Bhavan, 1958.

Banerjee, Madhulika. "Ayurveda in Modern India: Standardization and Pharmaceuticalization." In *Modern and Global Ayurveda: Pluralism and Paradigms*, edited by Dagmar Wujastyk and Frederick M Smith, 201–14. Albany: SUNY Press, 2008.

Banerjee-Dube, Ishita. "Caste, Race and Difference: The Limits of Knowledge and Resistance." *Current Sociology* 42:4 (2014): 512–30.

Banerji, Sures. *Tantra in Bengal: A Study in its Origin, Development, and Influence*. Calcutta: Naya Prokash, 1977.

Barman, Anandachandra. *Sarkaumudi*. Calcutta: Sudhasindhu Jantra, 1867 (1274 BE).

Barnes, David. *The Great Stink of Paris and Nineteenth-Century Struggle against Filth and Germs*, 107–8. Baltimore, MD: Johns Hopkins University Press, 2006.

Basalla, George. "The Spread of Western Science." *Science* 156: 3775 (1967): 611–22.

Basu, Nagendranath. "Gangadhar Kobiraj." In *Bishwokosh*, vol. 5. Calcutta: Sri Rakhalchandra Mitra, 1894 (1301 BE).

Basu, Shib Chunder. *The Hindoos as They Are*. Calcutta: Thacker, Spink & Co., 1883.

Bates, Don, ed. *Knowledge and the Scholarly Medical Tradition*. Cambridge: Cambridge University Press, 1995.

Baudrillard, Jean and Marc Guillaume. *Radical Alterity*. Los Angeles: Semiotexte/Smart Art, 2008.

Bennett, Jane. *Vibrant Matter: A Political Ecology of Things*. Durham, NC: Duke University Press, 2010.

Berger, Rachel. *Ayurveda Made Modern: Political Histories of Indigenous Medicine in North India, 1900–1955*. Basingstoke, UK: Palgrave Macmillan, 2013.

Berkowitz, Carin. "The Beauty of Anatomy: Visual Displays and Surgical Education in Nineteenth-Century London." *Bulletin of the History of Medicine* 85:2 (2011): 248–78.

Bettany, G. T. and rev. Richard Hankins. "Guy, William Augustus (bap. 1810, d. 1885)." *Oxford Dictionary of National Biography*. Oxford University Press, 2004 (online May 2009). Accessed May 28, 2015, http://dx.doi.org/10.1093/ref:odnb/11801.

Bhabha, Homi. *The Location of Culture*. London: Routledge, 1994.

Bharadwaj, Aditya. *Local Cells, Global Science: The Rise of Embryonic Stem Cells Research in India*. London: Routledge, 2009.

Bharatamallick. "Obotoronika." In *Ratnaprabha: Radhiya Baidyakuloponjika*, edited by Binodlal Sen, n.p. Calcutta, 1891 (1298 BE).

Bharatmullick, *Chandraprabha: Baidyakulapanjika*, edited by Binodlal Sen. Calcutta, 1892 (1299 BE).

Bhattacharya, Jayanta. "The First Dissection Controversy: Introduction to Anatomical Education in Bengal and British India." *Current Science* 101:9 (2011): 1227–32.

Bhattacharya, Upendranath. *Banglar Baul o Baul Gan*. Calcutta: Orient Book Company, 2001.

Bidyabhushan, Kalidas. *Chikitsatottwo o Bhaboushodh*. Calcutta: Ayurveda Bidya Mandir, 1912 (1319 BE).

Bidyabinod, Salimuddin Ahmad. *Podye Nadi-gyan*. Calcutta: Altafi Press, 1913 (1320BE).

Bidyaranta, Dwarakanath. *Chikitsaratna*. Calcutta: Sanskrita Jantra, 1909 (1316 BE).

Bidyaratna, Kaliprasanna. *Pachon Samgraha o Kobiraji Siksha*. Calcutta: n.p., 1906.

Bijker, Wiebe E. *Of Bicycles, Bakelite and Bulbs: Towards a Theory of Sociotechnical Change*. Cambridge, MA: MIT Press, 1997.

Bir, Atilla, Sinasi Acar, and Mustafa Kacar. "The Clockmaker Family Meyer and Their Watch Keeping the *Alla Turca* Time." In *Science between Europe and Asia: Historical Studies on the*

Transmission, Adoption and Adaptation of Knowledge, edited by Feza Gunergun and Dhruv Raina, 125–36. Dordrecht: Springer, 2011.

Biswas, Sailendra. *Samsad Bengali—English Dictionary*. Calcutta: Sahitya Samsad, 2000.

Bivins, Roberta. *Alternate Medicine? A History*. Oxford: Oxford University Press, 2007.

Blanes, Ruy, and Diana Espirito Santo, eds. *The Social Life of Spirits*. Chicago: University of Chicago Press, 2013.

Bode, Maarten. *Taking Traditional Knowledge to the Market: The Modern Image of the Ayurvedic and Unani Industry, 1980–2000*. Hyderabad: Orient Longman, 2008.

Bondyopadhyay, Probir K. "Sir J. C. Bose's Diode Detector Received Marconi's First Transatlantic Wireless Signal of December 1901 (The 'Italian Navy Coherer' Scandal Revisited)." *Proc. IEEE* 86:1 (1998): 259–85.

Bourdieu, Pierre "The Social Space and the Genesis of Groups." *Theory & Society* 14:6 (1985): 723–44.

Bourguet, Marie-Noelle, Christian Liccope, and H. Otto Sibum, eds. *Instruments, Travel and Science: Itineraries of Precision from the Seventeenth to the Twentieth Century*. New York: Routledge, 2014.

Brain, Robert. "Protoplasmania: Huxley, Haeckel, and the Vibratory Organism in Late Nineteenth-Century Art and Science." In *The Art of Evolution: Darwin, Darwinisms, and Visual Culture*, edited by B. Larson and E. Brauer, 92–103. Hanover, NH: University Press of New England, 2009.

Breckenridge, Carol A. ed. *Consuming Modernity: Public Culture in a South Asian World*. Minneapolis: University of Minnesota Press, 1995.

British Library, Indian Office Records & Private Papers, IOR/Z/E/4/37/H149.

Brown, Edward. "Neurology and Spiritualism in the 1870s." *Bulletin of the History of Medicine* 57 (1983): 563–77.

Buchanan, Francis. *An Account of the District of Purnea*. Patna: Bihar & Orissa Research Society, 1928.

Canales, Jimena. *A Tenth of a Second: A History*. Chicago: University of Chicago Press, 2009.

Capt. Everest. "On the Compensation Measuring Apparatus of the Great Trigonometrical Survey of India." *Asiatic Researches* 18 (1833): 189–214.

Cerulli, Anthony. *Somatic Lessons: Narrating Patienthood and Illness in Indian Medical Literature*. Albany: SUNY Press, 2014.

Chakrabarti, Pratik. "'Living Versus Dead': The Pasteurian Paradigm and Imperial Vaccine Research." *Bulletin of the History of Medicine* 84:3 (2010): 387–423.

Chakrabarty, Dipesh. "Community, State and the Body: Epidemics and Popular Culture in Colonial India." In *Medical Marginality in South Asia: Situating Subaltern Therapeutics*, edited by David Hardiman and Projit Bihari Mukharji, 36–58. Abingdon, UK: Routledge, 2012.

———. *Habitations of Modernity: Essays in the Wake of Subaltern Studies*. Chicago: University of Chicago Press, 2002.

———. "The Muddle of Modernity." *American Historical Review* 116:3 (2011): 663–75.

———. *Provincializing Europe: Postcolonial Thought and Historical Difference*. Princeton, NJ: Princeton University Press, 2000.

———. "Romantic Archives: Literature and the Politics of Identity in Bengal." *Critical Inquiry* 30:3 (2004): 654–83.

Chakrabarty, Ramakanta. *Bonge Vaishnav Dharma: Ekti Oitihasik Ebom Samajtattwik Odhyayon*. Calcutta: Ananda Publishers Pvt. Ltd., 1996.

Chakraborty, Sudhir. *Bangla Dehotottwer Gan*. Calcutta: Pustak Bipani, 1990.

Chakrapanidatta. *Charakchaturanan Srimach Chakrapanidatta Pranita Chikitshasamgraha.* Edited by Pyarimohan Sengupta. Calcutta: Bidyaratna Jantra, 1888 (1295 BE).

Chakravarti, Chintaharan. *Tantras: Studies on their Religion and Literature*. Calcutta: Punthi Pustak, 1963.

Chang, Hasok. *Inventing Temperature: Measurement and Scientific Progress*. New York: Oxford University Press, 2004.

Chatterjee, Kumkum. "King of Controversy: History and Nation-making in Late Colonial India." *American Historical Review* 110:5 (2005): 1454–75.

Chatterjee, Partha. *Our Modernity*. Rotterdam/Dakar: SEPHIS CODESRIA, 1997.

———. *The Nation and Its Fragments: Colonial and Postcolonial Histories*. Princeton, NJ: Princeton University Press, 1993.

Chattopadhyay, Ashwinikumar. *Pachon o Tahar Byabohar Siksha*. Calcutta: A. K. Chatterjee, 1927.

Chattopadhyay, Kishorimohan. *Swapnatattwa*. Calcutta: Metcalf Press, n.d.

Chattopadhyay, Sitalchandra. "Ayurveder Koshay Mahatmyo." *Ayurveda* 1:9 (1916–17 [1323 BE]): 406–15.

Chen, Mel Y. *Animacies: Biopolitics, Racial Mattering and Queer Affect*. Durham, NC: Duke University Press, 2012.

Chilson, Clark, and Peter Knecht, eds. *Shamans in Asia*. London: Routledge/Curzon, 2003.

Choubey, Khagendranath. "'Ambashtanang' Chikitsitam." *Baidya Hitoishi* 4:6 (1928 [1335 BE]): 192–95.

Clarke, Bruce. *Energy Forms: Allegory and Science in the Era of Classical Thermodynamics*. Ann Arbor: University of Michigan Press, 2001.

———. *Energy Forms: Allegory and Science in the Era of Classical Thermodynamics*. Ann Arbor: University of Michigan Press, 2001.

Cleetus, Burton. "Indigenous traditions and Practices in Medicine and the Impact of Colonialism in Kerala, 1900–1950." PhD diss. Jawaharlal Nehru University, 2007.

Cohen, Lawrence. "The Epistemic Carnival: Meditations on Disciplinary Intentionality and Ayurveda." In *Knowledge and the Scholarly Medical Tradition*, edited by Don Bates. Cambridge: Cambridge University Press, 1995, 320–43.

———. "Whodunit?" *Configurations* 2:2 (1994): 343–47.

———. *No Aging in India: Alzheimer's, the Bad Family, and Other Modern Things*. Berkeley: University of California Press, 1998.

Cook, Harold J. "Conveying Chinese Medicine to Seventeenth Century Europe." In *Science between Europe and Asia: Historical Studies on the Transmission, Adoption and Adaptation of Knowledge*, edited by Feza Gunergun and Dhruv Raina, 209–33. Dordrecht, Netherlands: Springer, 2011.

Cowan, Ruth Schwartz. *More Work for Mother*. New York: Basic Books, 1983.

Crenner, Christopher. "Race and Laboratory Norms: The Critical Insights of Julian Herman Lewis (1891–1989)." *Isis* 105:3 (2014): 477–507.

Cullingworth, Charles J. *The Nurse's Companion: A Manual of General and Monthly Nursing*. London: J & A Churchill, 1876.

Daechsel, Markus. "The Civilizational Obsessions of Ghulam Jilani Barq." In *Colonialism as Civilizing Mission: Cultural Ideology in British India*, edited by Harald Fischer-Tine and Michael Mann, 270–90. London: Anthem, 2004.

———. *The Politics of Self-Expression: The Urdu Middle-class Milieu in Mid-Twentieth Century India and Pakistan*. London: Routledge, 2006.

Daniel, E. Valentine. "The Pulse as an Icon in Siddha Medicine." *Contributions to Asian Studies* 18 (1984): 115–26.

Das, Lochan. *Chaitanyamangal*. Rangpur: Radharaman Jantra, 1892 (1299 BE).

Dasgupta, Amulyakumar ("Sambuddha"). "Torunayon." In *Dialektiks*, 33–64. Calcutta: Ranjan Publishing House, 1946 (1353BE).

Dasgupta, Krishnapriya. *Botoler Nanaprokar Koutuk ba Purono Kolkatar Dokandari ar Ponyo Pasarer Bigyapon: Oushudh-Ator-Toiladi*. Calcutta: Gangchil, 2012.

Dasgupta, Shashibhusan. *Obscure Religious Cults*. Calcutta: University of Calcutta Press, 1946.

Dash, Bhagwan, and Lalitesh Kashyap. *Basic Principles of Ayurveda*. New Delhi: Concept Publishing Company, 2003.

Datta, Amaresh. *Encyclopaedia of Indian Literature*. Vol. 1. New Delhi: Sahitya Akademi, 1987.

Datta, Atulchandra. *Gupto Mantra*. Calcutta, 1907 (1314 BE).

Datta, Dwarkanath. *Chikitsa Chakrasar*. Calcutta: Jyotishprakash Jantra, 1882 (1289 BE).

Deleuze, Gilles and Felix Guattari. *A Thousand Plateaus*. Translated by Brian Massumi. Minnesota: University of Minnesota Press, 1987.

Desjarlais, Robert. *Body and Emotions: The Aesthetics of Illness and Healing in the Nepal Himalayas*. Philadelphia: University of Pennsylvania Press, 1992.

Dey, Nabinchandra. "Drobyo Porichoy." *Ayurveda Bikash* 2:2 (1914–15 [1321 BE]): 61–62.

Dey, Rev. Lal Behari. *Govinda Samanta*. London: Macmillan, 1874.

Dikotter, Frank. *Things Modern: Material Culture and Everyday Life in China*. London: C. Hurst, 2006.

Dodson, Michael S. *Orientalism, Empire and National Culture: India, 1770–1880*. Basingstoke, UK: Palgrave Macmillan, 2007.

Donner, Henrike, ed. *Being Middle Class in India: A Way of Life*. Abingdon, UK: Routledge, 2011.

Dowse, Thomas Stretch. *The Brain and the Nerves*. London: Bailliere, Tyndall & Cox, 1884.

Dumit, Joseph. *Picturing Personhood: Brain Scans and Biomedical Identity*. Princeton, NJ: Princeton University Press, 2003.

Dutt, Udoy Chand. Preface to first edition. In *Nidana: A Sanskrit System of Pathology*. Calcutta: Ayurveda Yantra, 1880 (2nd ed.).

Dutta, Udaychand. *Nidan*. Calcutta: Ayurveda Yantra, 1880.

Edgerton, David. "Creole Technologies and Global Histories: Rethinking How Things Travel in Time and Space." *Journal of History of Science and Technology* 1 (2007): 75–112.

Edwards, Elizabeth, ed. *Anthropology and Photography, 1860–1920*. New Haven, CT: Yale University Press, 1994.

Elshakry, Marwa. *Reading Darwin in Arabic, 1860–1950*. Chicago: University of Chicago Press, 2013.

———. "When Science Became Western?: Historiographical Reflections." *Isis* 101:1 (2010): 146–58.

Ernst, Waltraud. *Plural Medicine: Tradition and Modernity, 1800–2000*. London: Routledge, 2002.

Faizullah, Munshi. *Goroksho Bijoy*. Edited by Abdul Karim. Calcutta: Bangiya Sahitya Parishat, 1917 (1324 BE).

Farquhar, Judith. *Appetites: Food and Sex in Post-Socialist China*. Durham, NC: Duke University Press, 2002.

Fazl-i-Allami, Abul. *Ain-i-Akbari*. Vol. 3. Translated by Colonel H. S. Jarrett. Calcutta: Baptist Mission Press, 1894.

Feierman, Steven. "Explanation and Uncertainty in the Medical World of the Ghaambo." *Bulletin of the History of Medicine* 74:2 (2000): 317–44.

———. *Peasant Intellectuals: Anthropology and History in Tanzania*. Madison: University of Wisconsin Press, 1990.

Feldhay, Rivka. "Religion." In *The Cambridge History of Science.* Volume 3, *Early Modern Science*, edited by Katharine Park and Lorraine Daston, 727–55. Cambridge: Cambridge University Press, 2006.

File Medl. 10–44, Proceedings 64–66, Local Self-Government Dept., Medical Branch Proceedings, July 1922. West Bengal State Archives.

———. 3M-11, Proceedings B 307 & 308, Local Self-Government Dept., Medical Branch Proceedings, February 1928. West Bengal State Archives.

Fisher, J. D., M. Rytting, and R. Heslin. "Hands Touching Hands: Affective and Evaluative Effects of an Interpersonal Touch." *Sociometry* 39:4 (1976): 416–21.

Fischer-Tine, Harald, and Michael Mann, eds. *Colonialism as Civilizing Mission: Cultural Ideology in British India*. London: Anthem, 2004.

Fleet. "The Ancient Indian Water-clock." *Journal of the Royal Asiatic Society of Great Britain and Ireland* 47:2 (1915): 213–30.

Flood, Gavin. *The Tantric Body: The Secret Tradition of Hindu Religion*. London: I. B. Tauris, 2006.

Fontana, Alessandro and Mauro Bertani. "Situating the Lectures." In Michel Foucault. *"Society Must Be Defended": Lectures at the College de France, 1975–76*. Edited by Alessandro Fontana and Mauro Bertani, and translated by David Macey. New York: Picador, 1997.

Frawley, David. *Ayurveda, Nature's Medicine*. Delhi: Motilal Banarsidass, 2004.

FRP. "Reviews." *American Journal of the Medical Sciences* 125:2 (1903): 342.

Frumer, Yulia. "Translating Time: Habits of Western-Style Timekeeping in Late Edo Japan." *Technology & Culture* 55:4 (2014): 785–820.

Furth, Charlotte. "Androgynous Males and Deficient Females: Biology and Gender Boundaries in Sixteenth- and Seventeenth-Century China." *Late Imperial China* 9:2 (1988): 1–31.

Gait, E. A. Census of India, 1901. Volume 6, The Lower Provinces of Bengal and Their Feudatories: Part 1. Calcutta: Bengal Secretariat Press, 1902.

———. *Report of The Census of India*. Vol. 1, part 1. Calcutta: Government Printing Press, 1903.

Gandhi, M. K. "Ayurvedic System." In *The Collected Works of Mahatma Gandhi*, vol. 31, 460–62. Delhi: Publications Division, Ministry of Information and Broadcasting, Government of India, 1969.

———. "Speech at Astanga Ayurveda Vidyalaya." In *The Collected Works of Mahatma Gandhi*, vol. 31, 280–83, Delhi: Publications Division, Ministry of Information and Broadcasting, Government of India, 1969.

Ganguly, T. N. *Svarnalata Or Scenes from Hindu Village Life in Bengal*. Translated by D. C. Roy. Calcutta: Sanyal, 1906.

Gaudilliere, Jean-Paul. "An Indian Path to Biocapital? The Traditional Knowledge Digital Library, Drug Patents, and the Reformulation Regime of Contemporary Ayurveda." *East Asian Science, Technology and Society: An International Journal* 8:4 (2014): 391–415.

———. "Biologics in the Colonies: Emile Perrot, Kola Nuts and the Industrial Reordering of Pharmacy." In *Biologics: A History of Agents Made from Living Organisms in the Twentieth Century*, edited by Alexander von Schwerin, Heiko Stoff, and Bettina Wahrig, 47–64. London: Pickering & Chatto, 2003.

———. "The Visible Industrialist: Standards and the Manufacture of Sex Hormones." In *Evaluating and Standardizing Therapeutic Agents, 1890–1950*, edited by Jonathan Simon and Christoph Gradmann, 174–201. Basingstoke, UK: Palgrave Macmillan, 2010.

Ghosh, Mrinalkanti. *Poroloker Kotha*. Calcutta: n.p., 1933.

Gilmartin, David, and Bruce Lawrence, eds. *Beyond Turk and Hindu: Rethinking Identities in Islamicate South Asia*. Gainesville: University Press Florida, 2000.

Gobindadas. *Bhaisajyaratnabali*. Translated by Binodlal Sen. Calcutta: Adi Ayurveda Machine Press, 1929 [1876]).

Golinski, Jan. "Is It Time to Forget Science? Reflections on a Singular Science and Its History." *Osiris* 27:1 (2012): 19–36.

Gordin, Michael. "How Lysenkoism Became Pseudoscience: Dobzhansky to Velikovsky." *Journal of the History of Biology* 45:1 (2012): 443–68.

Goswami, Bhagabat Kumar. Foreword. In *Ayurveda Ratnakarah*, Jogendranath Darshanshastri, unnumbered pages. Calcutta: Jyotirindra Bhattacharya, 1936 (1343 BE).

Goswami, Manu. "From Swadeshi to Swaraj: Nation, Economy and Territory in Colonial South Asia." *Comparative Studies in Society and History* 40:4 (1998): 609–36.

Goswami, Surendranath. *Ayurved o Malaria-jwor*. Calcutta: Janmabhumi Press, 1914 (1321 BE?).

———. "Bhut-bigyan." *Janmabhumi* 18:1 (1910 [1317 BE]): 2–7.

———. "Bhut Bigyan." *Janmabhumi* 18:3 (1911): 124–27.

———. "Vayur Boigyanik Byakhya." *Janmabhumi* 24:1–4 (1916 [1323 BE]): 22–26.

———. "Vayur Boigyanik Byakhya." *Janmabhumi* 24:2 (1916 [1323 BE]): 48–53.

Green, T. Henry. *Pathology and Morbid Anatomy*. Philadelphia: Lea Brothers, 1900.

Guenther, Katja. *A Body Made of Nerves: Reflexes, Body Maps and the Limits of the Self in Modern German Medicine*. PhD diss., Harvard University Press, 2009.

Guha, Ranajit. *Dominance without Hegemony: History and Power in Colonial India*. Cambridge, MA: Harvard University Press, 1997.

Gunergun, Feza and Dhruv Raina, eds. *Science between Europe and Asia: Historical Studies on the Transmission, Adoption and Adaptation of Knowledge*. Dordrecht, Netherlands: Springer, 2011.

Gupta Prakashak. *Bongiyo Bishwasyo Mantraboli*. Calcutta: Prakrita Jantralaya, 1863 (?).

Gupta, Amritalal "Agni," *Ayurveda* 1:2 (1916 [1323 BE]): 85–88

———. "Baidya Britti." *Ayurveda* 1:12 (1916–17 [1323 BE]): 539–42.

———. *Nadigyan Rohosyo*. Calcutta: Bandemataram Oushodhaloy, 1916 (1323 BE).

———. "Tej o Pitta." *Ayurveda Bikash* 3:1 (1915–16 [1322 BE]): 12–15.

———. "Tej O Pitta." *Ayurveda Bikash* 3:2/3 (1915–16 [1322 BE]): 33–38.

Gupta, Atulbihari. *Mrityur Pare o Punorjonmobad*. Benares: Sailendranath Gupta, 1941 (1348 BE): 117–24.

Gupta, Brahmananda. "Indigenous Medicine in Nineteenth and Twentieth Century Bengal." In *Asian Medical Systems: A Comparative Study*, edited by Charles Leslie, 368–82. Berkeley: University of California Press, 1976.

———. *Sutanati Parishad Smarak Grantha*. n.p., n.d.

Gupta, Charu. *Sexuality, Obscenity, Community: Women, Muslims and the Hindu Public in Colonial India*. New York: Palgrave, 2002.

Gupta, Haralal. *Nadi-siksha*. Calcutta: Rajendranath Sengupta, 1911 (1318 BE).

Gupta, Partha Sarathi. *Radio and the Raj, 1921–47*. Calcutta: K. P. Bagchi, 1995.

Gupta, Rasiklal. *Moharaj Rajballabh Sen o Totsomokalborti Banglar Itihaser Sthul Sthul Biboron*. Calcutta: Sakhi Press, n.d.

Gupta, Russick Lal. *Hindu Anatomy, Physiology, Therapeutics, History of Medicine and Practice of Physic*. Calcutta: S. C. Addy, 1892.

———. *Science of Sphygmica or Sage Kanad on Pulse*. Calcutta: S. C. Addy, 1891.

Gupta, S. V. *Units of Measurement: Past, Present and Future. International System of Units*. Dordrecht: Springer, 2010.

Gupta, Umeshchandra. *Jati-tottwo Baridhi*. Calcutta: Majumdar Library, 1902.

Guy, William A. "On the Effects Produced on the Pulse by Change of Posture." *Eclectic Journal of Medicine* 3:11 (1839): 403–6.

Guy, William A., and John Harley. *Hooper's Vade Mecum*. London: Henry Renshaw, 1869.

Haag, Pascale. "I Wanna Be a Brahmin Too: Grammar, Traditions and Mythology as Means for Social Legitimisation amongst the Vaidyas in Bengal." In *Samskrta-sadhuta 'Goodness of Sanskrit': Studies in Honour of Professor Ashok N. Aklujkar*, edited by Chikafumi Watanabe, Michele Marie Desmarais, and Yoshichika Honda, 226–49. New Delhi: D. K. Printworld, 2012.

Haeckel, Ernst. *Gesammelte populare Vortrage aus dem Gebiete der Entwicklungslehre*. Bonn, Germany: Emil Strauss, 1878.

Halliburton, Murphy. "Resistance or Inaction? Protecting Ayurvedic Medical Knowledge and Problems of Agency." *American Ethnologist* 38:1 (2011): 86–101.

———. "Rethinking Anthropological Studies of the Body: Manas and Bodham in Kerala." *American Anthropologist* 104:4 (2002): 1123–34.

Hamlin, Christopher. *More than Hot: A Short History of Fever*. Baltimore, MD: Johns Hopkins University Press, 2014.

Hanson, Marta. "Hand Mnemonics in Classical Chinese Medicine: Texts, Earliest Images and Arts of Memory." *Asia Major* 21:1 (2008): 325–57.

———. *Speaking of Epidemics in Chinese Medicine: Disease and the Geographic Imagination in Late Imperial China*. Abingdon, UK: Routledge, 2011.

Harder, Hans. *Sufism and Saint Veneration in Contemporary Bangladesh: The Maijbhandaris of Chittagong*. Abingdon, UK: Routledge, 2011.

Hardiman, David. "Indian Medical Indigeneity: From Nationalist Assertion to the Global Market." *Social History* 34:3 (2009): 263–83.

Hardiman, David, and Projit Bihari Mukharji, eds. *Medical Marginality in South Asia: Situating Subaltern Therapeutics*. Abingdon, UK: Routledge, 2012.

Harding, Sandra. *Sciences from Below: Feminisms, Postcolonialities, and Modernities*. Durham, NC: Duke University Press, 2008.

Harrison, Mark. "'The Tender Frame of Man': Disease, Climate and Racial Difference in India and the West Indies, 1760–1860." *Bulletin of the History of Medicine* 70:1 (1996): 68–93.

Hashimoto, Takehiko. "The Adoption and Adaptation of Mechanical Clocks in Japan." In *Science between Europe and Asia: Historical Studies on the Transmission, Adoption and Adaptation of Knowledge*, edited by Feza Gunergun and Dhruv Raina, 137–50. Dordrecht, Netherlands: Springer, 2011.

Hatley, Shaman. "Mapping the Esoteric Body in the Islamic Yoga of Bengal." *History of Religions* 46 (2007): 351–68.

Hausman, Gary J. "Siddhars, Alchemy and the Abyss of Tradition: 'Traditional' Tamil Medical Knowledge in 'Modern' Practice." PhD diss., University of Michigan, 2006.

Haynes, Douglas E., Abigail McGowan, Tirthankar Roy, and Haruki Yanagisawa, eds. *Towards a History of Consumption in South Asia*. New Delhi: Oxford University Press, 2009.

———. "Making the Ideal Home? Advertising of Electrical Appliances and the Education of the Middle Class Consumer in Bombay, 1925–40." Unpublished Paper.

———. "Masculinity, Advertising and the Reproduction of the Middle Class Family in Western India, 1918–1940." In *Being Middle Class in India: A Way of Life*, edited by Henrike Donner, 23–46. Abingdon, UK: Routledge, 2011.

———. "Selling Masculinity: Advertisements for Sex Tonics and the Making of Modern Conjugality in Western India, 1900–1945." *South Asia: Journal of South Asian Studies* 35:4 (2012): 787–831.

Heidegger, Martin. "The Question Concerning Technology." In *The Question Concerning Technology and Other Essays*. Translated by William Lovitt. New York: Garland, 1977.

Hess, Volker. "Standardizing Body Temperature: Quantification in Hospitals and Daily Life, 1850–1900." In *Body Counts: Medical Quantification in Historical and Sociological Perspective*, edited by Gerard Jorland, A. Opinel, and G. Weisz, 109–26. Montreal: McGill-Queen's University Press, 2005.

Hinojosa, Servando Z. "Bonesetting and Radiography in the Southern Maya Highlands." *Medical Anthropology* 23:4 (2004): 263–93.

Howell, Joel D. *Technology in the Hospital: Transforming Patient Care in the Early Twentieth Century*. Baltimore, MD: Johns Hopkins University Press, 1996.

Howell, William H. *A Textbook of Physiology for Medical Students and Physicians*. Philadelphia, 1907, 576–77.

Hsu, Elisabeth. *Pulse Diagnosis in Early Chinese Medicine: The Telling Touch*. Cambridge: Cambridge University Press, 2010.

Hughes, T. P. *Notes on Muhammadanism*. London: William H. Allen, 1875.

Hunt, Lynn. "Modernity: Are Modern Times Different?" *Historia Critica* 54 (2014): 107–24.

Hunt, Nancy Rose. *A Colonial Lexicon: Of Birth Ritual, Medicalization, and Mobility in the Congo*. Durham, NC: Duke University Press, 1999.

Ikuko, Nishimoto. "The 'Civilization' of Time: Japan and the Adoption of the Western Time System." *Time & Society* 6:2/3 (1997): 237–59.

Inden, Ronald B., and Ralph W Nicholas. *Kinship in Bengali Culture*. Chicago: Chicago University Press, 1977.

Inkster, Ian. "The West Had Science and the Rest Had Not? Queries of a Mindful Hand." *History of Technology* 29 (2009): 205–11.

Iyer, T. G. Ramamurthi. "A Scientific Study of Septic Cases Treated in Indigenous System." *Journal of Ayurveda* 7:4 (1930): 129–39.

Jaffrelot, Christophe, and Peter van der Veer, eds. *Patterns of Middle Class Consumption in India and China*. New Delhi: Sage, 2008.

James, Frank A. J. L. "Thermodynamics and the Sources of Solar Heat, 1846–1862." *British Journal of the History of Science* 15:2 (1982): 155–81.

Jarry, Alfred. *Exploits and Opinions of Dr Faustroll Pataphysician: A Neoscientific Novel*. Translated by Simon Watson Taylor. Boston: Exact Change, 1996.

Jorland, Gerard, A. Opinel, and G. Weisz eds. *Body Counts: Medical Quantification in Historical and Sociological Perspective*. Montreal: McGill-Queen's University Press, 2005.

Joshi, Prayagchandra. *Shrikanadmaharishipranit Nadivigyanam*. Varanasi: Chowkhamba Sanskrit Series Office, 1959.

Joshi, Sanjay, ed. *Middle Class in Colonial India*. New Delhi: Oxford University Press, 2010.

Juneja, Monica. "Objects, Frames, Practices: A Post Script on Agency and Braided Histories of Art." *Medieval History Journal* 15:2 (2012): 415–23.

Kabyatirtha, Ramsahay. "Arya Rishira Jibanutottwo Janiten Ki Na?" *Ayurveda* 2:7 (1917 [1324 BE]): 281–85.

Kalpagam, U. "Temporalities, Histories and Routines of Colonial Rule in India." *Time & Society* 8:1 (1999): 141–59.

Kapila, Shruti. "Race Matters: Orientalism and Religion, India and Beyond, c. 1770–1880." *Modern Asian Studies* 41:3 (2007): 471–513.

Katzir, Shaul. "On 'the Electromagnetic World-View': A Comment on an Article by Suman Seth." *Historical Studies in the Physical and Biological Sciences* 36:1 (2005): 189–92.

Kaviraj, Sudipta. "Modernity and Politics in India." *Daedalus* 129:1 (2000): 137–62.

Keswani, N. H. "Medical Education in India Since Ancient Times." In *History of Medical Education: A Symposium*, edited by Charles Donald O'Malley, 329–66. Berkeley: University of California Press, 1970.

Kinsley, David. *Tantric Visions of the Divine Feminine: The Ten Mahavidyas*. Berkeley: University of California Press, 1997.

Kobiraj, Gopinath. *Aspects of Indian Thought*. Burdwan: University of Burdwan, 1966.

———. *Notes on Religion and Philosophy*. Varanasi: Sampurnanand Sanskrit University, 1987.

Kobiraj, Krishnadas. *Chaitanya-charitamrita*. Edited by Sukumar Sen. New Delhi: Sahitya Akademi, 1936.

Kobirotno, Abinashchandra. "Jadihasti tadanyatra jannastih na tatkkachit." *Chikitsa Sammilani* 9:1 (1893): 30.

Kohler, Robert E. *From Medical Chemistry to Biochemistry: The Making of a Biomedical Discipline*. New York: Cambridge University Press, 1982.

———. "A Generalist's View." *Isis* 96:2 (2005): 224–29.

Kuriyama, Shighehisa. *The Expressiveness of the Body and the Divergence of Greek and Chinese Medicine*. New York: Zone, 1999.

Lahiri, Kshitishchandra. "Prachin Chikitsoker Totka o Mushtijog." *Ayurveda* 4:3 (1919–20 [1326 BE]): 141–42.

Lahiri Choudhury, Deep Kanta. *Telegraphic Imperialism: Crisis and Panic in the Indian Empire, c. 1830–1920*. Basingstoke, UK: Macmillan, 2010.

Lambert, Helen. "The Cultural Logic of Indian Medicine: Prognosis and Aetiology in Rajasthani Popular Therapeutics." *Social Science & Medicine* 34:10 (1992): 1069–76.

Langford, Jean M. *Fluent Bodies: Ayurvedic Remedies for Postcolonial Imbalance*. Durham, NC: Duke University Press, 2002.

Langwick, Stacey A. *Bodies, Politics and African Healing: The Matter of Maladies in Tanzania*. Bloomington: Indiana University Press, 2011.

Larson, B., and E. Brauer eds. *The Art of Evolution: Darwin, Darwinisms and Visual Culture*. Hanover, NH: University Press of New England, 2009.

Latour, Bruno. *Science in Action: How to Follow Scientists and Engineers through Society*. Cambridge, MA: Harvard University Press, 1987.

Laudan, Laurens. "The Clock Metaphor and Probabilism: The Impact of Descartes on English Methodological Thought, 1650–65." *Annals of Science* 22:2 (1966): 73–104.

Lawrie, Edward. "On the Temperature in Health." *Indian Medical Gazette* 7:10 (1873): 253–56.

Leach, Edmund, and S. N. Mukherjee, eds. *Elites in South Asia*. Cambridge: Cambridge University Press, 1970.

Lei, Sean Hsian-lin. *Neither Donkey Nor Horse: Medicine and the Struggle over China's Modernity*. Chicago: Chicago University Press, 2014.

Leichty, Mark. *Suitably Modern: Making Middle Class Culture in a New Consumer Society*. Princeton, NJ: Princeton University Press, 2003.

Leslie, Charles. "The Ambiguities of Medical Revivalism in Modern India." In *Asian Medical Systems: A Comparative Study*, edited by Charles Leslie, 356–67. Berkeley: University of California Press, 1976.

———. *Asian Medical Systems: A Comparative Study*. Berkeley: University of California Press, 1976.

———. "Interpretations of Illness: Syncretism in Modern Ayurveda." In *Paths to Asian Medical Knowledge*, edited by Charles Leslie and Allan Young, 177–208. Berkeley: University California Press.

Letter from J. Donald, Secretary to the Government of Bengal. December 18, 1916. Financial Dept., Medical No. 3340 Med. File 3M-5/19. West Bengal State Archives.

Liebeskind, Claudia. "Arguing Science: Unani Tibb, Hakims and Biomedicine in India." In *Plural Medicine: Tradition and Modernity, 1800–2000*, edited by Waltraud Ernst, 58–75. London: Routledge, 2002.

Livingston, Julie. "Productive Misunderstandings and the Dynamism of Plural Medicine in Mid-Century Bechuanaland." *Journal of Southern African Studies* 33:4 (2007): 801–10.

Local Self Govt. Dept., Proceedings, 7, 1921, File Med. 2R-28(1). West Bengal State Archives.

Local Self Govt. Dept., Proceedings, 16–21, 1921, File: Med. 2R-12 (1). West Bengal State Archives.

Local Self Govt. Dept., Proceedings, 16–21, 1921, File: Med. 2R-12 (4). West Bengal State Archives.

Local Self Govt. Dept., Proceedings, 215, 1922, File: Med. 2R-17. West Bengal State Archives.

Lock, Margaret and Vinh-Kim Nguyen. *An Anthropology of Biomedicine*. Chichester, UK: Wiley, 2010.

Loh, Jogendrakishore. "Chikitsoker Kortobyo." *Ayurveda* 2:12 (1917–18 [1324 BE]): 492–96.

Long, James. *A Descriptive Catalogue of Bengali Works*. Calcutta: Sanders, Cones and Co., 1855.

Lopez Jr., Donald S, ed. *Religions of India in Practice*. Princeton, NJ: Princeton University Press, 1995.

Lorenzen, David, and Adrian Muñoz. *Yogi Heroes and Poets: Histories and Legends of the Naths*. Delhi: Oxford University Press, 2011.

Lovitt, William, trans. *The Question Concerning Technology and Other Essays*. New York: Garland, 1977.

Magee, Karl. "Graves, Robert James (1796–1853)." *Oxford Dictionary of National Biography*. Oxford University Press, 2004. Accessed on May 28, 2015. http://dx.doi.org/10.1093/ref:odnb/11317.

Maguire, H. F. J. T. *Report on the Census of Calcutta taken on the 26th February 1891*. Calcutta: Bengal Secretariat Press, 1891.

Maitra, Prasannakumar. "Dhatu-byakhya." *Chikitsa Sammilani* 6 (1889 [1296 BE]): 161–64.

Majumdar, Haramohan. *Ayurveda Gurhotottwo Prokash*. Calcutta: Bangabasi Ltd.., n.d.

———. "Sharir Vayu." *Ayurveda* 1:9 (1917 [1324 BE]): 399–406.

Mandal, Panchanan. *Chithi Potre Somaj Chitro*. Vol. 2. Santiniketan: Vishwabharati, 1952 (1359 BE).

———. *Punthi Porichoy*. Vol. 3. Santiniketan, India: Vishwabharati, 1963.

Martin, Emily. "Towards an Anthropology of Immunology: The Body as Nation State." *Medical Anthropology Quarterly* 4:4 (1990): 410–26.

Matilal, Bimal Krishna. "Causality in the Nyaya-Vaisesika School." *Philosophy East and West* 25:1 (1975): 41–48.

Maurice, Graham. "Round the World—And Some Gas Works." *Journal of Gas Lighting, Water Supply Etc.* 103 (1908): 26–29.

McDaniel, June. *Offering Flowers, Feeding Skulls: Popular Goddess Worship in Bengal*. Oxford: Oxford University Press, 2004.

McDermott, Rachel Fell. *Singing to the Goddess: Poems to Kali and Uma from Bengal*. Oxford: Oxford University Press, 2001.

McHugh, JA. "Blattes de Byzance in India: Mollusk Opercula and the History of Perfumery." *Journal of the Royal Asiatic Society* 23:1 (2013): 53–67.

McLeod, Kenneth. "Summary of Meteorological Observations Taken at the Office of the Civil Assistant Surgeon of Jessore for the Month of November 1867," *Indian Medical Gazette*, 1 January 1868, 7.

McMahon, Darrin and Sam Moyn, eds. *Rethinking Modern European Intellectual History*. New York: Oxford University Press, 2014.

Mehta, Uday Singh. *Liberalism and Empire: A Study in Nineteenth-Century British Liberal Thought*. Chicago: University of Chicago Press, 1999.

Meister, Michael. "Regional Variations in Matrika Conventions." *Artibus Asia* 47:3–4 (1986): 233–62.

Mendelsohn, J. Andrew. "Lives of the Cell." *Journal of the History of Biology* 36:1 (2003): 1–37.

Meulenbeld, G. J. *A History of Indian Medical Literature*. Vols. 1–3. Groningen: Egbert Forsten, 1999.

Meulenbeld, G. J. *History of Indian Medical Literature*. Vol. 2B. Groningen: Egbert Forsten, 2000.

Meulenbeld, Gerrit Jan, trans. "Introduction." In *The Madhavanidana and Its Chief Commentaries: Chapters 1—10*, 1–28. Leiden: Brill, 1974.

Mignolo, Walter. "The Many Faces of Cosmo-Polis: Border Thinking and Critical Cosmopolitanism." *Public Culture* 12:3 (2000): 721–48.

Mines, Diane P. *Caste in India*. Key Issues in Asian Studies 3, Association of Asian Studies, 2009.

Minsky, Lauren Nauta. "Pursuing Protection from Disease: The Making of Smallpox Prophylactic Practice in Colonial Punjab." *Bulletin of the History of Medicine* 83:1 (2009): 163–89.

Mishra, Jayakanta. "Kaviraj Ji as I knew Him." In *Gopinath Kaviraja Commemoration Volume*, iii–vii, Allahabad: Ganganatha Jha Kendriya Sanskrit Vidyapeetha, 1976.

Mitchell, Silas Weir. *Doctor and Patient*. Philadelphia: J. B. Lippincott, 1888.

Mitra, Rajendralal. *Notices of Sanskrit MSS*. Vol. 1. Calcutta: Baptist Mission Press, 1871.

Mol, Annemarie. *The Body Multiple: Ontology in Medical Practice*. Durham, NC: Duke University Press, 2002.

Mooney, Graham. "The Material Consumptive: Domesticating the Tuberculosis Patient in Edwardian England." *Journal of Historical Geography* 42 (2013): 152–66.

Morus, Iwan Rhys. *Shocking Bodies: Life, Death and Electricity in Victorian England*. Stroud: History Press, 2011.

Mottin, Jean. "A Hmong Shaman's Séance." In *Shamans in Asia*, edited by Clark Chilson and Peter Knecht, 86–96. London: Routledge/Curzon, 2003.

Mouat, F. J. "Hindu Medicine." In *Calcutta Review*, vol. 8, 379–433. Calcutta: Sanders, Cones and Co., 1847.

Mouat. T. "Epidemic Diseases Which Occurred at Bangalore during the Year 1833." *Transactions of the Medical and Physical Society of Calcutta: Volume the Seventh*. Calcutta: Baptist Mission Press, 1835, 282–343.

Mukerjee, Girindra Nath. "Human Parasites in the Atharva Vedas." *Journal of Ayurveda* 4:5 (1927): 165–86.

Mukharji, Projit Bihari. "Cultures of Fear: Technonationalism and the Postcolonial Responsibilities of STS." *East Asian Science, Technology and Society* 6 (2012): 267–74.

———. "In-Disciplining Jwarasur: The Folk/Classical Divide and the Transmateriality of Fevers in Colonial Bengal." *Indian Economic & Social History Review* 50:3 (2013): 261–88.

———. "Lokman, Chholeman and Manik Pir: Multiple Frames for Institutionalising Islamic Medicine in Modern Bengal." *Social History of Medicine* 24:3 (2011): 720–38.

———. *Nationalizing the Body: The Medical Market, Print and Daktari Medicine*. London: Anthem, 2009.

Mukherjee, K. P. "Preface." In *Science of Sphygmica or Sage Kanad On Pulse*, edited by Russick Lall Gupta. Calcutta: S. C. Addy, 1891.

Mukherjee, S. N. "Caste, Class and Politics in Calcutta, 1815–38." In *Elites in South Asia*, edited by Edmund Leach and S. N. Mukherjee, 33–78. Cambridge: Cambridge University Press, 1970.

Mukherjee, Umacharan. *Tailanga Swamir Jibonchorit o Tottwopodesh*. Calcutta: Jogendranath Mukhopadhyay, 1918 (1325 BE).

Mukhopadhyaya, Girindranath. *Ancient Hindu Surgery: Surgical Instruments of the Hindus: With a Comparative Study of the Surgical Instruments of the Greek, Roman, Arab, and the Modern European Surgeons*. New Delhi: Cosmo, 1994.

Mukhopadhyay, Girindranath. *History of Indian Medicine*. Vol. 2. New Delhi: Oriental Books Reprint Corp., 1974.

Mukhopadhyaya, Girindranath. *History of Indian Medicine: Containing Notices, Biographical, of the Ayurvedic Physicians and their Works on Medicine, from the Earliest Ages to the Present Time*. New Delhi: Munshiram Manoharlal Publishers Private Limited, 1994.

Mullick, Kalyani. *Nath Somprodayer Itihas, Darshan o Sadhonpronali*. Calcutta: Calcutta University Press, 1946.

Mullick, Kumudnath. *Nadia Kahini*. Ranaghat: n.p., 1910.

Mullick, Ramchandra. *Bishwo Chikitsa*. Calcutta: Ramayana Press, 1889 (1296 BE).

Murray, John C. "On the Nervous, Bilious, Lymphatic and Sanguine Temperaments: Their Connection with Races in England, and their Relative Longevity." *Anthropological Review* 8 (1870): 14–28.

Nanda, Meera. *The God Market: How Globalization is Making India More Hindu*. New York: New York University Press, 2011.

———. *Prophets Facing Backwards: Postmodern Critiques of Science and Hindu Nationalism in India*. New Brunswick, NJ: Rutgers University Press, 2003.

Nandy, Ashis, and Shiv Visvanathan. "Modern Medicine and Its Non-Modern Critics: A Study in Discourse." In *Dominating Knowledge: Development, Culture and Resistance*, edited by Frederique Apffel Marglin and Stephen A. Marglin, 145–84. Oxford: Clarendon, 1990.

Naraindas, Harish. "Care, Welfare and Treason: The Advent of Vaccination in the 19th Century." *Contributions to Indian Sociology* 32:1 (1998): 67–96.

———. "Of Relics, Body Parts and Laser Beams: The German Heilpraktiker and his Ayurvedic Spa." *Anthropology & Medicine* 18:1 (2011): 67–86.

———. "Of Spineless Babies and Folic Acid: Evidence and Efficacy in Biomedicine and Ayurvedic Medicine." *Social Science & Medicine* 62:11 (2006): 268–69.

———. "Poisons, Putrescence and the Weather: A Genealogy of the Advent of Tropical Medicine." *Contributions to Indian Sociology* 30:1 (1996): 1–35.

———. "Preparing for the Pox: A Theory of Smallpox in Bengal and Britain." *Asian Journal of Social Science* 31:2 (2003): 304–39.

Narwekar, Sanjit. *Films Division and the Indian Documentary*. New Delhi: Publications Division, Ministry of Information & Broadcasting, Government of India, 1992.

Newell, James. "Unseen Power: Aesthetic Dimensions of Symbolic Healing in *Qawwali*." *Muslim World* 97:4 (2007): 640–56.

Nicholson, Malcolm. "The Art of Diagnosis: Medicine and the Five Senses." In *Companion Encyclopaedia of the History of Medicine*, vol. 2, edited by W. F. Bynum and Roy Porter, 801–25. London: Routledge, 1993.

Obeyesekere, Gananath. "Impact of Ayurvedic Ideas on the Culture and the Individual in Sri Lanka." In *Asian Medical Systems: A Comparative Study*, edited by Charles Leslie, 201–26. Berkeley: University of California Press, 1976.

Obituary notice for Mrs. Henrietta Eliza Peters. In *Calcutta Monthly Journal and General Register*, 111. Calcutta: Samuel Smith, 1840.

O'Connell, J. T. "Chaitanya Vaishnava Devotion (bhakti) and Ethics as Socially Integrative in Sultanate Bengal." *Bangladesh e-Journal of Sociology* 8:1 (2011): 51–63.

O'Malley, Charles Donald. *History of Medical Education: A Symposium*. Berkeley: University of California Press, 1970.

Otis, Laura. *Networking: Communicating Between Bodies and Machines in the Nineteenth Century*. Ann Arbor: University of Michigan Press, 2001.

Otter, Chris. *The Victorian Eye: A Political History of Light and Vision in Britain, 1800–1910*. Chicago: University of Chicago Press, 2008.

Oudshoorn, Nelly. *Beyond the Natural Body: An Archaeology of Sex Hormones*. London: Routledge, 1994.

Packard, Francis Randolph, ed. *Annals of Medical History*. Vol. 4. New York: Paul B. Hoeber, 1922.

Pahari, Subrata. *Unish Shotoker Banglaye Sonatoni Chikitsa Byabosthar Sworup*. Calcutta: Progressive Publishers, 2001.

Panikkar, K. N. "Indigenous Medicine and Cultural Hegemony: A Study of the Revitalization Movement of Kerala." *Studies in History* 8:2 (1992): 283–308.

Parasuram. "*Chikitsa Sonkot*." In *Goddalika*, 40–67, Calcutta: M. C Sarkar & Sons, 1951 (1358 BE).

Pati, Biswamoy, and Mark Harrison, eds. *Health, Medicine, and Empire: Perspectives on Colonial India*. Hyderabad: Orient Longman, 2001.

Patwardhan, Bhushan. "Let's Plan for National Health." *Journal of Ayurveda and Integrated Medicine* 2:3 (2011): 103–4.

Pearce, J. M. S. "A Brief History of the Clinical Thermometer." *Quarterly Journal of Medicine* 95 (2002): 251–52.

Pickering, Andrew. *The Mangle of Practice: Time, Agency and Science*. Chicago: University of Chicago Press, 1995.

Pilloud, Severine, and Micheline Louis-Courvoisier. "The Intimate Experience of the Body in the Eighteenth Century: Between Interiority and Exteriority." *Medical History* 47:4 (2003): 451–72.

Pingree, David Edwin. *Census of Exact Sciences in Sanskrit, series A*. Vol. 3. Montpelier, VT: Beckman and Beckman, 1992.

———. *Census of the Exact Sciences in Sanskrit*. Vols. 1–5. Philadelphia: American Philosophical Society, 1970.

Pinkerton, Alasdair. "Radio and the Raj: Broadcasting in British India (1920–1940)." *Journal of the Royal Asiatic Society of Great Britain & Ireland* 18:2 (2008): 167–91.

Pinney, Christopher. "Creole Europe: A Reflection of a Reflection." *Journal of New Zealand Literature* 20 (2003): 125–61.

———. "The Parallel Histories of Anthropology and Photography." In *Anthropology and Photography, 1860–1920*, edited by Elizabeth Edwards, 74–95. New Haven, CT: Yale University Press, 1994.

Plofker, Kim. *Mathematics in India*. Princeton, NJ: Princeton University Press, 2009.

Pollock, Sheldon. "The Theory of Practice and the Practice of Theory in Indian Intellectual History." *Journal of the American Oriental Society* 105:3 (1985): 499–519.

Pordie, Laurent, ed. *Tibetan Medicine in the Contemporary World: Global Politics of Medical Knowledge and Practice*. London: Routledge, 2008.

Porter, Roy. *Blood and Guts: A Short History of Medicine*. New York: Norton, 2004.

Prakash, Gyan. *Another Reason: Science and the Imagination of Modern India*. Princeton, NJ: Princeton University Press, 1999.

Prakash, Gyan. "Review of David Arnold. Everyday Technology: Machines and the Making of India's Modernity." Isis 105:2 (2014): 445–46.

Pramanik, Basantakumar. "Nadigyan Lobhopay." *Chikitsa Sammilani* 12:2–3 (1907): 149–51.

Qibla, Baba Jan Sharif Sureswari. *Nur-e-Haqq, Ganj-e-Nur*. Shariatpur: Syed Nur-e-Iqbal Shah Sharif al Sureswari, 2002.

Quaiser, Ahsan Jan. *The Indian Response to European Technology & Culture, AD 1498–1707*. New Delhi: Oxford University Press, 1982.

Quaiser, Neshat. "Politics, Culture, and Colonialism: Unani's Debate with Doctory." In *Health, Medicine, and Empire: Perspectives on Colonial India*, edited by Biswamoy Pati and Mark Harrison, 317–55. Hyderabad: Orient Longman, 2001.

———. "Unani Medical Culture: Memory, Representation, and the Literate Critical Anticolonial Public Sphere." In *Contesting Colonial Authority: Medicine and Indigenous Responses in 19th and 20th Century India*, edited by Poonam Bala, 115–36. Lanham, MD: Lexington Books, 2012.

Rabinbach, Anson. *The Human Motor: Energy, Fatigue, and the Origins of Modernity*. Berkeley: University of California Press, 1992.

Raj, Kapil. "Beyond Postcolonialism . . . and Post-positivism: Circulation and the Global History of Science." *Isis* 104:2 (2013): 337–47.

———. "The Historical Anatomy of a Contact Zone: Calcutta in the Eighteenth Century." *Indian Economic and Social History Review* 48:1 (2011): 55–82.

Ramanna, Mridula. *Health Care in Bombay Presidency, 1896–1930*. Delhi: Primus, 2012.

Ray, Binodbihari. "Jwor." *Chikitsak* 1:1 (1889 [1296 BE]): 13–17.

———. *Podye Ayurveda Siksha*. Calcutta: Lila Printing Works Yantra, 1908 (1315 BE).

———. "Vayu Pitta Kapha." *Chikitsak* 1:1 (1889–1890 [1296–97 BE]): 5–9.

———. "Vayu Pitta Kapha." *Chikitsak* 1:2 (1889–90 [1296–97 BE]): 32–35.

———. "Vayu Pitta Kapha." *Chikitsak* 1:3 (1889 [1296 BE]): 52–56.

———. "Vayu Pitta Kapha." *Chikitsak* 1:4 (1889–90 [1296–97 BE]): 77–80.

Ray, Brajaballabh. "Ayurveda." *Ayurveda* 1:1 (1916 [1323 BE]): 5–6.

———. "Jwor." *Ayurveda* 1:6 (1916 [1323 BE]): 254–58.

Ray, D. N. "Modern Problems in Ayurveda." *Journal of Ayurveda* 7:4 (1930): 125–29.

Ray, Jaminibhushan. "Boidya Sommelone Sobhapotir Obhibhashon." *Ayurveda* 1:5 (1916 [1323 BE]): 189–90.

Ray, Srikanta. "Gangadhar Sen." In *Bengal Celebrities*, p. 27. Calcutta: City Book Society, 1906.

Ray, Tryambakeshwar. "Acharjyo Gangadharer Jiboni." *Ayurveda Bikash* 2:1 (1914 [1321 BE]): 19–29.

Reiser, Stanley Joel. *Medicine and the Reign of Technology*. Cambridge: Cambridge University Press, 1978.

Rele, Vasant G. *The Mysterious Kundalini: The Physical Basis of the 'Kundalini (Hatha) Yoga' in terms of Western Anatomy and Physiology*. Bombay: D. B. Taraporevala Sons, 1931.

Report of the Commission Appointed by the Govt. of Bengal to Enquire into the Excise of Country Spirit in Bengal, 1883–84. Vol. 2. Calcutta: Bengal Secretariat Press, 1884.

Richardson, Ruth. *Death, Dissection and the Destitute*. Chicago: University of Chicago Press, 1987.

Risley, H. H. *The Tribes and Castes of Bengal: Ethnographic Glossary*. Vol. 1. Calcutta: Bengal Secretariat Press, 1892.

Roberts, Lissa. "The Death of the Sensuous Chemist: The 'New' Chemistry and the Transformations of Sensuous Technology." *Studies in the History & Philosophy of Science* 26:4 (1995): 503–29.

Roberts, Lissa, and Simon Schaffer. Preface. *The Mindful Hand: Inquiry and Invention from the Late Renaissance to Early Industrialisation*, edited by Lissa Roberts, Simon Schaffer, and Peter Dear, xiii–xxvii. Amsterdam: Koninklijke Nederlandse Akademie van Wetenschappen, 2007.

Rosenberg, Charles. *Explaining Epidemics and Other Studies in the History of Medicine*. Cambridge: Cambridge University Press, 1992.

Rosenbloom, Jacob. "The History of Pulse Timing with Some Remarks on Sir John Floyer and His Physicians' Pulse Watch." In *Annals of Medical History*, vol. 4, edited by Francis Randolph Packard, 97–114. New York: Paul B. Hoeber, 1922.

Roy, Ashutosh. "Astrology in Hindu Medicine." *Journal of Ayurveda* 7:3 (1930): 113–16.

———. "Astrology in Hindu Medicine." *Journal of Ayurveda* 7:4 (1930): 139–46.

———. "The Nervous System of the Ancient Hindus." *Journal of Ayurveda* 6:9 (1930): 327–33.

———. *Pulse in Ayurveda*. Calcutta: Journal of Ayurveda, 1929.

Royle, John Forbes. *An Essay on the Antiquity of Hindoo Medicine*. London: W. H. Allen, 1837.

Sahlins, Marshall. *Islands of History*. Chicago: University of Chicago Press, 1987.

Salisbury, Laura, and Andrew Shail, eds. *Neurology and Modernity: A Cultural History of Nervous Systems 1800–1950*. Basingstoke, UK: Palgrave Macmillan, 2010.

Sandelowski, Margarete. *Devices & Desires: Gender, Technology and American Nursing*. Chapel Hill: University of North Carolina Press, 2000.

Sanyal, Hitesranjan. *Bangla Kirtoner Itihas*. Calcutta: K. P. Bagchi, 1989.

———. "Continuities of Social Mobility in Traditional and Modern Society in India." *Journal of Asian Studies* 30:2 (1971): 315–39.

Sanyal, Pulinbihari, "Dhatu-Allopathy-mote." *Chikitsa Sammilani* 5:10/11/12 (1888 [1295 BE]): 197–212.

Sanyal, Ram Brahma. *A Handbook of the Management of Animals in Captivity in Lower Bengal*. Calcutta: Bengal Secretariat Press, 1892.

Sappol, Michael. *A Traffic of Dead Bodies: Anatomy and Embodied Social Identity in Nineteenth Century America*. Princeton, NJ: Princeton University Press, 2002.

Sarangadhar. *Sarangadhar Samhita*. Edited by Peary Mohan Sengupta. Calcutta: Benimadhab Dey, 1889.

Sarkar, Jadunath. *Chaitanya's Life and Teachings: From His Contemporary Bengali Biography the Chaitanya-Charit-Amrita*. Calcutta: M. C. Sarkar & Sons, 1922.

Sarkar, Jagadish Narayan. *Thoughts on Trends of Cultural Contact in Medieval India*. Calcutta: OPS Publishers, 1984.

Sarkar, Jashodanandan trans. *Susruta Samhita (Mul o Bonganubad)*. Calcutta: Bangabasi Steam Machine Press, 1894.

Sarkar, Sumit. *Beyond Nationalist Frames: Postmodernism, Hindu Fundamentalism, History*. Delhi: Permanent Black, 2002.

———. *Writing Social History*. Oxford: Oxford University Press, 1997.

Sarma, S. R. *The Archaic and the Exotic: Studies in the History of Indian Astronomical Instruments*. Delhi: Manohar, 2008.

Sarton, George. "Second Preface to Volume XXIII: History of Science versus the History of Medicine." *Isis* 23:2 (1935): 313–20.
Satz, Aura. "'The Conviction of Its Existence': Silas Wier Mitchell, Phantom Limbs and Phantom Bodies in Neurology and Spiritualism." In *Neurology and Modernity: A Cultural History of Nervous Systems 1800–1950*, edited by Laura Salisbury and Andrew Shail, 113–29. Basingstoke, UK: Palgrave Macmillan, 2010.
Schaffer, Simon. "The Asiatic Enlightenments of British Astronomy." In *The Brokered World: Go-Betweens and Global Intelligence, 1770–1820*, 49–104. Sagamore Beach, MA: Science History Publications, 2009.
———. "Godly Men and Mechanical Philosophers: Souls and Spirits in Restoration Natural Philosophy." *Science in Context* 1:1 (1987): 53–85.
Schaffer, Simon, Lissa Roberts, Kapil Raj, and James Delbourgo, eds. *The Brokered World: Go-Betweens and Global Intelligence, 1770–1820*. Sagamore Beach, MA: Science History Publications, 2009.
Scheid, Volker. *Chinese Medicine in Contemporary China: Plurality and Synthesis*. Durham, NC: Duke University Press, 2002.
Schipper, Kristofer Marinus. "The Taoist Body." *History of Religion* 17:3/4 (1978): 355–86.
———. *The Taoist Body*. Translated by Karen C. Duval. Berkeley: University of California Press, 1993.
Secord, James A. "The Big Picture." *British Journal for the History of Science* 26 (1993): 387–483.
Selby, Martha Ann. "Narratives of Conception, Gestation and Labour in Sanskrit Ayurvedic Texts." *Asian Medicine: Tradition and Modernity* 1:2 (2005): 254–75.
Sen Gupta, Dhurmo Dass. *Nari Vijnana, Or An Exposition of the Pulse*. Calcutta: Union Printing Works, 1893.
Sen Gupta, Nagendra Nath. "Jib Yantre Ayurveder Bisheshotyo." *Ayurveda Bikash* 3:4 (1915–16): 64–69.
Sen Gupta, Sankar. *Folklorists of Bengal*. Calcutta: Indian Publications, 1965.
Sen, Binodlal. *Arya Grihachikitsa*. Calcutta: n.p., 1879 (1285 BE).
———. *Ayurveda Bigyan*. Vol. 1. Calcutta: Adi Ayurveda Machina Yantra, 1930 (1337 BE])
Sen, Dineshchandra. "Therapeutics," *Ayurveda* 6:2 (1921–22 [1328 BE]): 57–58.
Sen, Gananath. *Ayurveda Porichoy*. Calcutta: Vishwabharati, 1944.
———. *Ayurveda Samhita*. Calcutta: Gobardhan Press, 1924 (1331 BE).
———. "Baidyer Jatiyo Obonotir Karon (Oitihasik Tottwo)." *Baidya Hitoishi* 4:11 (1928 [1335 BE]): 368–72.
———. "Bharotborshe Ayurveder Jagriti." *Ayurveda Bikash* 1:12 (1913 [1320 BE]): 352–61.
———. "Sharir Bidya." *Ayurveda* 4:5 (1919 [1326 BE]): 193–201.
———. *Hindu Medicine: Address Delivered at Ceremony for Foundation of Benares Hindu University*. Calcutta: Sushilkumar Sen, 1916.
Sen, Hemchandra. "Nadichakra." *Chikitsa Sammilani* 11:11/12 (1907): 33–50.
———. *Original Researches in the Treatment of Tropical Diseases*. Calcutta: Record Press, 1902.
———. "Pitter Byabohar." *Chikitsha Sammilani* 11:10 (1906): 17–21.
———. "Swabhabik Rogbadhok Shokti." *Chikitsa Sammilani* 12:2/3 (1907): 137–43.
———. *A Thesis on Tropical Abscess of the Liver*. Calcutta: Record Press, 1902.
Sen, Pitambar. *Nadiprokash*. Calcutta: Chaitanyachandrodaya Jantra, 1865.
Sen, P. K. Advertisement. *Baidya Protibha* 2:1 (1925 [1332 BE]).
———. Advertisement. *Baidya Protibha* 2:2 (1925 [1332 BE]).
———. Advertisement. *Baidya Protibha* 2:3 (1925 [1332 BE]).

Sen, Rathindra Nath. *Life and Times of Deshbandhu Chittaranjan Das*. New Delhi: Northern Book Centre, 1989.

Sen, Saradacharan. "Hookworm ba Krimi." *Ayurveda* 3:10 (1919 [1326]): 376–78.

Sengoopta, Chandak. "'Dr Steinach Coming to Make Old Young!': Sex Glands, Vasectomy and the Quest for Rejuvenation in the Roaring Twenties." *Endeavour* 27:3 (2003): 122–26.

———. *The Most Secret Quintessence of Life: Sex, Glands and Hormones, 1850–1950*. Chicago: University of Chicago Press, 2006.

Sengupta, Debendranath, and Upendranath Sengupta. *Ayurved Samgraha*. Vol. 1. Calcutta: Dipayan, n.d. [reprint 1892].

———. *Pachon Samgraha*. Calcutta: n.p., 1895.

Sengupta, Gopalchandra. *Ayurveda Sarsamgraha*. Vol. 1. Calcutta: Columbian Press, 1871 [1278 BE].

———. "Grahokdiger Nam." *Ayurveda Sarsamgraha*. Vol. 2. Calcutta: Columbian Press, 1871 [1278 BE], unnumbered page.

Sengupta, Indubhushan. "Ayurveder Bonoushodhi." *Ayurveda* 8: 10/11 (1923–24 [1330 BE]): 246–51.

———. "Prachin Chikitsoker Porikshito Mushtijog," *Ayurveda* 6:11 (1921–22 [1327 BE]): 407–8.

Sengupta, Nagendranath. *The Ayurvedic System of Medicine*. Vol. 1. Delhi: Neeraj Publishing House, 1984.

———. *Pachon o Mushtijog*. Calcutta: Nagendra Steam Printing Works, 1911.

———. *Pachon o Mushtijog*. 4th ed. Calcutta: Nagendranath Sengupta, 1920.

———. *Sochitro Kobiraji Siksha*. Vol. 1. Calcutta: Nagendra Steam Printing Works, 1930.

Sengupta, Sarachchandra. "Pitta." *Ayurveda* 8:3 (1923 [1330 BE]): 69–72.

Sengupta, Satyacharan. "Chikitsoker Kotha," *Ayurveda* 8:4 (1923–24 [1330 BE]): 89–92.

———. "*Onukorone Amader Obostha.*" *Ayurveda* 1:1 (1916 [1323 BE]): 428–37.

Sengupta, Shyamacharan. *Ayurvedarthachandrika*. Kashipur: Pallibikashini Yantra, 1887 (1294 BE). Mukhopadhyay, Surendramohan. *Jonmantor Rohosyo*. n.p., 1899 (1306 BE).

Sensharma, Madhusudan. "Baidya-Brahmin Somiti Bhabanipur Kendrer Barshik Biboron." *Baidya Hitoishi* 3:4 (1926 [1333 BE]): 121–24.

Sensharma, Surendralal. "*Dasham Graha.*" *Baidya Protibha* 1:1 (1925 [1332 BE]): 62–63.

Seth, Suman. "Quantam Theory and the Electromagnetic World-view." *Historical Studies in the Physical and Biological Sciences* 35:1 (2004): 67–93.

Shapin, Steven, and Simon Schaffer. *The Leviathan and the Air-Pump: Hobbes, Boyle and the Experimental Life*. Princeton, NJ: Princeton University Press, 2011.

Sharif, Ahmed. *Banglar Sufi Sahityo*. Dhaka: Samay Prakashan, 1969.

Sharma, Acharya Priyavrat. *Ayurved ka Vaigyanik Itihas*. Varanasi: Chaukhamba Orientalia, 2009.

Sharma, Haradhan. *Ayurvedokto Nuton Chikitsa Darshan*. Calcutta: Bishwambhar Laha, 1883 (1290 BE).

Sharma, Madhuri. *Indigenous and Western Medicine in Colonial India*. New Delhi: Foundation Books, 2012.

Sharma, Priya vrat, trans. *Cakradatta*. Delhi: Chaukhamba Orientalia, 2007.

Sharma, Satishchandra, trans. *Charaka Samhita*. Calcutta: Bhaisajya Steam Machine Jantra, 1904 (1311 BE).

Sharma, Shiv. *Ayurvedic Medicine: Past and Present*. Calcutta: Dabur Publications, 1975.

———. *The System of Ayurveda*. Bombay: Khemraj Shrikrishnadas Shri Venkateshwar Steam Press, 1929.

Shefer-Mossensohn, Miri. *Ottoman Medicine: Healing and Medical Institutions 1500–1700*. Albany: SUNY Press, 2009.

Sibum, Heinz Otto. "Reworking the Mechanical Value of Heat: Instruments of Precision and Gestures of Accuracy in Early Victorian England." *Studies in the History & Philosophy of Science* 26:1 (1995): 73–106.

Sigerist, Henry E. "The History of Medicine *and* the History of Science: An Open Letter to George Sarton, Editor of *Isis*." *Bulletin of the Institute of the History of Medicine* 4 (1936): 1–13.

Sinding, Christiane. "Les multiples usages de la quantification en medicine: Le cas du diabète sucré." In *Body Counts: Medical Quantification in Historical and Sociological Perspective*, edited by Gerard Jorland, A. Opinel, and G. Weisz, 127–44. Montreal: McGill-Queen's University Press, 2005.

Singh Jee, Bhagvat. *A Short History of Aryan Medical Science*. London: Macmillan, 1896.

Sircar, Dines Chandra. "The Ambashtha Jati." *Journal of the United Provinces Historical Society* 17:1 (1944): 148–61.

Sivaramakrishnan, Kavita. *Old Potions, New Bottles: Recasting Indigenous Medicine in Colonial Punjab, 1850–1945*. Hyderabad: Orient Longman, 2006.

Sivin, Nathan. *Traditional Medicine in Contemporary China*. Ann Arbor: University of Michigan Press, 1987.

Smajic, Srdjan. *Ghost-Seers, Detectives and Spiritualists: Theories of Vision in Victorian Literature and Science*. Cambridge: Cambridge University Press, 2010.

Smith, Pamela. *The Body of the Artisan: Art and Experience in the Scientific Revolution*. Chicago: University of Chicago Press, 2004.

Sointu, Eeva. *Theorizing Complementary and Alternative Medicines: Wellbeing, Self, Gender, Class*. Basingstoke, UK: Palgrave Macmillan, 2012.

Srinivas, M. N. "A Note on Sanskritization and Westernization." *Far Eastern Quarterly* 15:4 (1956): 481–96.

Staal, Frits. "Indian Bodies." In *Self as Body in Asian Theory and Practice*, edited by Thomas Kasulis, Roger T. Amis, and Wimal Dissanayake, 59–102. Albany: SUNY Press, 1993.

Stewart, Duncan "Observations on the Fever which Prevailed at Howrah, During the months of June and July, 1834." *Transactions of the Medical and Physical Society of Calcutta: Volume the Seventh*. Calcutta: Baptist Mission Press, 1835, 363–88.

Stewart, Tony K. *The Final Word: The Caitanya Caritamrita and the Grammar of Religious Tradition*. Oxford: Oxford University Press, 2010.

———. "Replicating Vaisnava Worlds: Organizing Devotional Space through the Architectonics of the Mandala." *South Asian History and Culture* 2:2 (2011): 300–336.

———. "Satya Pir: Muslim Holy Man and Hindu God." In *Religions of India in Practice*, edited by Donald S. Lopez Jr, 578–98. Princeton, NJ: Princeton University Press, 1995.

Stolz, Daniel A. "'By Virtues of Your Knowledge': Scientific Materialism and the Fatwas of Rashid Rida." *Bulleting of the School of Oriental and African Studies* 75:2 (2011): 223–47.

———. "The Lighthouse and the Observatory: Islam, Authority, and Cultures of Astronomy in Late Ottoman Egypt." PhD diss., Princeton University, 2013.

Tagore, Rabindranath. *Chithi Potro*. Vol. 1. Santiniketan: Vishwabharati Granthabibhag, 1993 [1942].

———. *Farewell Song*. Translated by Radha Chakravarty. New Delhi: Penguin, 2011.

Tarkabhushan, Pramathanath. "Ayurveder Unnoti ki Obonoti?" *Ayurveda* 1:11 (1916–17 [1323 BE]): 474–77.

Tarkaratna, Jadabeshwar. "Prachin Bharote Kitanu Tottwo." *Ayurveda* 3:12 (1919 [1326 BE]): 442–45.

Taylor, Kathleen. *Sir John Woodroffe, Tantra and Bengal: An Indian Soul in a European Body?* London: Routledge, 2001.

Thapar, Romila. *Time as a Metaphor of History: Early India*. Delhi: Oxford University Press, 1996.

Thomas, Nicholas. *Entangled Objects: Exchange, Material Culture, and Colonialism in the Pacific*. Cambridge, MA: Harvard University Press, 1991.

Tilley, Helen. "Global Histories, Vernacular Science, and African Genealogies; Or is the History of Science Really for the World?" *Isis* 101:1 (2010): 110–19.

Tiwari, Premavati. *Kasyapa-samhita Or Vriddhajivakiya Tantra*. Benares: Caukhamba Visvabharati, 1996.

Tomes, Nancy J and John Harley Warner. "Introduction to the Special Issue on Rethinking the Reception of the Germ Theory of Disease: Comparative Perspectives." *Journal of the History of Medicine and Allied Sciences* 52 (1997): 7–16.

Trautmann, Thomas R. *Aryans and British India*. Berkeley: University of California Press, 1997.

Tresch, John. "Cosmologies Materialized: History of Science and the History of Ideas." In *Rethinking Modern European Intellectual History*, edited by Darrin McMahon and Sam Moyn, 153–72. New York: Oxford University Press, 2014.

———. *The Romantic Machine: Utopian Science and Technology after Napoleon*. Chicago: University of Chicago Press, 2012.

Upadhyay, Sarva Dev. *Nadi Vijnana: Ancient Pulse Science*. Benares: Chaukhamba Vidyabhawan, 1986.

Upjohn, A. "Calcutta in the Olden Time—Its Localities." In *Calcutta Review* 18, 275–320. Calcutta: Sanders, Cones and Co., 1852.

van der Eijk, Philip. "Medicine and Health in the Graeco-Roman World." In *The Oxford Handbook of the History of Medicine*, edited by Mark Jackson, 21–39. Oxford: Oxford University Press, 2011.

van der Veer, Peter. *Imperial Encounter: Religion and Modernity in India and Britain*. Princeton, NJ: Princeton University Press, 2001.

Vasu, Srisa Chandra, trans. *Siva Samhita*. Allahabad: Indian Press, 1914.

Vaughan, Megan. *Curing Their Ills: Colonial Power and African Illness*. Stanford, CA: Stanford University Press, 1991.

Vink, Markus. *Dutch Sources on South Asia c. 1600–1825*. Vol. 4. Delhi: Manohar, 2012.

von Schwerin, Alexander, Heiko Stoff, and Bettina Wahrig, eds. *Biologics: A History of Agents Made from Living Organisms in the Twentieth Century*. London: Pickering & Chatto, 2003.

Wadley, Susan S. "Sitala: The Cool One." *Asian Folklore Studies* 39:1 (1980): 33–62.

Waghorne, Joanne Punzo. *Diaspora of the Gods: Modern Hindu Temples in an Urban Middle Class World*. Oxford: Oxford University Press, 2004.

Walhouse, M. J. "A Hindu Prophetess." *Journal of the Anthropological Institute of Great Britain and Ireland* 14 (1885): 187–92.

Waliullah, Syed. *Lal Salu*. Dhaka: Nouroj Kitabistan, 2006.

Ward, William. *Account of the Writings, Religion and Manners of the Hindoos*. Vol. 4. Serampore: Mission Press, 1811.

———. *A View of the History, Literature and Mythologies of the Hindoos*. Vol. 4. 3rd ed. London: Black, Kingsbury, Parbury & Allen, 1820.

Warner, John Harley. "'Exploring the Labyrinths of Creation': Popular Microscopy in Nineteenth Century America." *Journal of the History of Medicine and Allied Sciences* 37:1 (1982): 7–33.

———. "'The Nature-Trusting Heresy': American Physicians and the Concept of the Healing Power of Nature in the 1850s and 1860s." *Perspectives in American History* 11 (1977–78): 291–324.

———. "Science in Medicine." *Osiris* 1 (1985): 37–85.

Warner, John Harley, and Lawrence J. Rizzolo. "Anatomical Instruction and Training for Professionalism: From the 19th to the 21st Centuries." *Clinical Anatomy* 19:5 (2006): 403–14.

Webb, Allan. *Elephentiasis Orientalis, and Especially Elephentiasis Genitalis in Bengal*. Calcutta: Bengal Military Orphan Press, 1855.

———. *The Historical Relations of Ancient Hindu with Greek Medicine*. Calcutta: Military Orphan Press, 1850.

Weiss, Richard S. *Recipes for Immortality: Medicine, Religion, and Community in South India*. Oxford: Oxford University Press, 2009.

Werner, Michael, and Benedicte Zimmermann. "Beyond Comparison: Histoire Croisee and the Challenge of Reflexivity." *History & Theory* 45:1 (2006): 30–50.

Westermeyer, Joseph. "Folk Medicine in Laos: A Comparison between two Ethnic Groups." *Social Science & Medicine* 27:8 (1988): 769–78.

Whitcher, Sheryle J., and Jeffrey D Fisher. "Multidimensional Reaction to Therapeutic Touch in a Hospital Setting." *Journal of Personality and Social Psychology* 37:1 (1979): 87–96.

White, David Gordon. *The Alchemical Body: The Siddha Tradition in Medieval India*. Chicago: University of Chicago Press, 1996.

———. "On the Magnitude of the Yogic Body." In *Yogi Heroes and Poets: Histories and Legends of the Naths*, edited by David Lorenzen and Adrian Muñoz, 79–90. Delhi: Oxford University Press, 2011.

Wilson, H. H. "Medical and Surgical Sciences of the Hindus." In *Works of the Late Horace Hayman Wilson*, vol. 3, edited by Reinhold Rost, 269–76, 380–93. London: Trubner, 1864.

Winawer, J., N. Witthoft, M. C. Frank, L. Wu, A. R. Wade, and L. Boroditsky. "Russian Blues Reveal Effects of Language on Color Discrimination." *Proceedings of the National Academy of the Sciences* 104:19 (2007): 7780–85.

Winner, Langdon. "Do Artifacts Have Politics?" *Daedalus* 109:1 (1980): 121–36.

Wise, James. *Notes on the Races, Castes and Trades of Eastern Bengal*. London: Harrison and Sons, 1883.

Wise, T. A. *Commentary on the Hindu System of Medicine: New Issue*. London: Trubner, 1860.

Wise, T. A. *Review of the History of Medicine*. Vols. 1–2. London: J. Churchhill, 1867.

Wishnitzer, Avner. "The Transformation of Ottoman Temporal Culture During the Long Nineteenth Century." PhD diss., Tel Aviv University, 2009.

Worboys, Michael. "The Emergence of Tropical Medicine: A Study in the Establishment of a Scientific Speciality." In *Perspectives on the Emergence of Scientific Disciplines*, edited by G. Lemaine, R. Macleod, M. Mulkay, and P. Weingart, 75–98. Hague: Mouton, 1976.

———. *Spreading Germs: Disease Theories and Medical Practice in Britain, 1865–1900*. Cambridge: Cambridge University Press, 2000.

———. "Was There a Bacteriological Revolution in Late Nineteenth-Century Medicine?" *Studies in History and Philosophy of Biological and Biomedical Sciences* 38:1 (March 2007): 20–42.

Wright, David, Nathan Flis, and Mona Gupta, "The 'Brain-Drain' of Physicians: Historical Antecedents to an Ethical Debate, c. 1960–79." *Philosophy, Ethics and Humanities in Medicine* 3:24 (2008).

Wujastyk, Dominik. "Change and Creativity in Early Modern Indian Medical Thought." *Journal of Indian Philosophy* 33:1 (2005): 95–118.

———. "Interpreting the Image of the Human Body in Premodern India." *International Journal of Hindu Studies* 13:2 (2009): 189–228.

———. *Well-Mannered Medicine: Medical Ethics and Etiquette in Classical Ayurveda*. New York: Oxford University Press, 2012.

Wujastyk, Dominik, ed. and trans. *The Roots of Ayurveda: Selections from Sanskrit Medical Writings*. New Delhi: Penguin, 1998.

Wujastyk, Dagmar, and Frederick M. Smith, eds. *Modern and Global Ayurveda: Pluralism and Paradigms*. Albany: SUNY Press, 2008.

Zimmermann, Francis. *The Jungle and the Aroma of the Meats*. Delhi: Motilal Banarsidass, 2011.

Zysk, Kenneth. *Asceticism and Healing in Ancient India: Medicine in the Buddhist Monastery*. Delhi: Motilal Banarsidass, 1998.

———. "Mantra in Ayurveda: A Study of the Use of Magico-Religious Speech in Ancient Indian Medicine." In *Understanding Mantras*, edited by Harvey P Alper, 123–43. Albany: SUNY Press, 1989.

INDEX